First Edition

THE MARS ATLAS

Topography & Places

RAY & RAY

RedMapper — Subsidiary of Planetary Modeling, Inc.
PO Box 3421
Kirkland, WA 98083
U.S.A.

THE MARS ATLAS – TOPOGRAPHY & PLACES, FIRST EDITION

ISBN 978-0-578-23165-5

Author & cartographer: *Eian Ray*
Co-author & cartographic editor: *Tyler Ray*

For more information visit: RedMapper.com

INTRODUCTION

Geography is a field of study that seeks to describe and provide meaning to the spatial relationships within the universe. As such the domain of geography is comprised of geology, hydrology, climate and geomorphology. It is also concerned with the ways in which place, space and environment are the consequences of human activities.

Geographers, and more specifically cartographers, design maps to present this spatial information in graphical form in order to adequately convey a deeper message about the places and objects that make up the world (or worlds) we live in. A map, even a very complex map, is simply a discreet symbol that represents an infinitely complex set of circumstances about a place. Maps assist people as a tool of intuition, indicating more about place than would otherwise be possible with other forms of media, written narratives, or even a firsthand visit to a place. They are an instrument of inquisition and create as many new questions as they provide answers to.

SPACE EXPLORATION & MAPS

Human exploration of Mars has been underway since the 1960's through national and multinational programs, most notably those of the United States, United Kingdom, European Union, and the former Soviet Union. More recently many of the associated costs of space exploration have been driven down through improvement in technology, economies of scale, and the global generation of unprecedented societal wealth. Many private, corporate, and non-profit organizations have picked up the torch of Martian and other space exploration activities. Some have begun collaborating with government programs while others are operating independently. Remote exploration of Mars has also entered the public sphere in the form of satellite imagery, online mapping platforms, and citizen efforts afforded by the collaborative nature of the internet.

There are numerous challenges associated with mapping planetary surfaces. Data and imagery for extraterrestrial places at a spatial and spectral resolution sufficient for mapping purposes is difficult to capture. Developing national space programs and public support to collect this data and imagery requires substantial amounts of funding. The planning endeavors and time scales required for the development of these programs, as well as the inherent risk involved in mission failures adds layers of logistical complexity and financial burden to governments and corporations. This has resulted in a waxing and waning of public interest over the last half century as organizations strive to justify their activities.

One of the key elements of a successful mission, regardless of scale or scope, is how to best communicate the geospatial aspects of the mission planning process to the general public. Notable imagery and customized maps may be presented as illustrations on the news or written material for specific locations. These may not adequately convey the scope, scale, or spatial relationships of off-Earth places to individuals reading about them. The maps portrayed within this book assist in this effort by using existing spatial data and traditional, static cartographic methods. By applying the existing resources extra-terrestrial geographic discoveries can be made and disseminated to the public in an attractive and profound manner.

THE MARS ATLAS

The Mars Atlas contains detailed topography, geology, named places, historical landing sites and human activity. This compilation of maps provides a baseline for understanding the spatial relationships present among numerous geographic features and the places our species has laid claim to within the Martian landscape. It is designed to create a sense of place in the minds of those who peruse these maps with a goal of transcending Mars from that of an abstract idea to that of a real place. Like all places, the use of maps supports us to create a sense of understanding, belonging and attachment. In short, this atlas is a bridge that serves to connect the places and spaces familiar to our species to new, similar landscapes from another world.

The Mars Atlas portrays a cartographically accurate representation of the planet Mars in order to inspire and initiate conversation. Early in this work it was decided the best way to convey the spatial arrangement of places on Mars was to provide a comprehensive set of standardized maps that cover every part of the Martian globe and not just those areas which are most conspicuous, interesting, or news-worthy. The atlas covers the famed Olympus Mons and Valles Marineris. It also covers many lesser known places, such as the Kasei Valley, the Ophir Mesa, and the Ulysses Hills. Most prominently however, the atlas shows in detail the multitude of places as yet unnamed and mostly unexplored. These are the places we wish to bring to the forefront of people's imagination. To show Mars explorers and space enthusiasts alike that Earth does not stand alone in the Cosmos. It is but one of many worlds with a variety of landscapes, many of which are very similar to our own.

MARS PORTAL

To further circulate Martian geographic information beyond the limitations of a book, a web map featuring the same mapping data used to develop The Mars Atlas has been co-developed along with this publication. While The Mars Atlas is designed to be used as a stand-alone reference tool, it can also be used seamlessly with RedMapper's Mars Portal. This web map can be accessed at RedMapper.com. This portal interface allows users to freely and dynamically explore the surface of Mars, functioning similarly to other web-based map services. Each map page in the atlas refers to grid-cell that overlays the web-map, and vice-versa. If a user wishes to learn more about a particular area within the atlas, they can visit Mars Portal to explore that same area in greater detail.

METADATA

Most data used in The Mars Atlas and Mars Portal are based on two publicly available base datasets sourced from the USGS Planetary GIS Data Server (PIGDA). These two datasets are the IAU nomenclature database and the Mars Orbital Laser Altimeter (MOLA) digital elevation model (DEM).

The USGS IAU nomenclature database contains a comprehensive list of named features and places across the planet Mars that have been approved by the International Astronomical Union (IAU). It is a GIS layer file that contains nearly 2000 named places on Mars that were extracted and incorporated into this atlas. Most named features on the planet were acquired from this source. For more information on this dataset and the IAU please visit the Gazetteer of Planetary Nomenclature and the official IAU website.

The DEM was created by the Mars Orbital Laser Altimeter (MOLA) which flew onboard the Mars Global Surveyor (MGS) from 1997 to 2006. The accuracy of this dataset approximately 100 meters vertically and three meters of horizontal accuracy. The DEM was used to generate secondary datasets such as a hill shade raster image and contour features using geospatial analysis. Since these datasets are both comprehensive and georeferenced, they represent the most accurate global topography of Mars available.

ADDITIONAL ATLAS DETAILS

The Mars Atlas has broken down the Martian surface into 182 distinct map pages representing six mid-latitude series of 27 pages, two series of nine pages representing sub-polar regions and a single page for the north and south poles. Each page within the mid-latitude series refers to a subsection of the planet's surface equal to approximately 13° longitude and 18 ° latitude while the sub-polar series are equal to approximately 40° longitude and 18 ° latitude. This resulted in an ideal scale for presenting the Martian landscape with adequate detail to be both interesting and visually appealing. Great care was taken to individually place labels to not overcrowd each map with excessive labeling and geographic features.

The purpose of a projection is to "project" (display) three-dimensional objects (planets) onto a two-dimensional plane. All projections, however, come at a cost in the form of spatial distortion. A coordinate system called *World From Space* using an orthographic projection were chosen to represent the atlas's maps in an effort to reduce the distortion that is inherent in all projected maps.

During research into Martian place names to include in The Mars Atlas it was discovered that many places have been given "unofficial" names which have entered the public lexicon through popular usage. These places are not officially recognized by the IAU but do represent an appropriation of these places by other interested parties. As such these places are culturally significant and many were included in the scope of The Mars Atlas and Mars Portal.

Labeling existing features and places with established and approved nomenclature was a critical part of the development of this atlas. Proper labeling standards were reviewed, and existing named places were examined from the USGS nomenclature GIS layer file.

Labels for features from the IAU nomenclature database show both the international scientific naming designation as well as an English translation for said feature. Both names are included in the atlas's index.

LEGEND

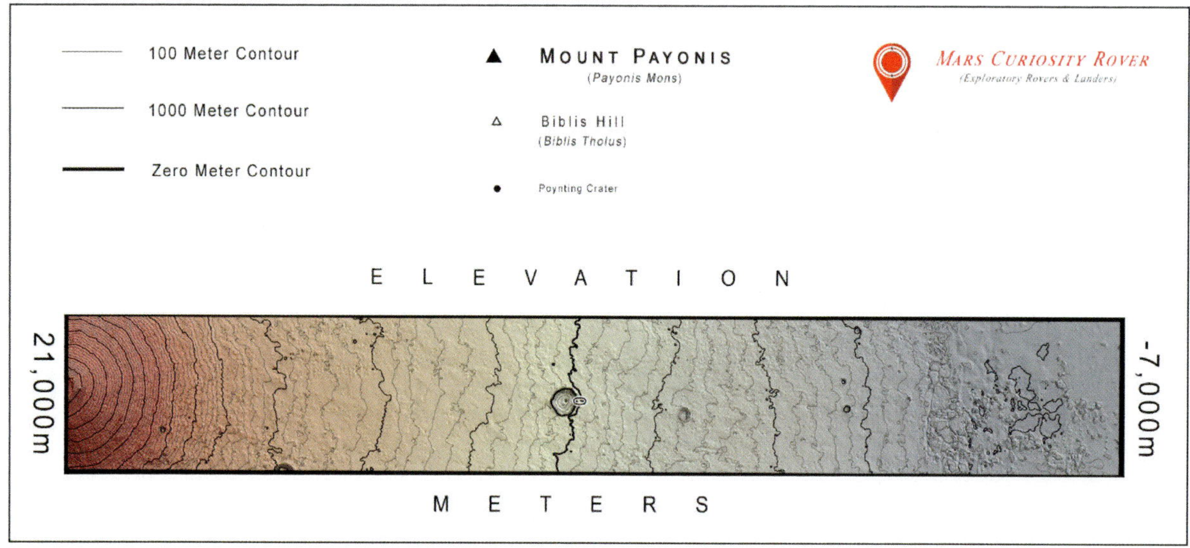

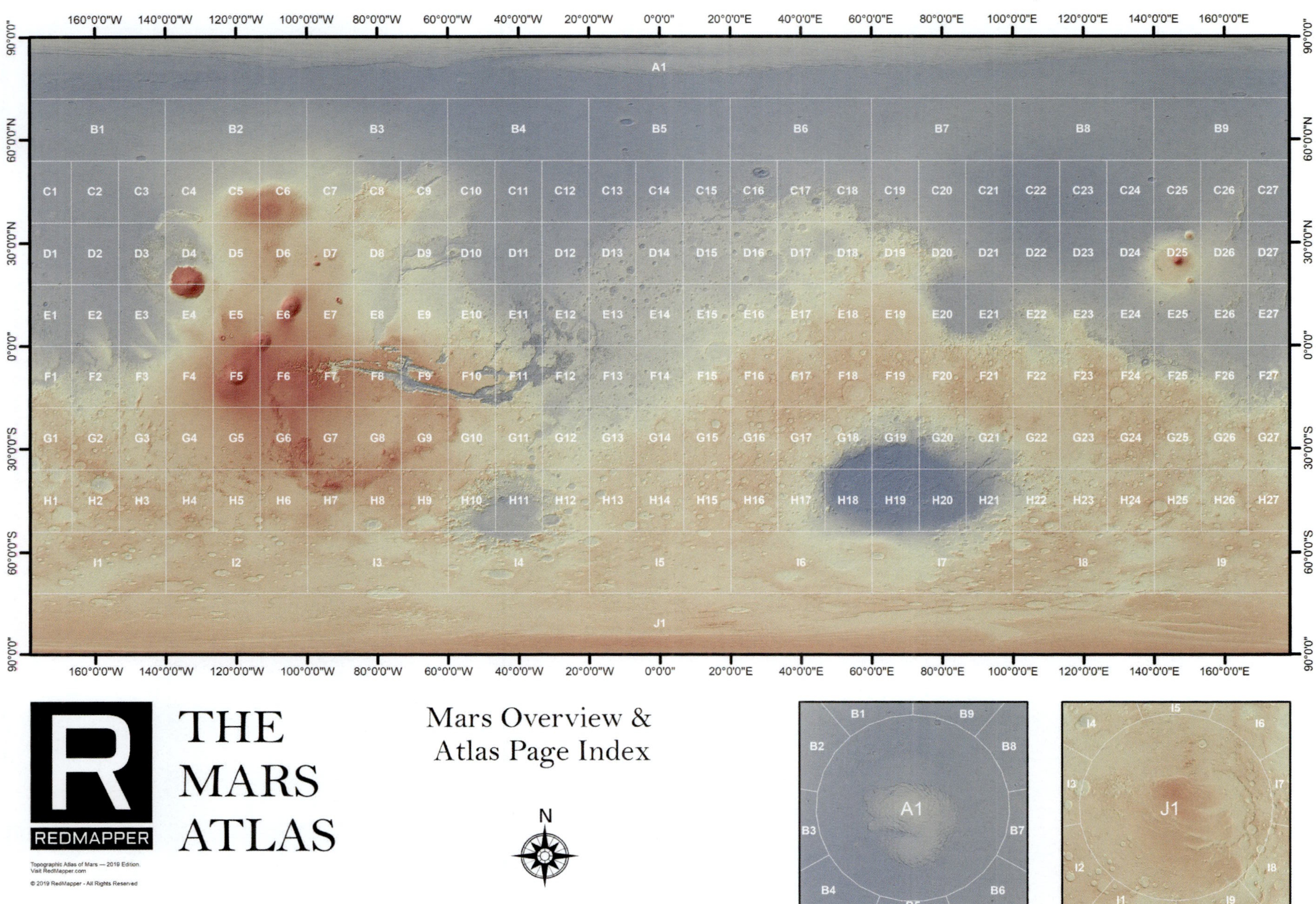

THE MARS ATLAS

REDMAPPER

Topographic Atlas of Mars — 2019 Edition.
Visit RedMapper.com

Mars Overview & Atlas Page Index

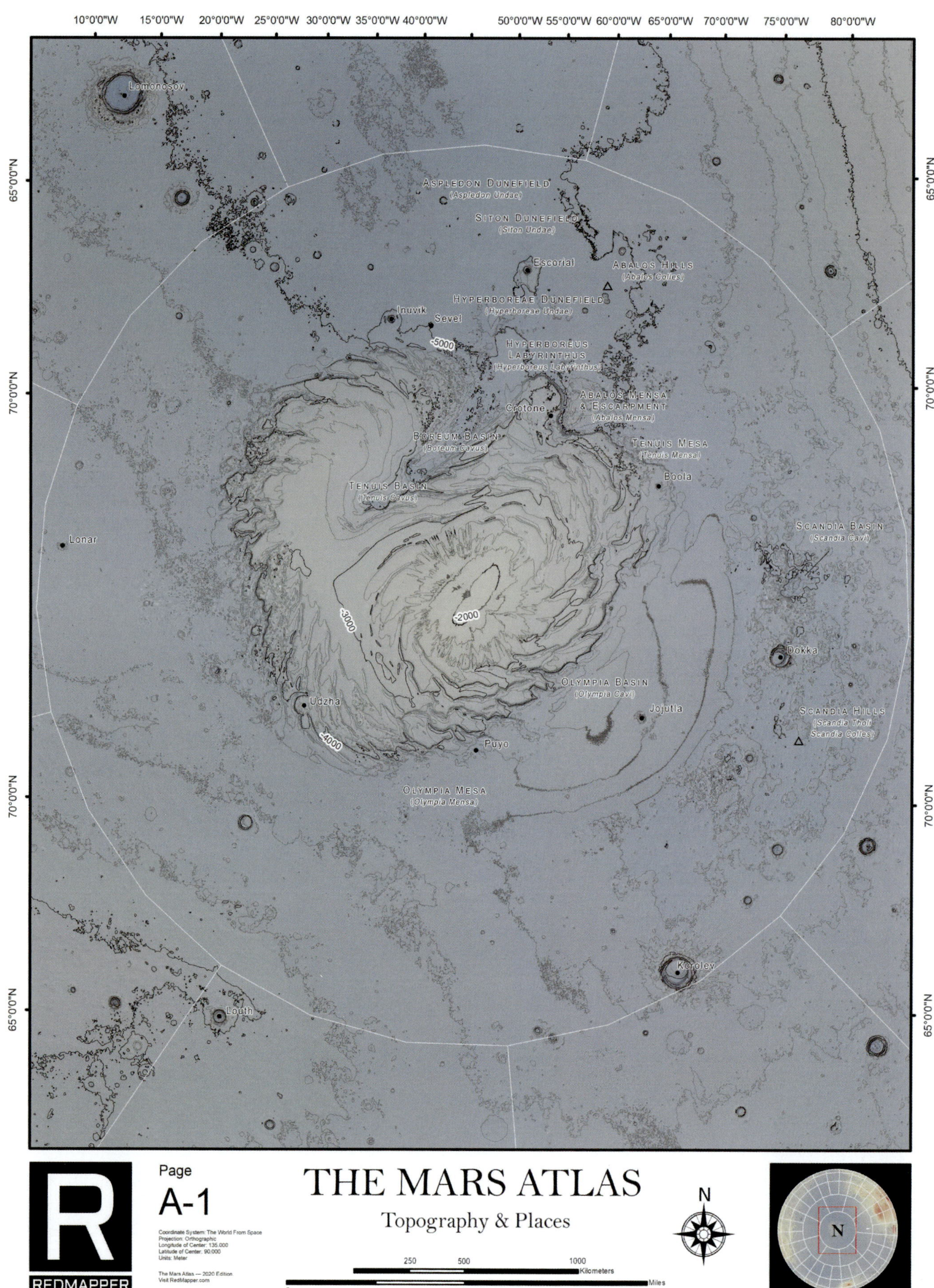
10°0'0"W
15°0'0"W
20°0'0"W
25°0'0"W
30°0'0"W
35°0'0"W
40°0'0"W
50°0'0"W
55°0'0"W
60°0'0"W
65°0'0"W
70°0'0"W
75°0'0"W
80°0'0"W
65°0'0"N
70°0'0"N
Lomonosov
ASPLEDON DUNEFIELD
(Aspledon Undae)
SITON DUNEFIELD
(Siton Undae)
Escorial
ABALOS HILLS
(Abalos Colles)
HYPERBOREAE DUNEFIELD
(Hyperboreae Undae)
Inuvik
Sevel
-5000
HYPERBOREUS LABYRINTHUS
(Hyperboreus Labyrinthus)
ABALOS MENSA & ESCARPMENT
(Abalos Mensa)
Crotone
BOREUM BASIN
(Boreum Cavus)
TENUIS MESA
(Tenuis Mensa)
Boola
TENUIS BASIN
(Tenuis Cavus)
Lonar
SCANDIA BASIN
(Scandia Cavi)
-3000
-2000
Dokka
OLYMPIA BASIN
(Olympia Cavi)
Udzha
Jojutla
SCANDIA HILLS
(Scandia Tholi
Scandia Colles)
-4000
Puyo
OLYMPIA MESA
(Olympia Mensa)
Korolev
Louth
R
REDMAPPER
Page
A-1
Coordinate System: The World From Space
Projection: Orthographic
Longitude of Center: 135.000
Latitude of Center: 90.000
Units: Meter
The Mars Atlas — 2020 Edition
Visit RedMapper.com
© 2019-2020 RedMapper - All Rights Reserved
THE MARS ATLAS
Topography & Places
N
0
250
500
1000
Kilometers
Miles
N

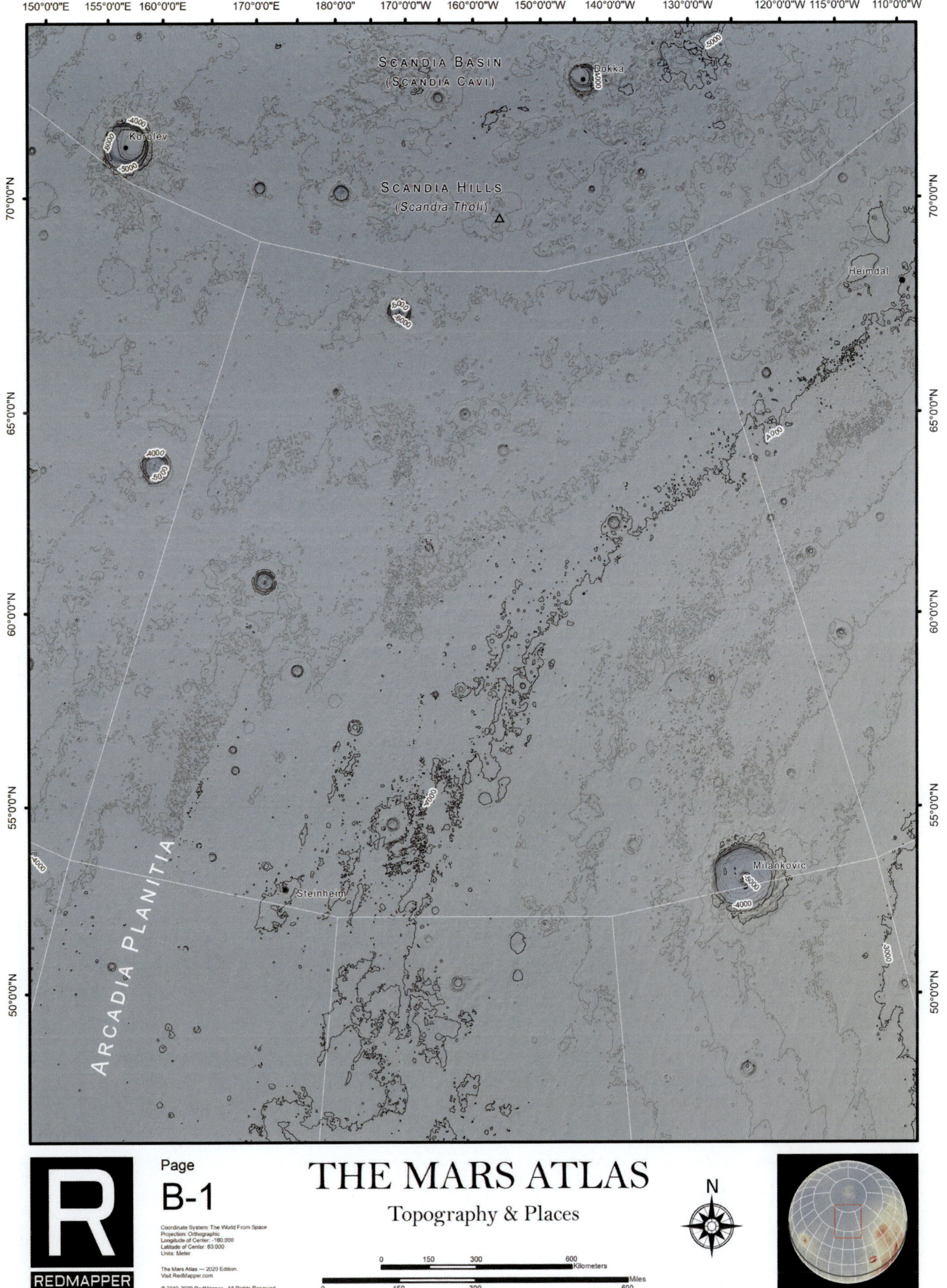

Page

B-1

Coordinate System: The World From Space
Projection: Orthographic
Longitude of Center: -180.000
Latitude of Center: 63.000
Units: Meter

The Mars Atlas — 2020 Edition.
Visit RedMapper.com

THE MARS ATLAS

Topography & Places

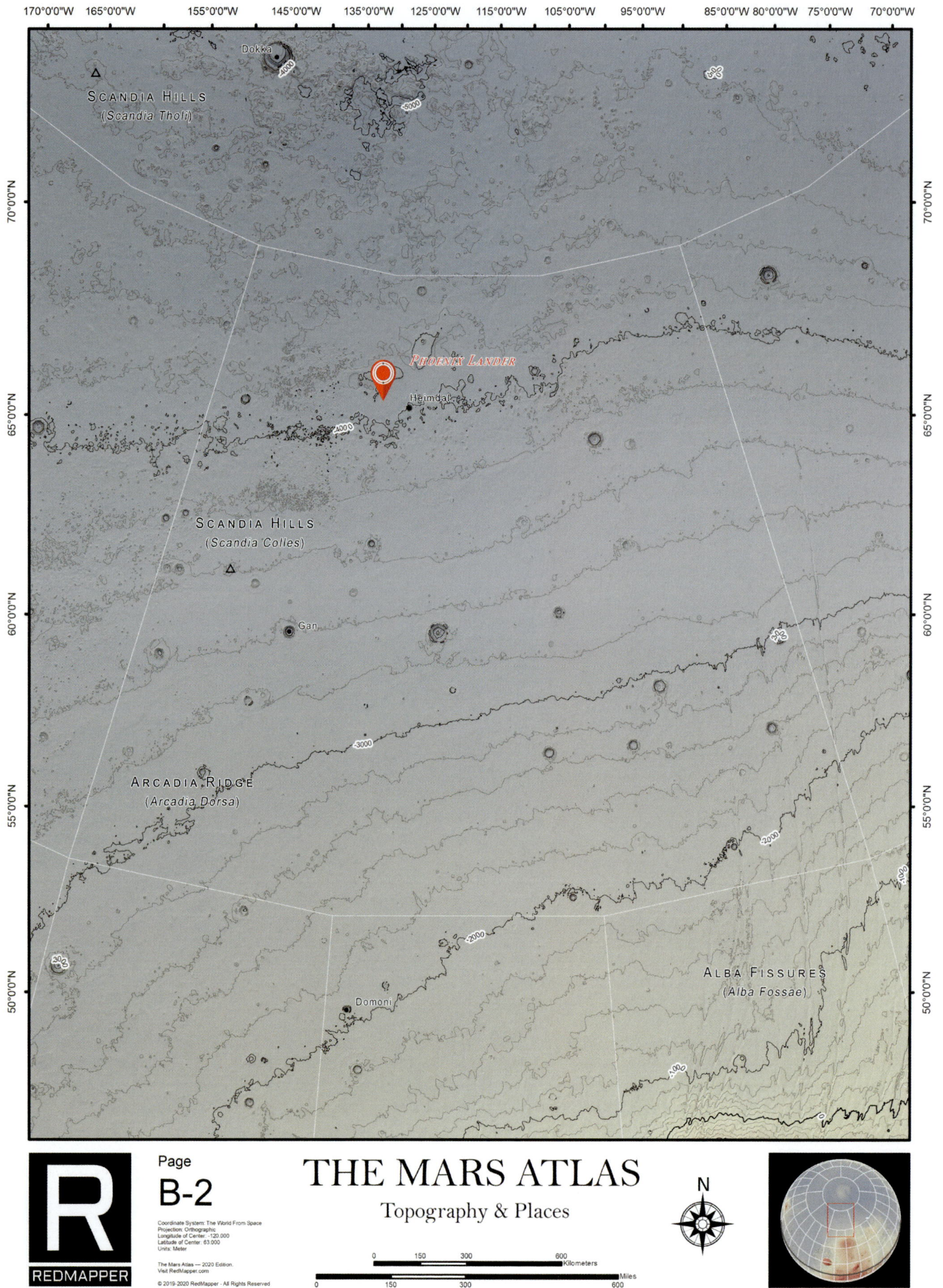
170°0'0"W
165°0'0"W
155°0'0"W
145°0'0"W
135°0'0"W
125°0'0"W
115°0'0"W
105°0'0"W
95°0'0"W
85°0'0"W
80°0'0"W
75°0'0"W
70°0'0"W
70°0'0"N
65°0'0"N
60°0'0"N
55°0'0"N
50°0'0"N
Dokka
-4000
-5000
SCANDIA HILLS
(Scandia Tholi)
PHOENIX LANDER
Heimdal
-4000
SCANDIA HILLS
(Scandia Colles)
Gan
-3000
ARCADIA RIDGE
(Arcadia Dorsa)
-2000
-2000
-1000
ALBA FISSURES
(Alba Fossae)
Domoni
-1000
0
REDMAPPER
Page
B-2
Coordinate System: The World From Space
Projection: Orthographic
Longitude of Center: -120.000
Latitude of Center: 63.000
Units: Meter
The Mars Atlas — 2020 Edition.
Visit RedMapper.com
© 2019-2020 RedMapper - All Rights Reserved
THE MARS ATLAS
Topography & Places
N
0 150 300 600 Kilometers
0 150 300 600 Miles

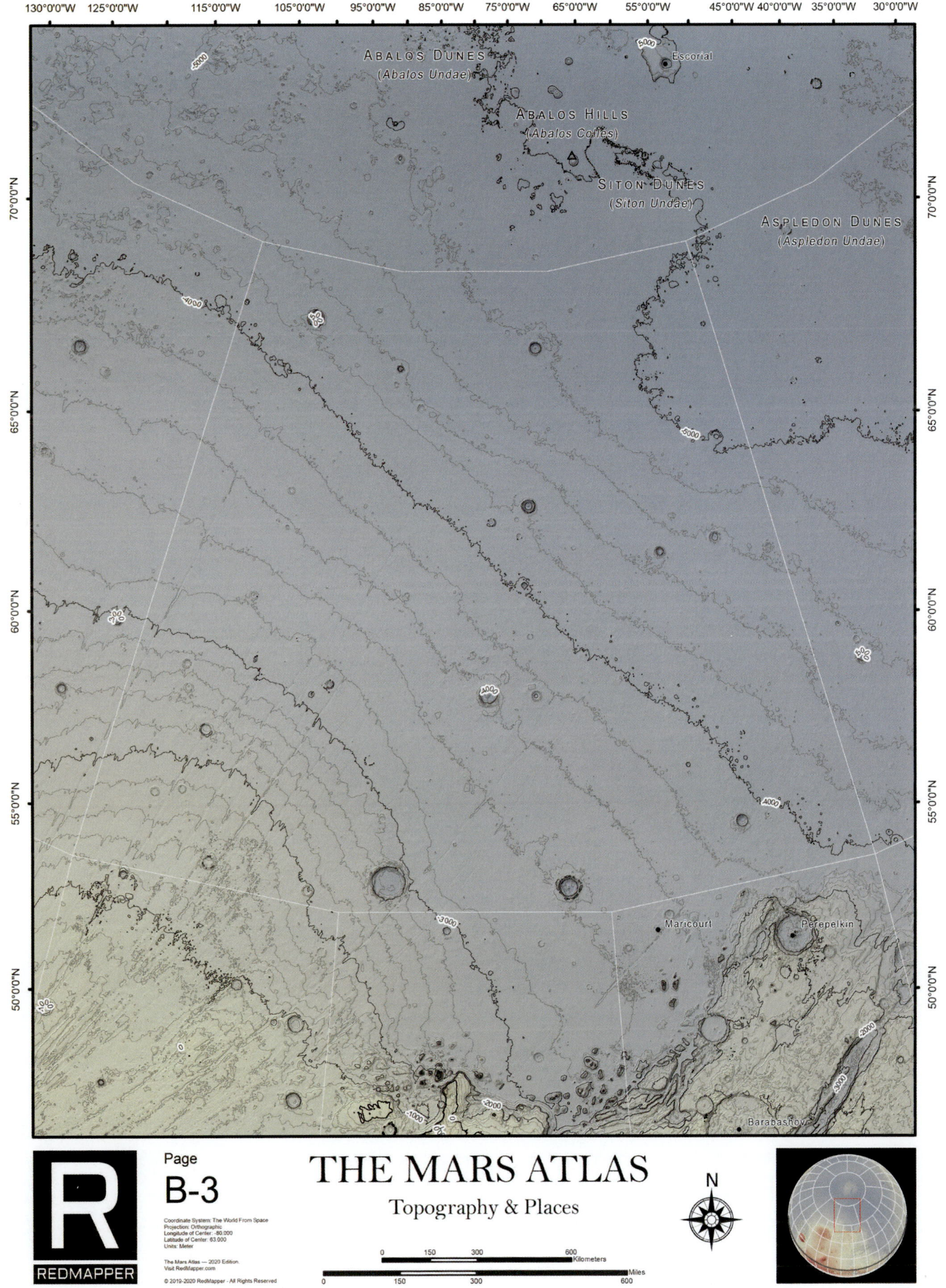
130°0'0"W
125°0'0"W
115°0'0"W
105°0'0"W
95°0'0"W
85°0'0"W
75°0'0"W
65°0'0"W
55°0'0"W
45°0'0"W
40°0'0"W
35°0'0"W
30°0'0"W
70°0'0"N
65°0'0"N
60°0'0"N
55°0'0"N
50°0'0"N
ABALOS DUNES
(Abalos Undae)
Escorial
ABALOS HILLS
(Abalos Colles)
SITON DUNES
(Siton Undae)
ASPLEDON DUNES
(Aspledon Undae)
Maricourt
Perepelkin
Barabashov
REDMAPPER
Page
B-3
Coordinate System: The World From Space
Projection: Orthographic
Longitude of Center: -80.000
Latitude of Center: 63.000
Units: Meter
The Mars Atlas — 2020 Edition.
Visit RedMapper.com
© 2019-2020 RedMapper - All Rights Reserved
THE MARS ATLAS
Topography & Places
N
0 150 300 600 Kilometers
0 150 300 600 Miles

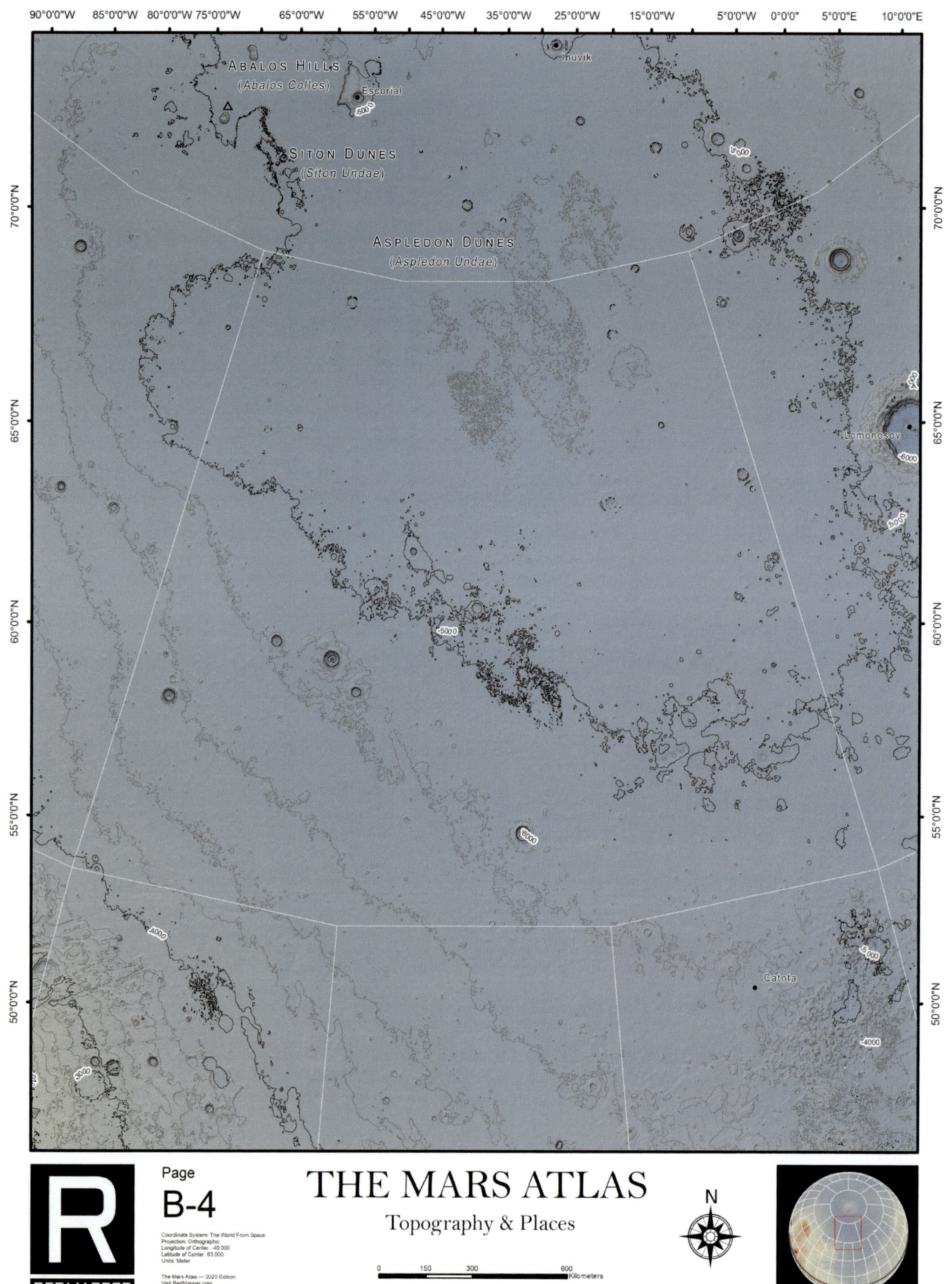

REDMAPPER

Page
B-4

Coordinate System: The World From Space
Projection: Orthographic
Longitude of Center: -40.000
Latitude of Center: 83.000
Units: Meter

The Mars Atlas — 2020 Edition.
Visit RedMapper.com

THE MARS ATLAS

Topography & Places

N

0 150 300 600 Kilometers

0 150 300 600 Miles

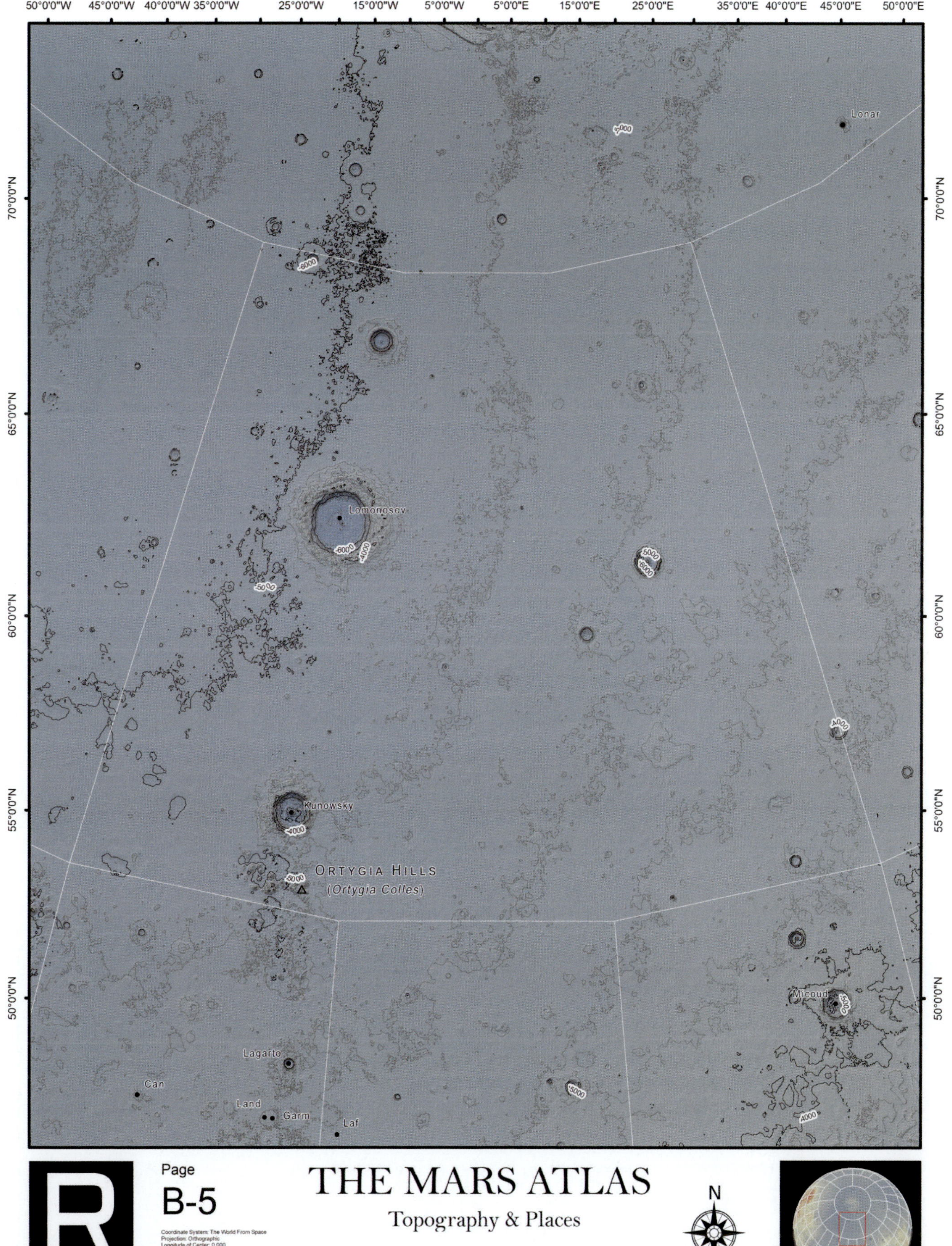

R
REDMAPPER

Page
B-5

Coordinate System: The World From Space
Projection: Orthographic
Longitude of Center: 0.000
Latitude of Center: 63.000
Units: Meter

The Mars Atlas — 2020 Edition.
Visit RedMapper.com

THE MARS ATLAS

Topography & Places

N

0 150 300 600 Kilometers

0 150 300 600 Miles

10°0'0"W
5°0'0"W
0°0'0"
5°0'0"E
15°0'0"E
25°0'0"E
35°0'0"E
45°0'0"E
55°0'0"E
65°0'0"E
75°0'0"E
80°0'0"E
85°0'0"E
90°0'0"E
70°0'0"N
65°0'0"N
60°0'0"N
55°0'0"N
50°0'0"N
Lonar
Lyot
-4000
R
REDMAPPER
Page
B-6
Coordinate System: The World From Space
Projection: Orthographic
Longitude of Center: 40.000
Latitude of Center: 63.000
Units: Meter
The Mars Atlas — 2020 Edition.
Visit RedMapper.com
© 2019-2020 RedMapper - All Rights Reserved
THE MARS ATLAS
Topography & Places
N
0 150 300 600 Kilometers
0 150 300 600 Miles

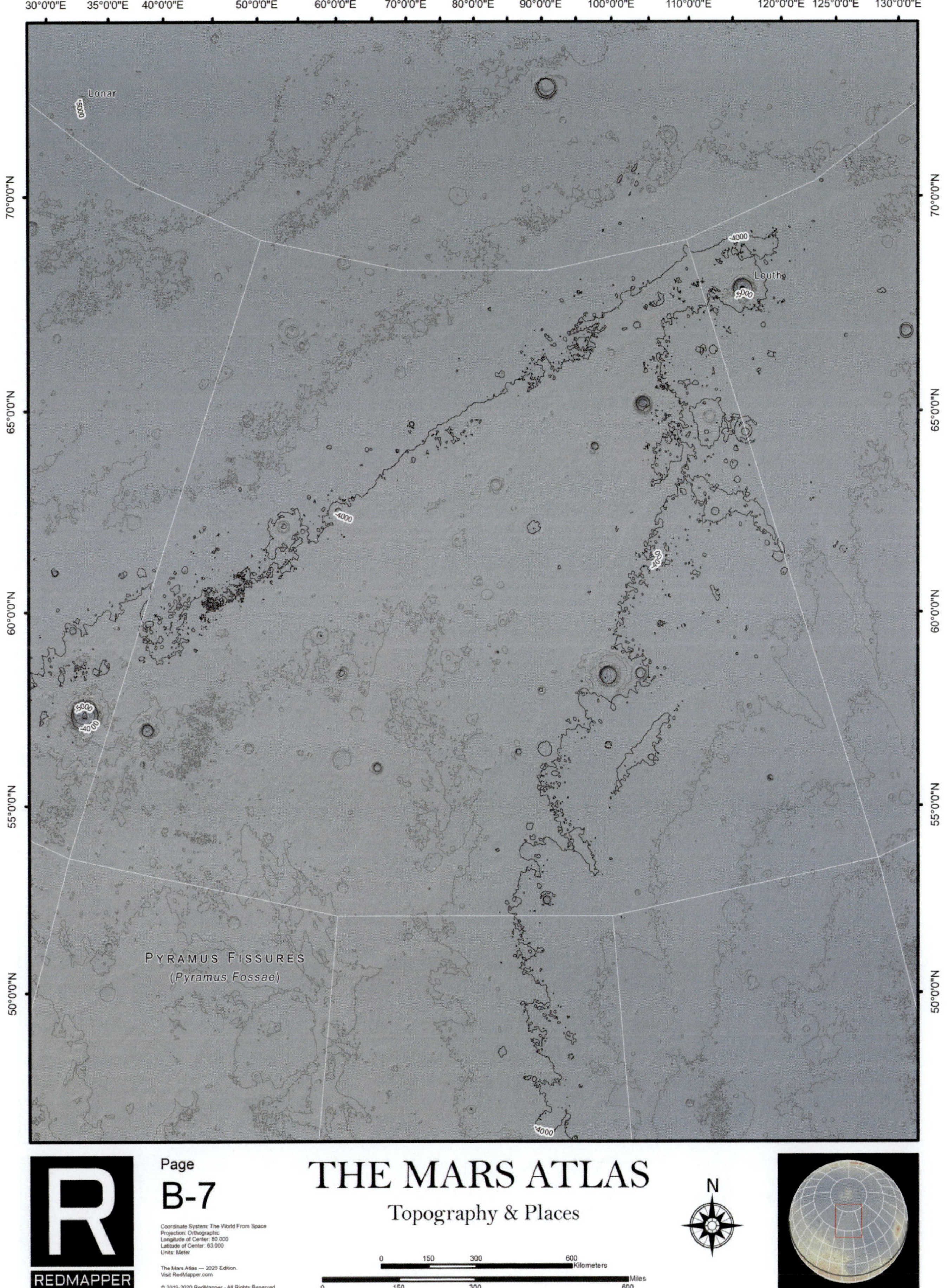
30°0'0"E
35°0'0"E
40°0'0"E
50°0'0"E
60°0'0"E
70°0'0"E
80°0'0"E
90°0'0"E
100°0'0"E
110°0'0"E
120°0'0"E
125°0'0"E
130°0'0"E
70°0'0"N
65°0'0"N
60°0'0"N
55°0'0"N
50°0'0"N
Lonar
Louth
-4000
-5000
PYRAMUS FISSURES
(Pyramus Fossae)
R
REDMAPPER
Page
B-7
Coordinate System: The World From Space
Projection: Orthographic
Longitude of Center: 80.000
Latitude of Center: 63.000
Units: Meter
The Mars Atlas — 2020 Edition.
Visit RedMapper.com
© 2019-2020 RedMapper - All Rights Reserved
THE MARS ATLAS
Topography & Places
0 150 300 600 Kilometers
0 150 300 600 Miles
N

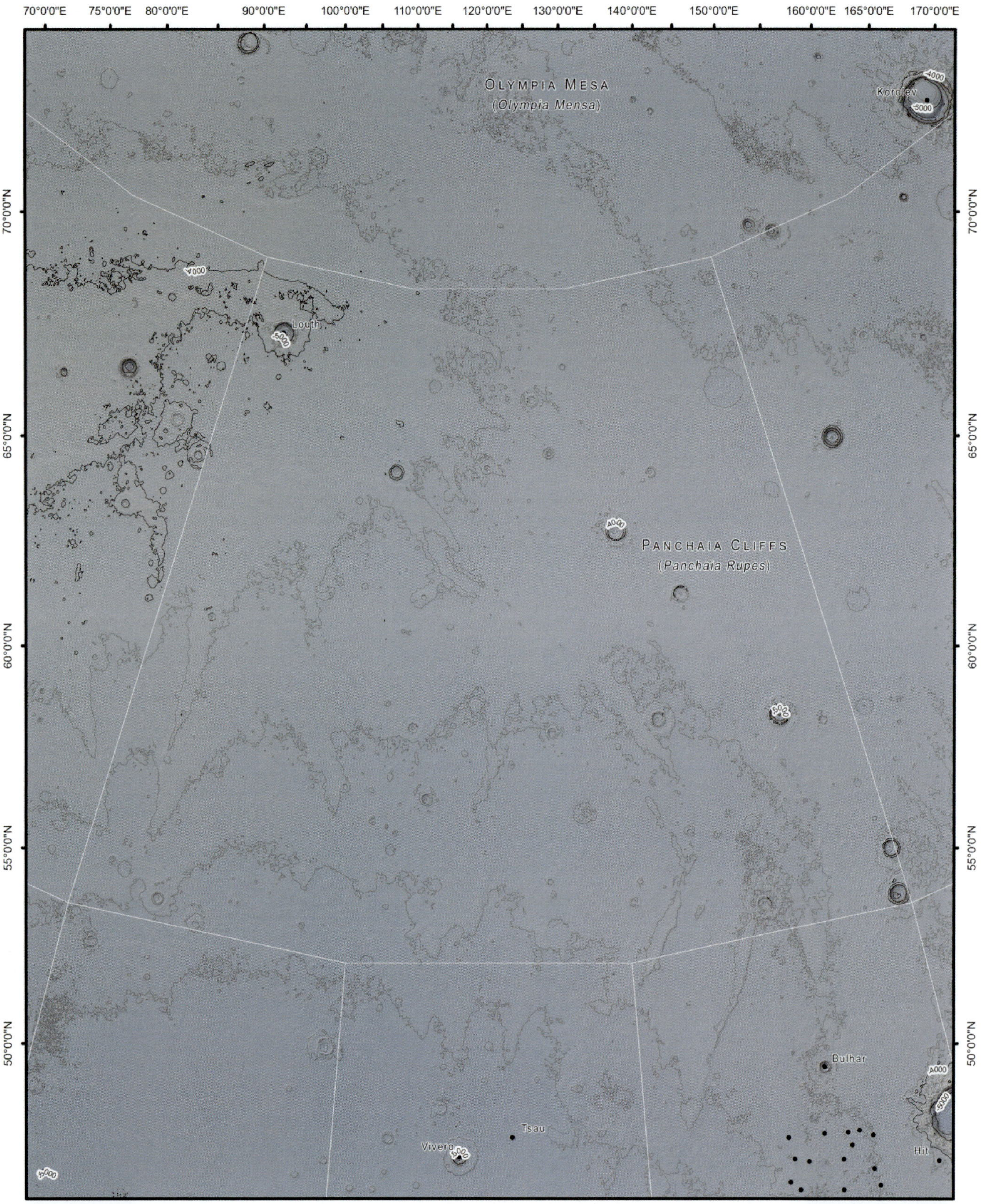

70°0'0"E 75°0'0"E 80°0'0"E 90°0'0"E 100°0'0"E 110°0'0"E 120°0'0"E 130°0'0"E 140°0'0"E 150°0'0"E 160°0'0"E 165°0'0"E 170°0'0"E

70°0'0"N
65°0'0"N
60°0'0"N
55°0'0"N
50°0'0"N

OLYMPIA MESA
(*Olympia Mensa*)

Korolev
-4000
-5000

-4000
Louth
-5000

-4000
PANCHAIA CLIFFS
(*Panchaia Rupes*)

-5000

Bulhar
-4000
-5000
Tsau
Vivero
-5000
Hit
-5000

Page
B-8

Coordinate System: The World From Space
Projection: Orthographic
Longitude of Center: 120.000
Latitude of Center: 63.000
Units: Meter

The Mars Atlas — 2020 Edition.
Visit RedMapper.com

THE MARS ATLAS
Topography & Places

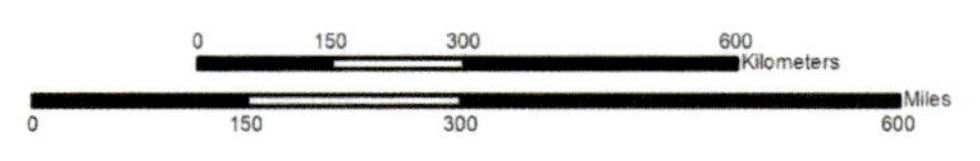

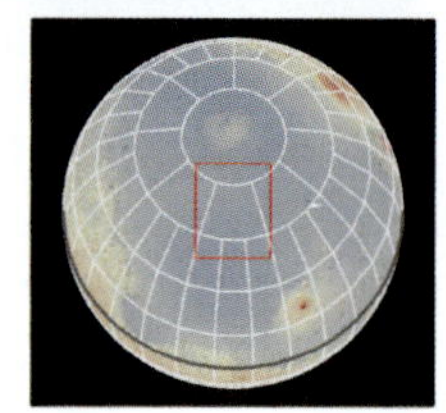

Page
B-9

Coordinate System: The World From Space
Projection: Orthographic
Longitude of Center: 160.000
Latitude of Center: 63.000
Units: Meter

The Mars Atlas — 2020 Edition.
Visit RedMapper.com

THE MARS ATLAS

Topography & Places

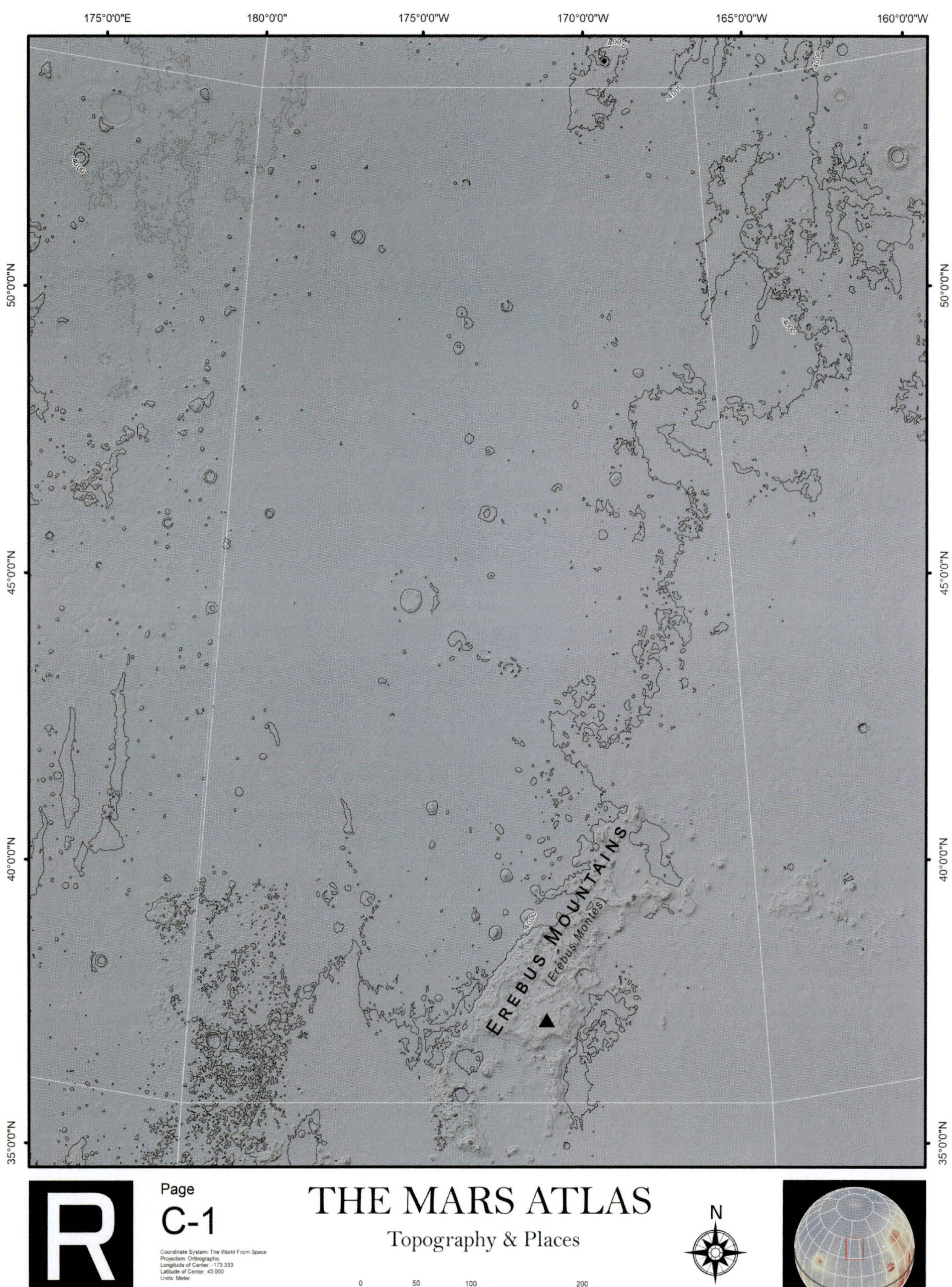

175°0'0"E
180°0'0"
175°0'0"W
170°0'0"W
165°0'0"W
160°0'0"W
50°0'0"N
45°0'0"N
40°0'0"N
35°0'0"N
-4000
EREBUS MOUNTAINS
(Erebus Montes)
R
REDMAPPER
Page
C-1
Coordinate System: The World From Space
Projection: Orthographic
Longitude of Center: -173.333
Latitude of Center: 45.000
Units: Meter
The Mars Atlas — 2020 Edition.
Visit RedMapper.com
© 2019-2020 RedMapper - All Rights Reserved
THE MARS ATLAS
Topography & Places
N
0
50
100
200
Kilometers
Miles

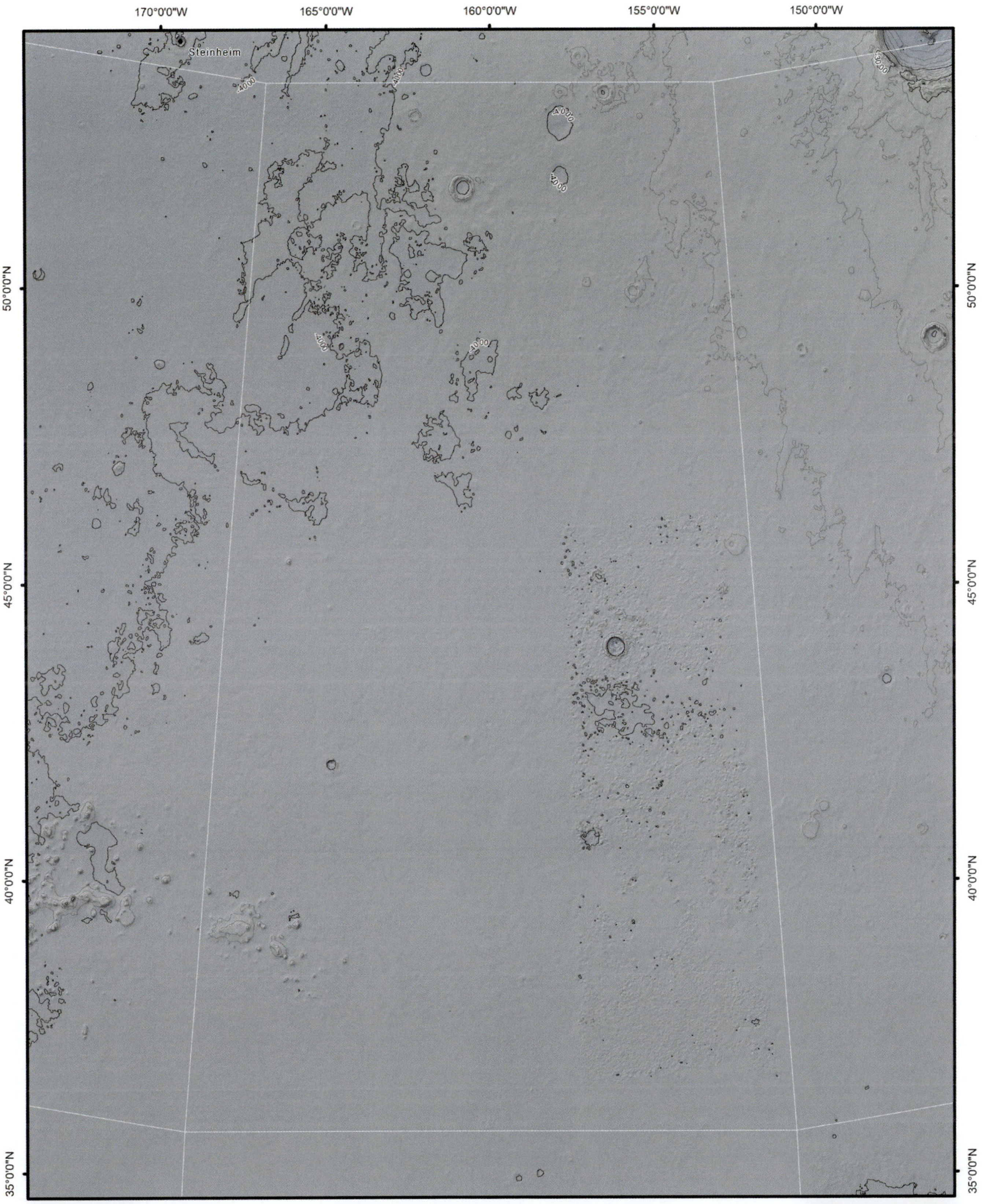

170°0'0"W 165°0'0"W 160°0'0"W 155°0'0"W 150°0'0"W

Steinheim

50°0'0"N

45°0'0"N

40°0'0"N

35°0'0"N

Page

C-2

Coordinate System: The World From Space
Projection: Orthographic
Longitude of Center: -160.000
Latitude of Center: 45.000
Units: Meter

The Mars Atlas — 2020 Edition.
Visit RedMapper.com

THE MARS ATLAS

Topography & Places

0 50 100 200 Kilometers

0 50 100 200 Miles

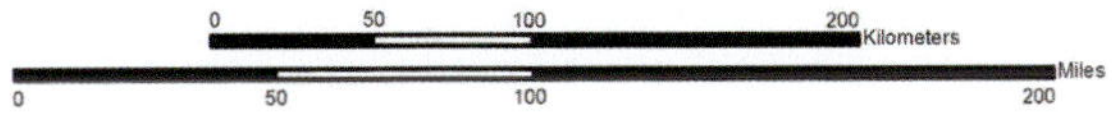

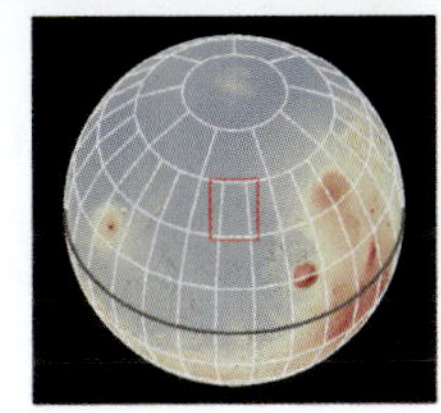

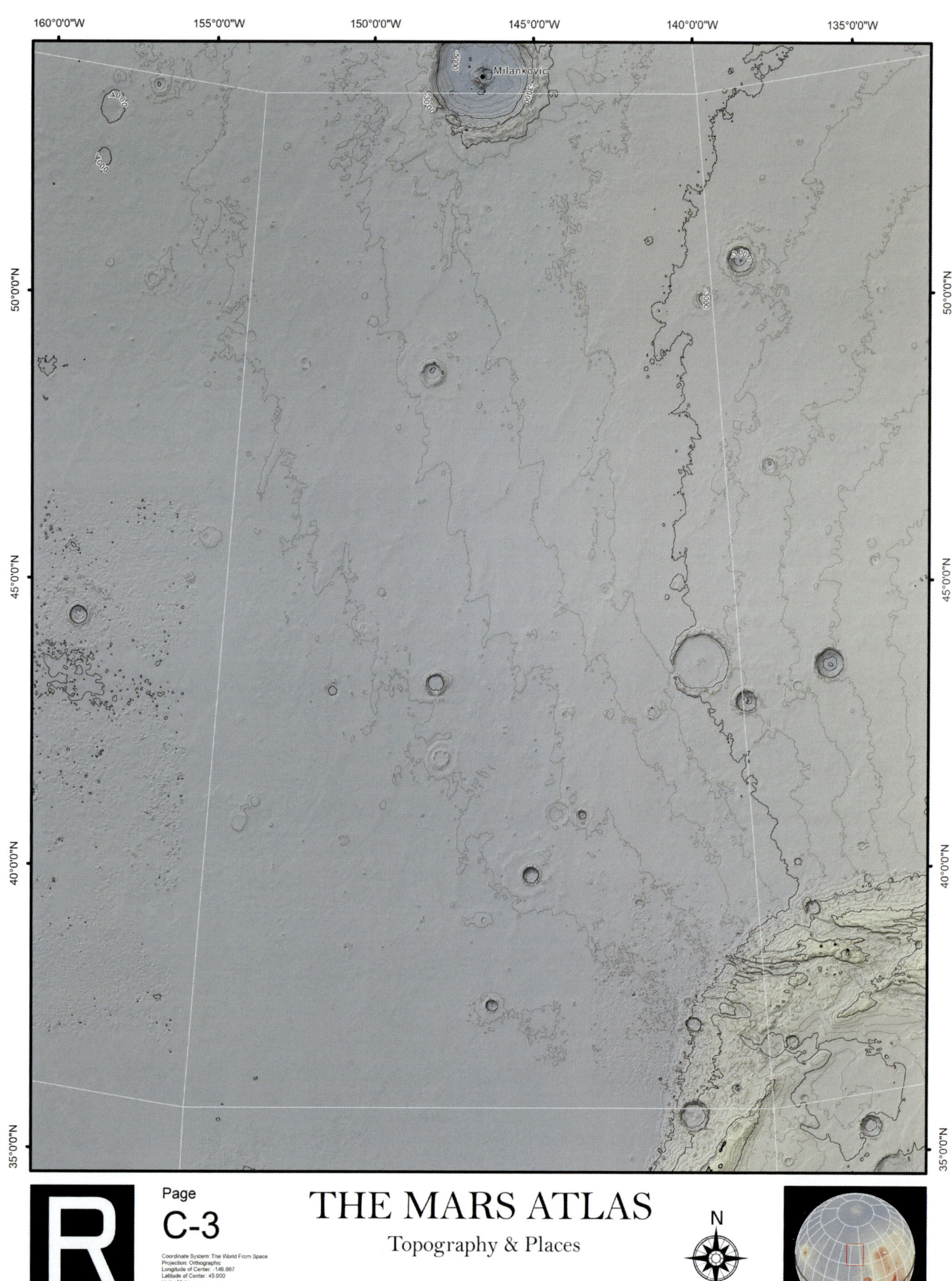

REDMAPPER

Page

C-3

Coordinate System: The World From Space
Projection: Orthographic
Longitude of Center: -146.667
Latitude of Center: 45.000
Units: Meter

The Mars Atlas — 2020 Edition
Visit RedMapper.com

THE MARS ATLAS

Topography & Places

N

0 50 100 200 Kilometers

0 50 100 200 Miles

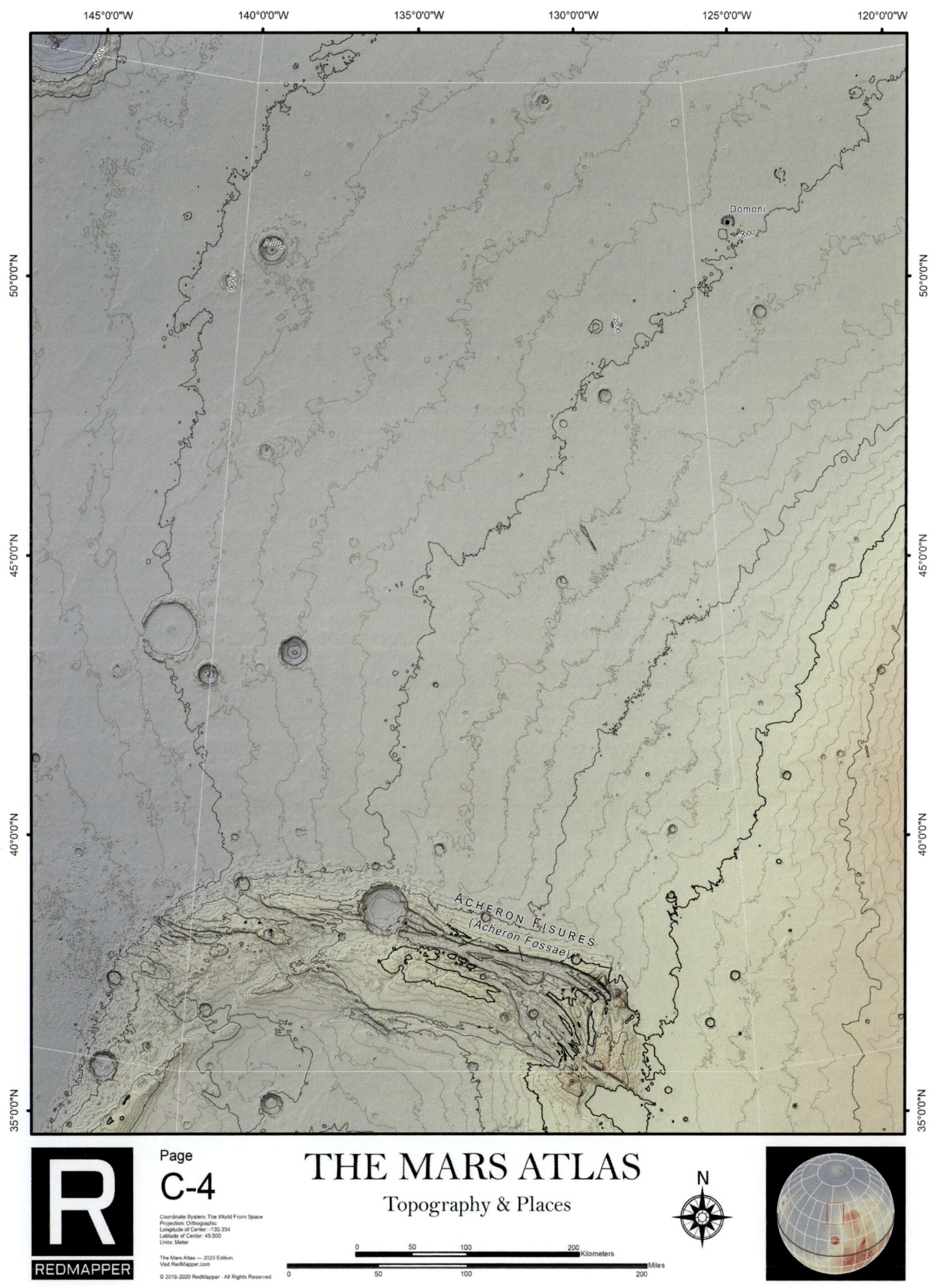
145°0'0"W
140°0'0"W
135°0'0"W
130°0'0"W
125°0'0"W
120°0'0"W
50°0'0"N
45°0'0"N
40°0'0"N
35°0'0"N
Domoni
ACHERON FISURES
(Acheron Fossae)
R
REDMAPPER
Page
C-4
Coordinate System: The World From Space
Projection: Orthographic
Longitude of Center: -133.334
Latitude of Center: 45.000
Units: Meter
The Mars Atlas — 2020 Edition.
Visit RedMapper.com
© 2019-2020 RedMapper - All Rights Reserved
THE MARS ATLAS
Topography & Places
N
0
50
100
200
Kilometers
Miles

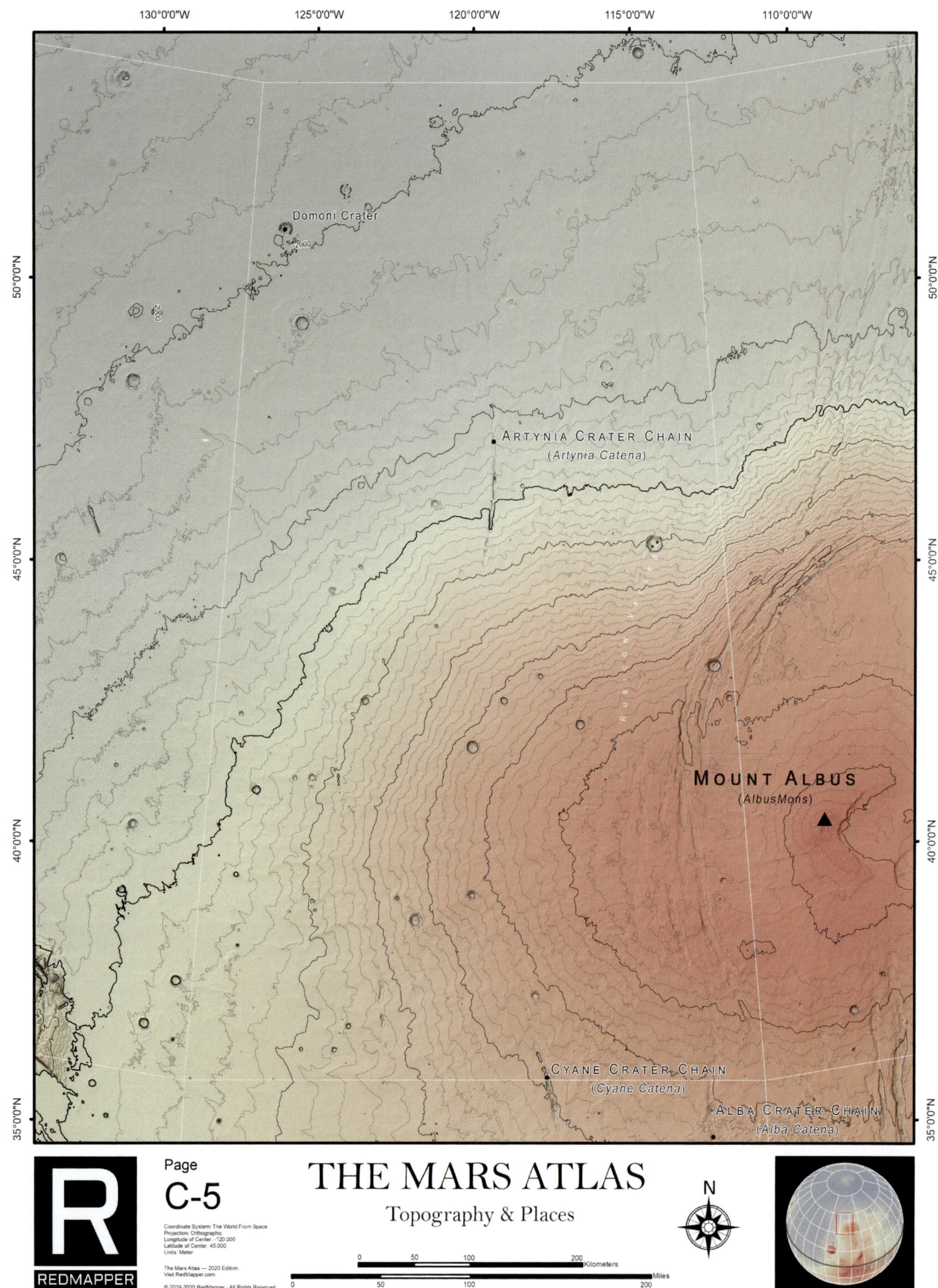
130°0'0"W
125°0'0"W
120°0'0"W
115°0'0"W
110°0'0"W
50°0'0"N
45°0'0"N
40°0'0"N
35°0'0"N
Domoni Crater
-2000
Artynia Crater Chain
(Artynia Catena)
Rubicon Valles
Mount Albus
(AlbusMons)
Cyane Crater Chain
(Cyane Catena)
Alba Crater Chain
(Alba Catena)
R
REDMAPPER
Page
C-5
Coordinate System: The World From Space
Projection: Orthographic
Longitude of Center: -120.000
Latitude of Center: 45.000
Units: Meter
The Mars Atlas — 2020 Edition.
Visit RedMapper.com
© 2019-2020 RedMapper - All Rights Reserved
THE MARS ATLAS
Topography & Places
N
0 50 100 200 Kilometers
0 50 100 200 Miles

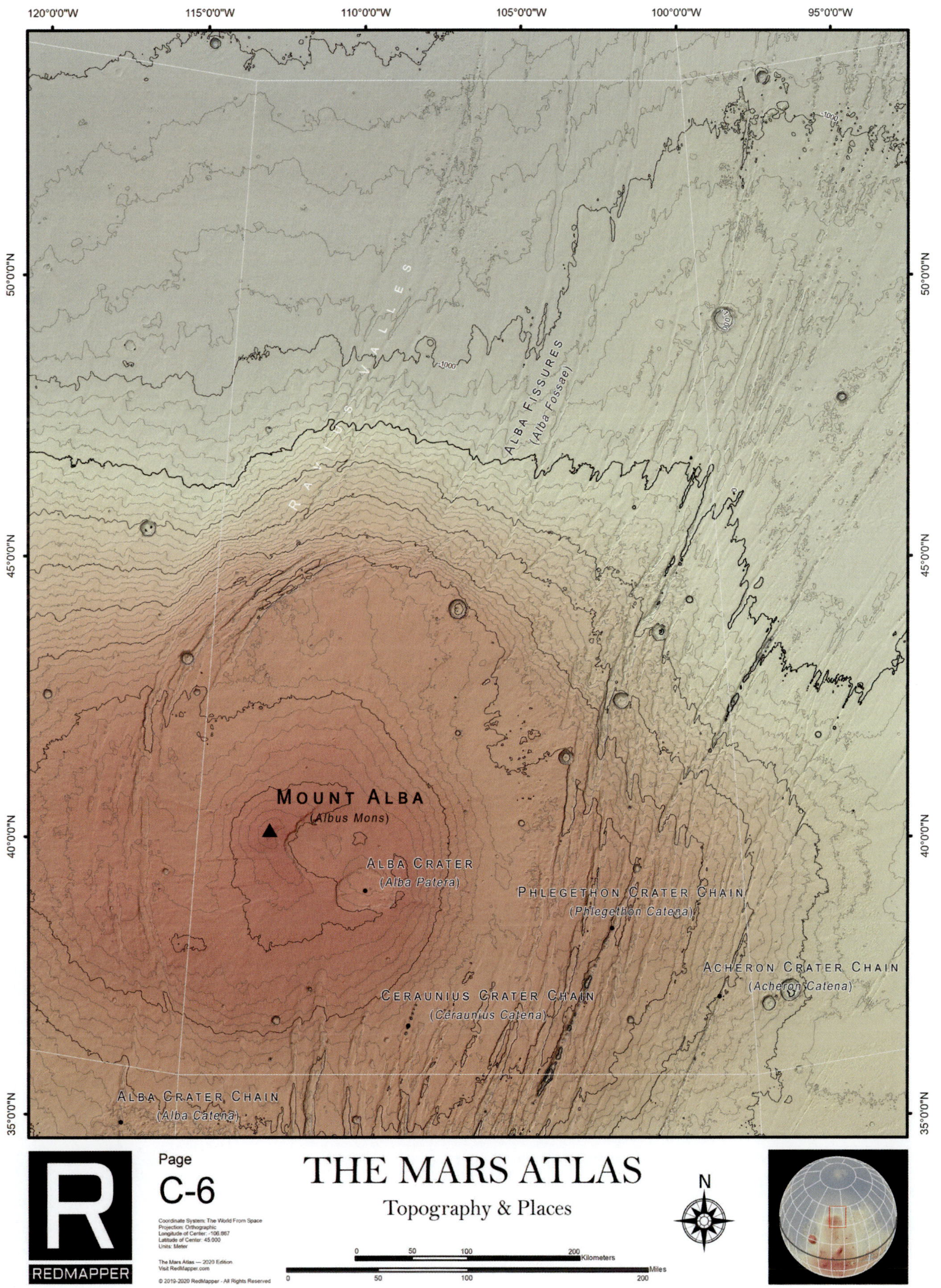

120°0'0"W
115°0'0"W
110°0'0"W
105°0'0"W
100°0'0"W
95°0'0"W
50°0'0"N
45°0'0"N
40°0'0"N
35°0'0"N
-1000
RAVIUS VALLES
ALBA FISSURES
(Alba Fossae)
MOUNT ALBA
(Albus Mons)
ALBA CRATER
(Alba Patera)
PHLEGETHON CRATER CHAIN
(Phlegethon Catena)
ACHERON CRATER CHAIN
(Acheron Catena)
CERAUNIUS CRATER CHAIN
(Ceraunius Catena)
ALBA CRATER CHAIN
(Alba Catena)
R
REDMAPPER
Page
C-6
Coordinate System: The World From Space
Projection: Orthographic
Longitude of Center: -106.867
Latitude of Center: 45.000
Units: Meter
The Mars Atlas — 2020 Edition
Visit RedMapper.com
© 2019-2020 RedMapper - All Rights Reserved
THE MARS ATLAS
Topography & Places
N
0
50
100
200
Kilometers
Miles

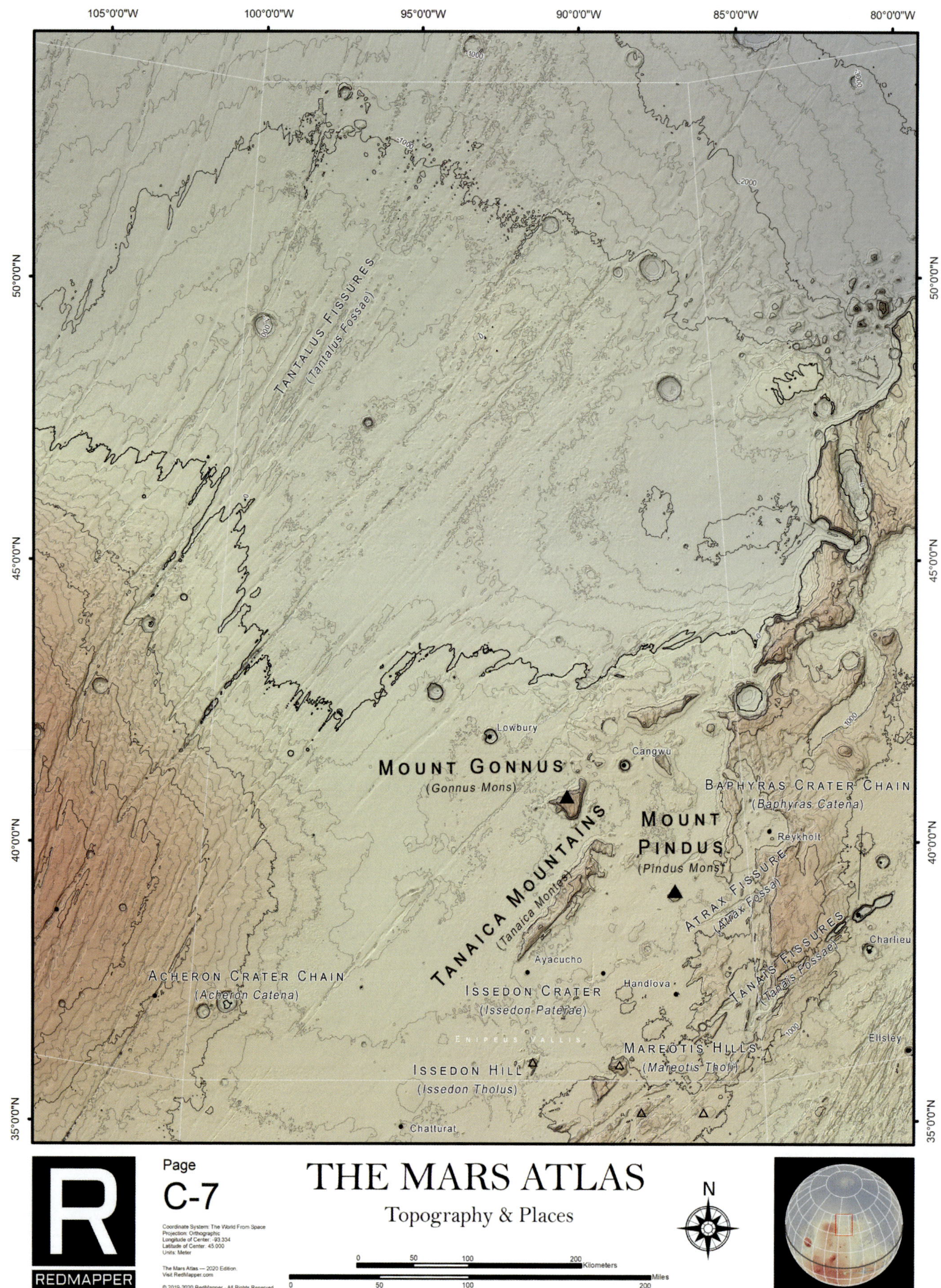

Page

C-7

Coordinate System: The World From Space
Projection: Orthographic
Longitude of Center: -93.334
Latitude of Center: 45.000
Units: Meter

The Mars Atlas — 2020 Edition.
Visit RedMapper.com

THE MARS ATLAS

Topography & Places

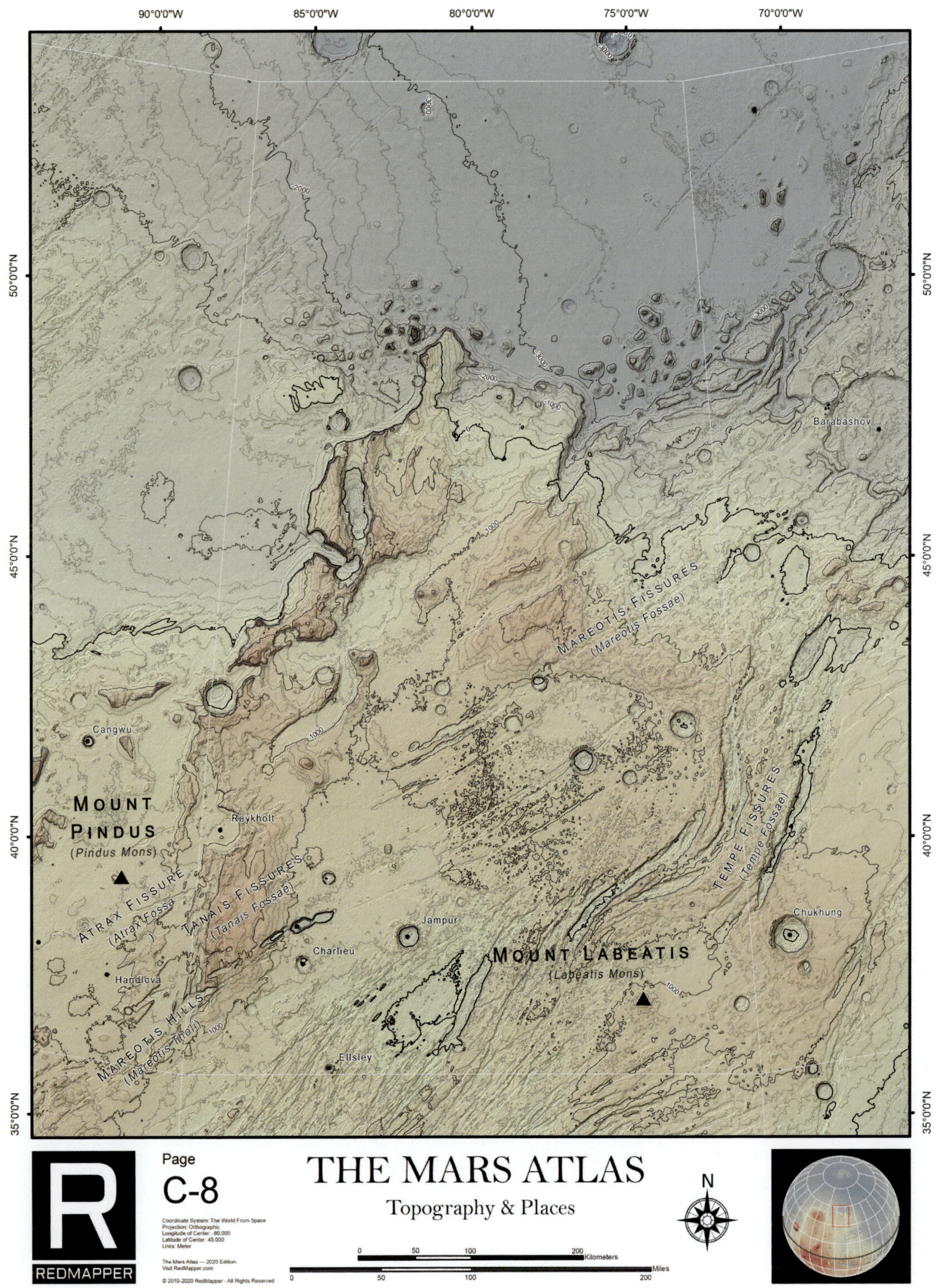
90°0'0"W
85°0'0"W
80°0'0"W
75°0'0"W
70°0'0"W
50°0'0"N
45°0'0"N
40°0'0"N
35°0'0"N
-2000
-3000
1000
Barabashov
MAREOTIS FISSURES
(Mareotis Fossae)
Cangwu
MOUNT PINDUS
(Pindus Mons)
Reykholt
ATRAX FISSURE
(Atrax Fossa)
TANAIS FISSURES
(Tanais Fossae)
Jampur
Charlieu
Handlova
MAREOTIS HILLS
(Mareotis Tholi)
Ellsley
MOUNT LABEATIS
(Labeatis Mons)
TEMPE FISSURES
(Tempe Fossae)
Chukhung
REDMAPPER
Page
C-8
Coordinate System: The World From Space
Projection: Orthographic
Longitude of Center: -80.000
Latitude of Center: 45.000
Units: Meter
The Mars Atlas — 2020 Edition.
Visit RedMapper.com
© 2019-2020 RedMapper - All Rights Reserved
THE MARS ATLAS
Topography & Places
0 50 100 200 Kilometers
0 50 100 200 Miles
N

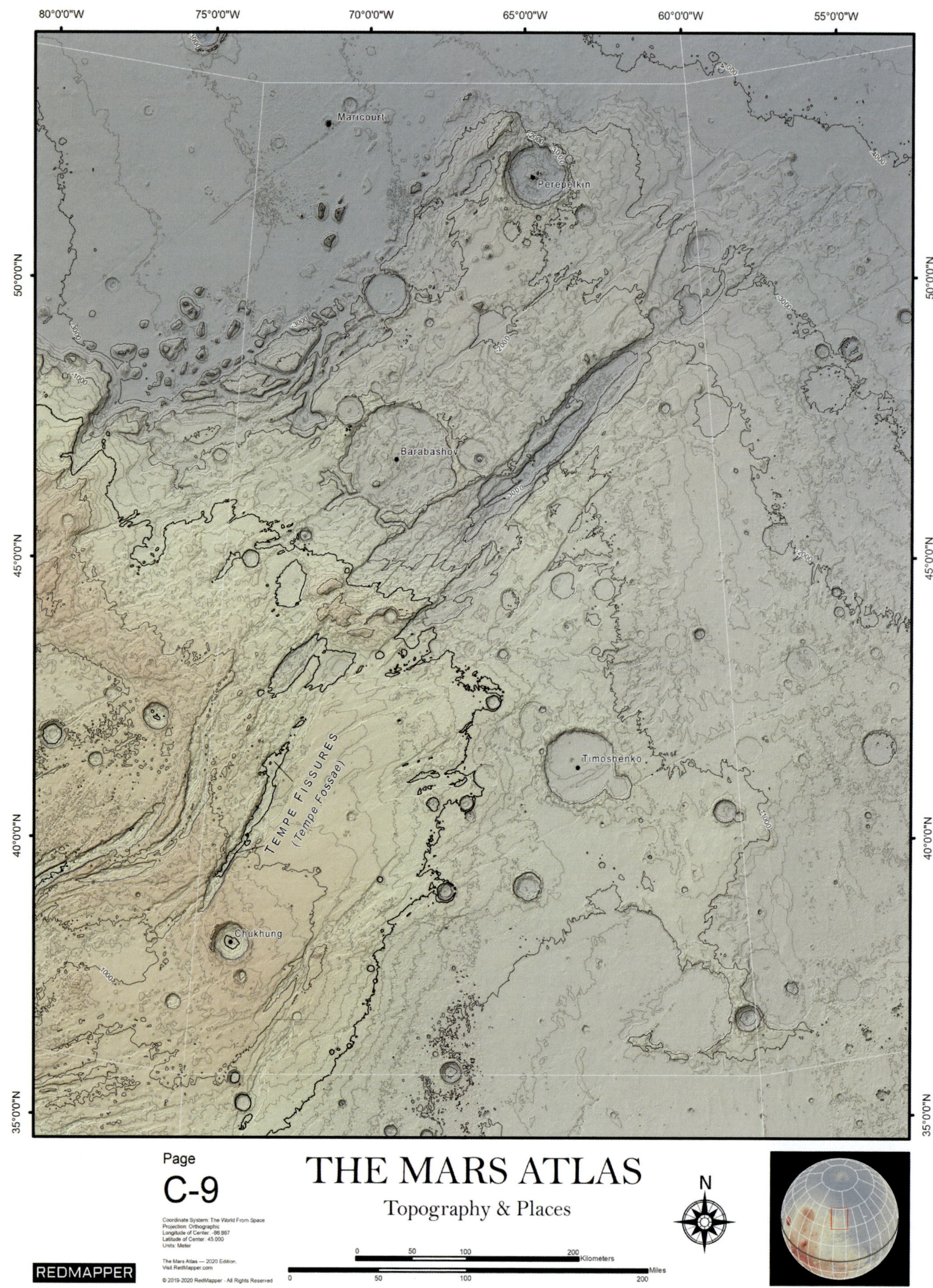

Page

C-9

THE MARS ATLAS

Topography & Places

Coordinate System: The World From Space
Projection: Orthographic
Longitude of Center: -66.667
Latitude of Center: 45.000
Units: Meter

The Mars Atlas — 2020 Edition.
Visit RedMapper.com

REDMAPPER

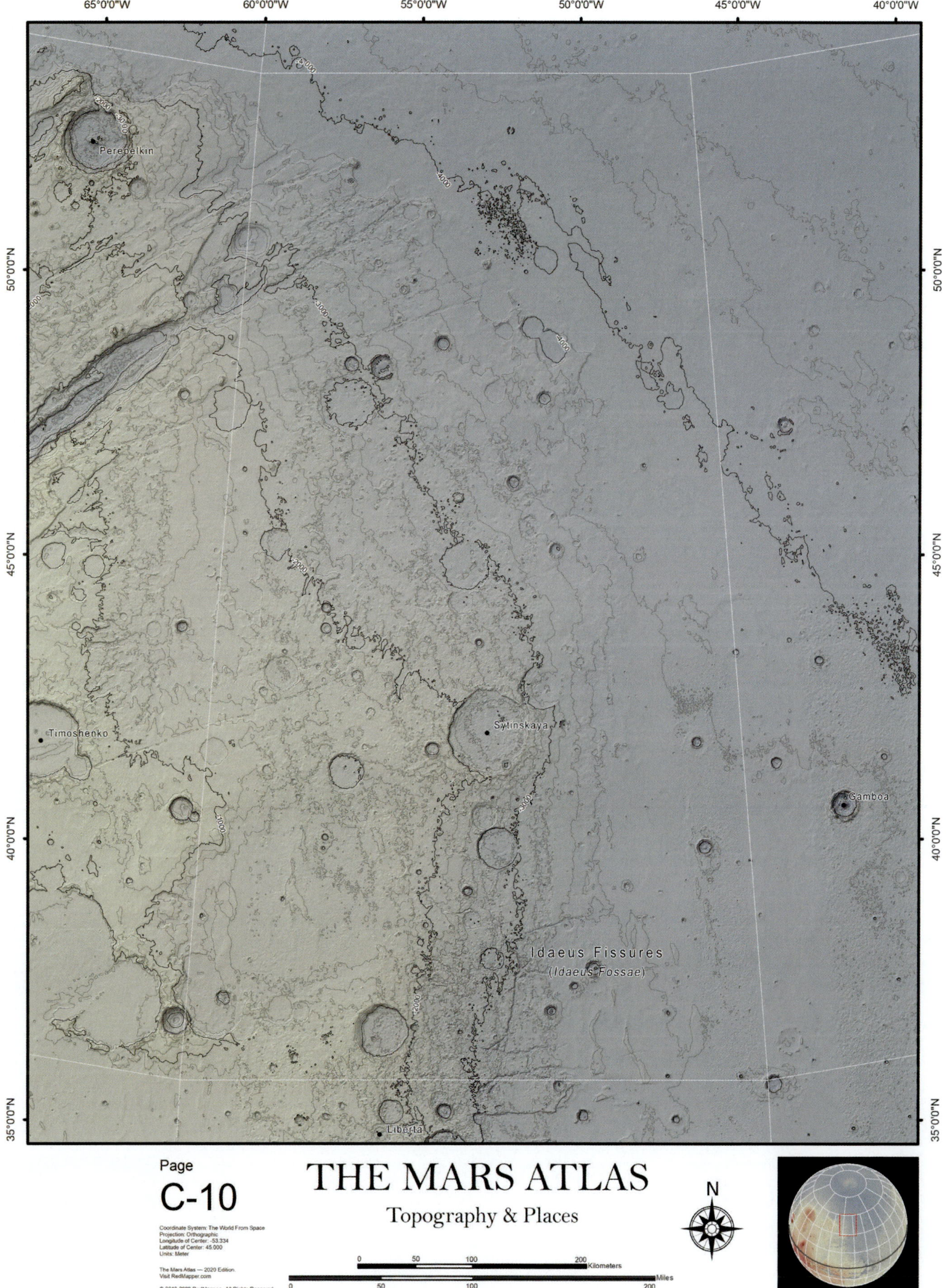

Page
C-10

Coordinate System: The World From Space
Projection: Orthographic
Longitude of Center: -53.334
Latitude of Center: 45.000
Units: Meter

The Mars Atlas — 2020 Edition.
Visit RedMapper.com

THE MARS ATLAS

Topography & Places

N

0 50 100 200 Kilometers

0 50 100 200 Miles

R
REDMAPPER

Page
C-11

Coordinate System: The World From Space
Projection: Orthographic
Longitude of Center: -40.000
Latitude of Center: 45.000
Units: Meter

The Mars Atlas — 2020 Edition.
Visit RedMapper.com

THE MARS ATLAS

Topography & Places

N

0 50 100 200 Kilometers
0 50 100 200 Miles

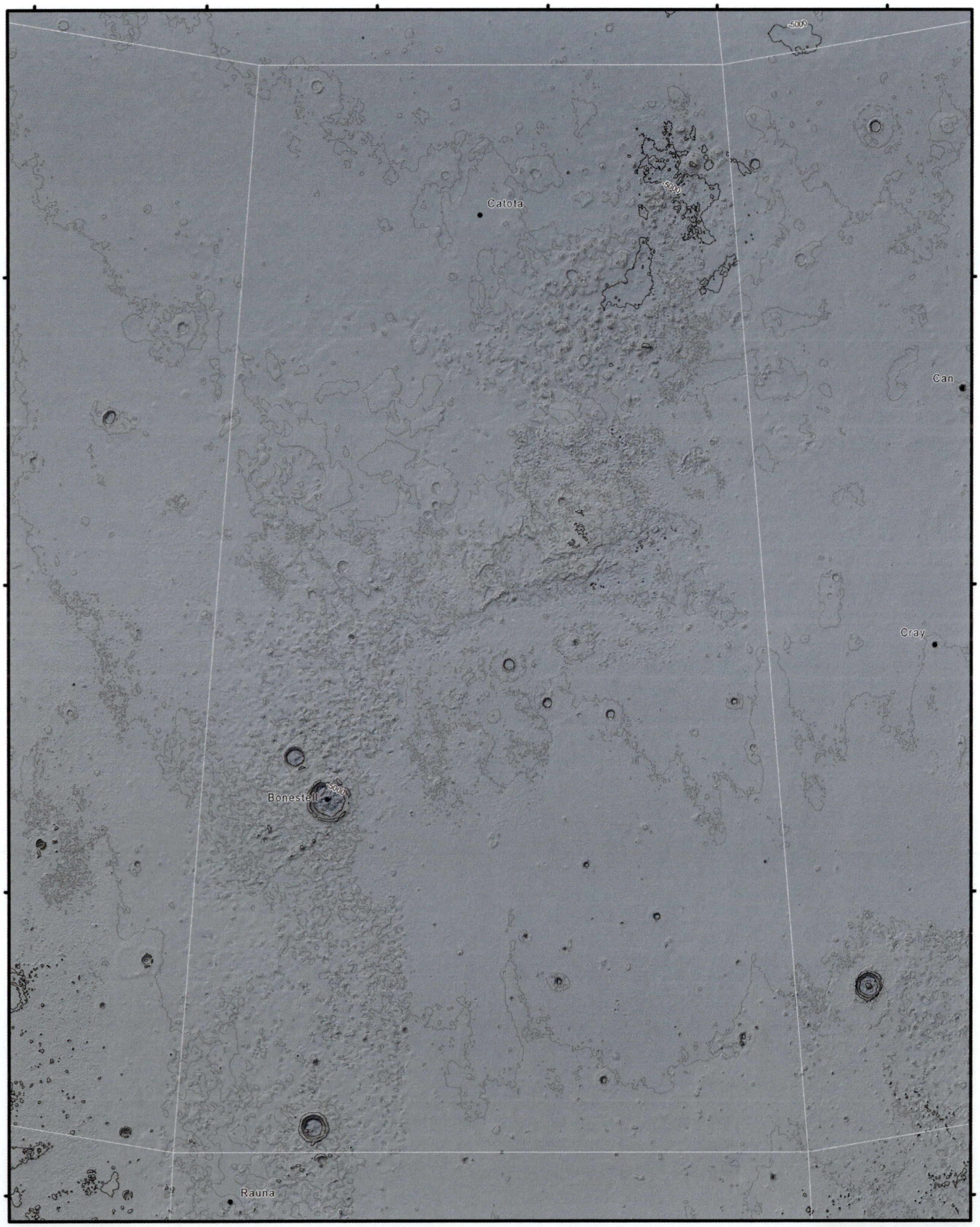

40°0'0"W 35°0'0"W 30°0'0"W 25°0'0"W 20°0'0"W 15°0'0"W

50°0'0"N

45°0'0"N

40°0'0"N

35°0'0"N

-5000

Catota

Can

Cray

Bonestell

Rauna

Page

C-12

Coordinate System: The World From Space
Projection: Orthographic
Longitude of Center: -26.667
Latitude of Center: 45.000
Units: Meter

The Mars Atlas — 2020 Edition.
Visit RedMapper.com

THE MARS ATLAS

Topography & Places

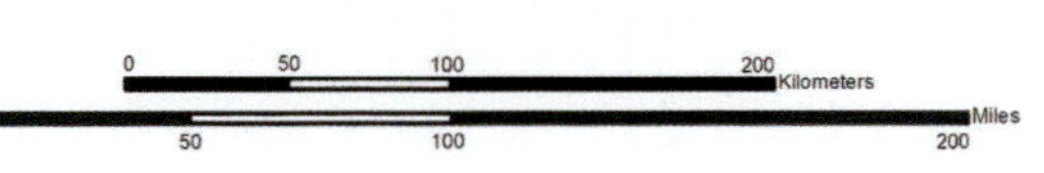

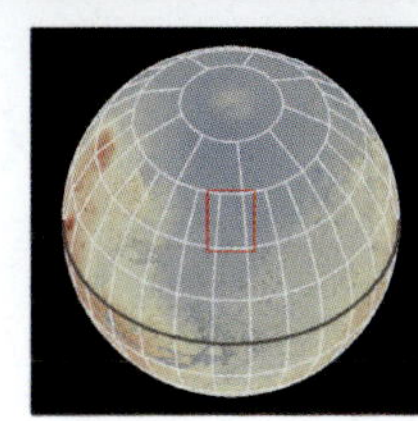

REDMAPPER

Page

C-13

Coordinate System: The World From Space
Projection: Orthographic
Longitude of Center: -13.334
Latitude of Center: 45.000
Units: Meter

The Mars Atlas — 2020 Edition.
Visit RedMapper.com

THE MARS ATLAS

Topography & Places

0 50 100 200 Kilometers

0 50 100 200 Miles

N

10°0'0"W
5°0'0"W
0°0'0"
5°0'0"E
10°0'0"E
50°0'0"N
45°0'0"N
40°0'0"N
35°0'0"N
-5000
-4000
Lagarto
Land
Garm
Laf
Gol
Lota
Wer
Esk
Hope
Davies
Eagle
Freedom
Mohawk
Yakima
Tem
Eil
Semeykin
OKAVANGO
Jen
Ant
Bamberg
OXUS CRATER
(Oxus Patera)
Gwash
Lutsk
Gaan
Chom
Cruz
Dein
ISMENIA CRATER
(Ismenia Patera)
EUPHRATES CRATER
(Euphrates Patera)
SILOE CRATER
(Siloe Patera)
R
REDMAPPER
Page
C-14
THE MARS ATLAS
Topography & Places
N
Coordinate System: The World From Space
Projection: Orthographic
Longitude of Center: 0.000
Latitude of Center: 45.000
Units: Meter
The Mars Atlas — 2020 Edition.
Visit RedMapper.com
© 2019-2020 RedMapper - All Rights Reserved
0 50 100 200 Kilometers
0 50 100 200 Miles

R
REDMAPPER

Page
C-15

THE MARS ATLAS

Topography & Places

Coordinate System: The World From Space
Projection: Orthographic
Longitude of Center: 13.333
Latitude of Center: 45.000
Units: Meter

The Mars Atlas — 2020 Edition.
Visit RedMapper.com

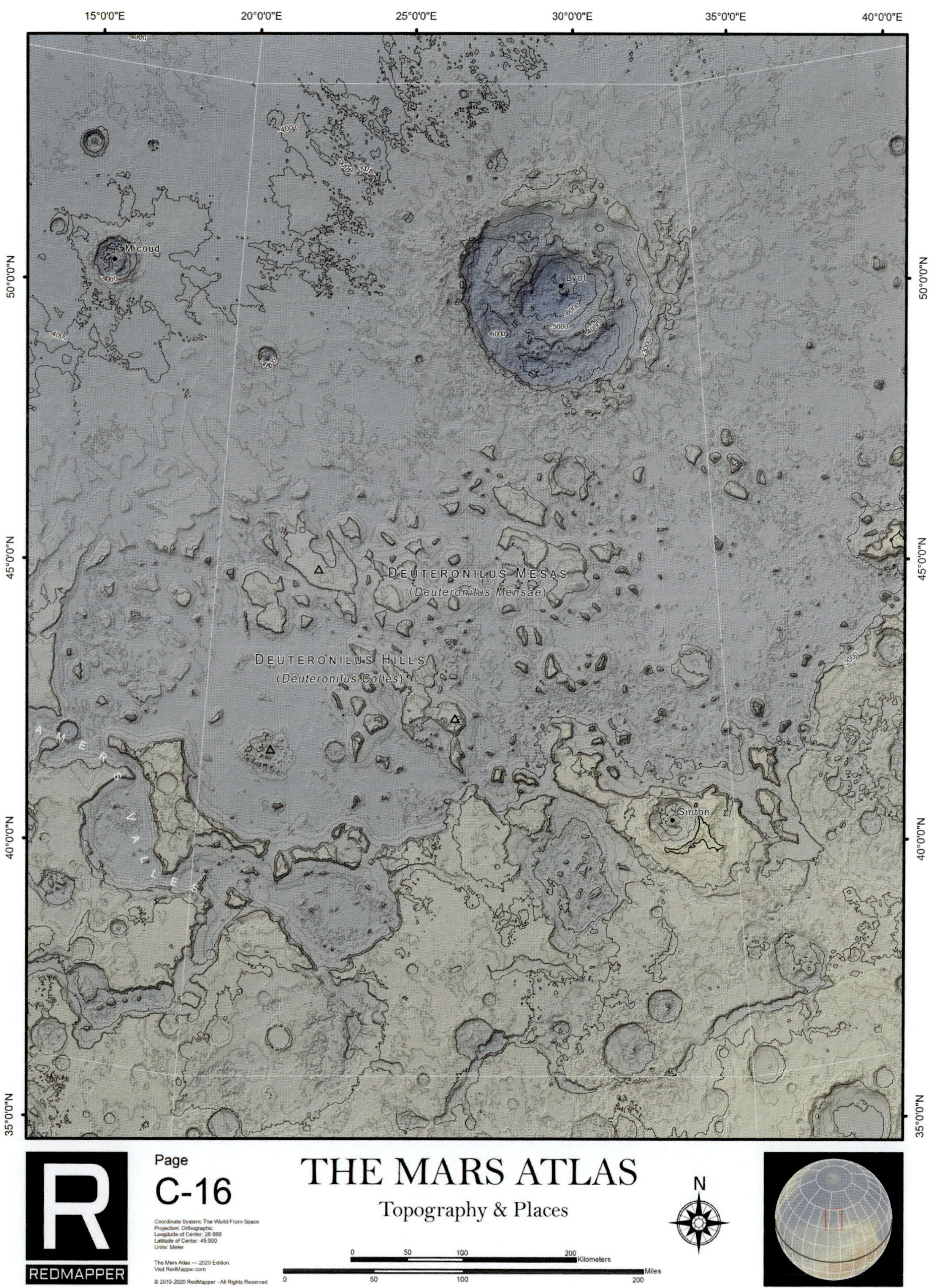

Page
C-16

Coordinate System: The World From Space
Projection: Orthographic
Longitude of Center: 26.666
Latitude of Center: 45.000
Units: Meter

The Mars Atlas — 2020 Edition.
Visit RedMapper.com

THE MARS ATLAS

Topography & Places

REDMAPPER

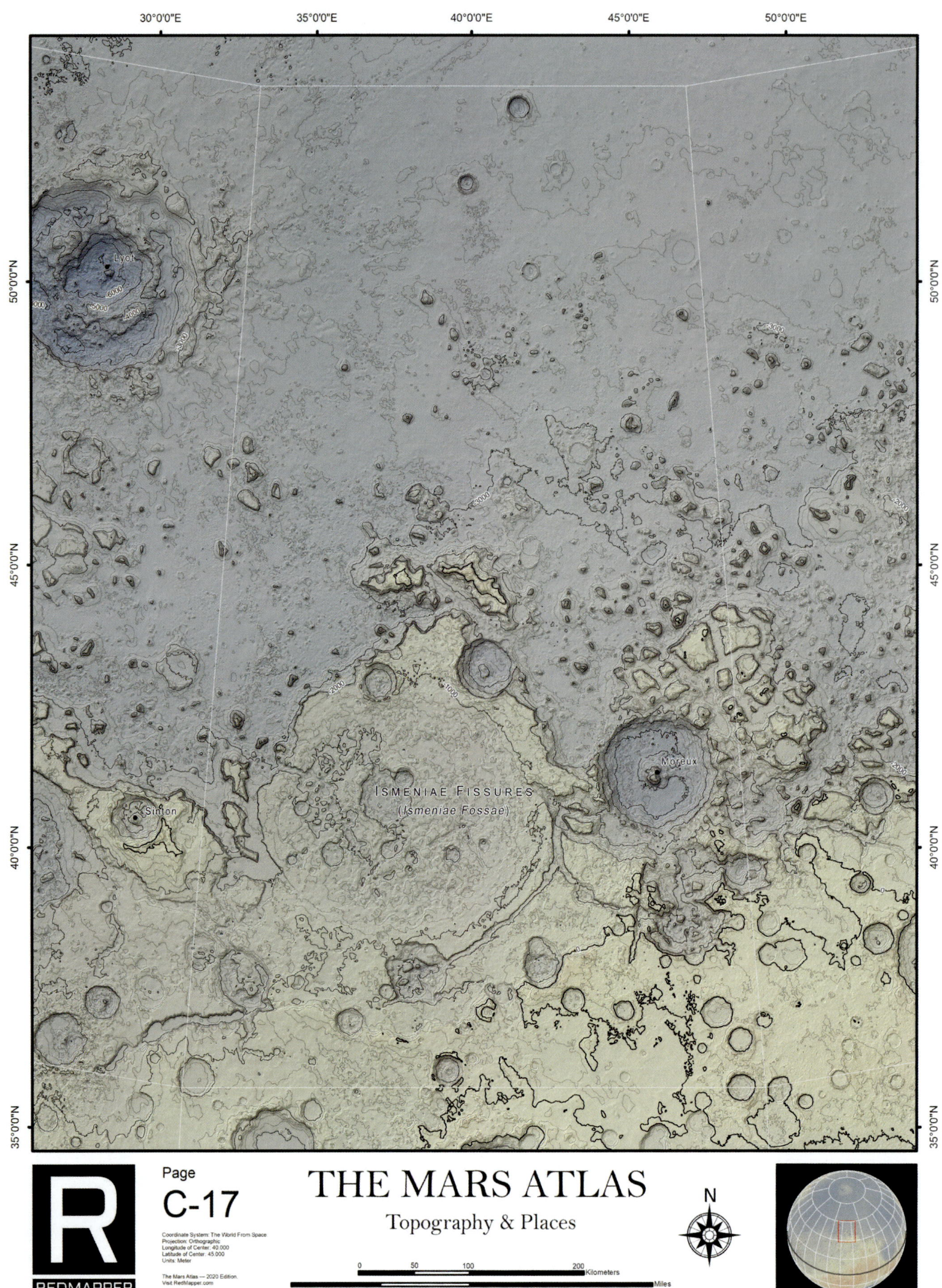

Page

C-17

Coordinate System: The World From Space
Projection: Orthographic
Longitude of Center: 40.000
Latitude of Center: 45.000
Units: Meter

The Mars Atlas — 2020 Edition.
Visit RedMapper.com

THE MARS ATLAS

Topography & Places

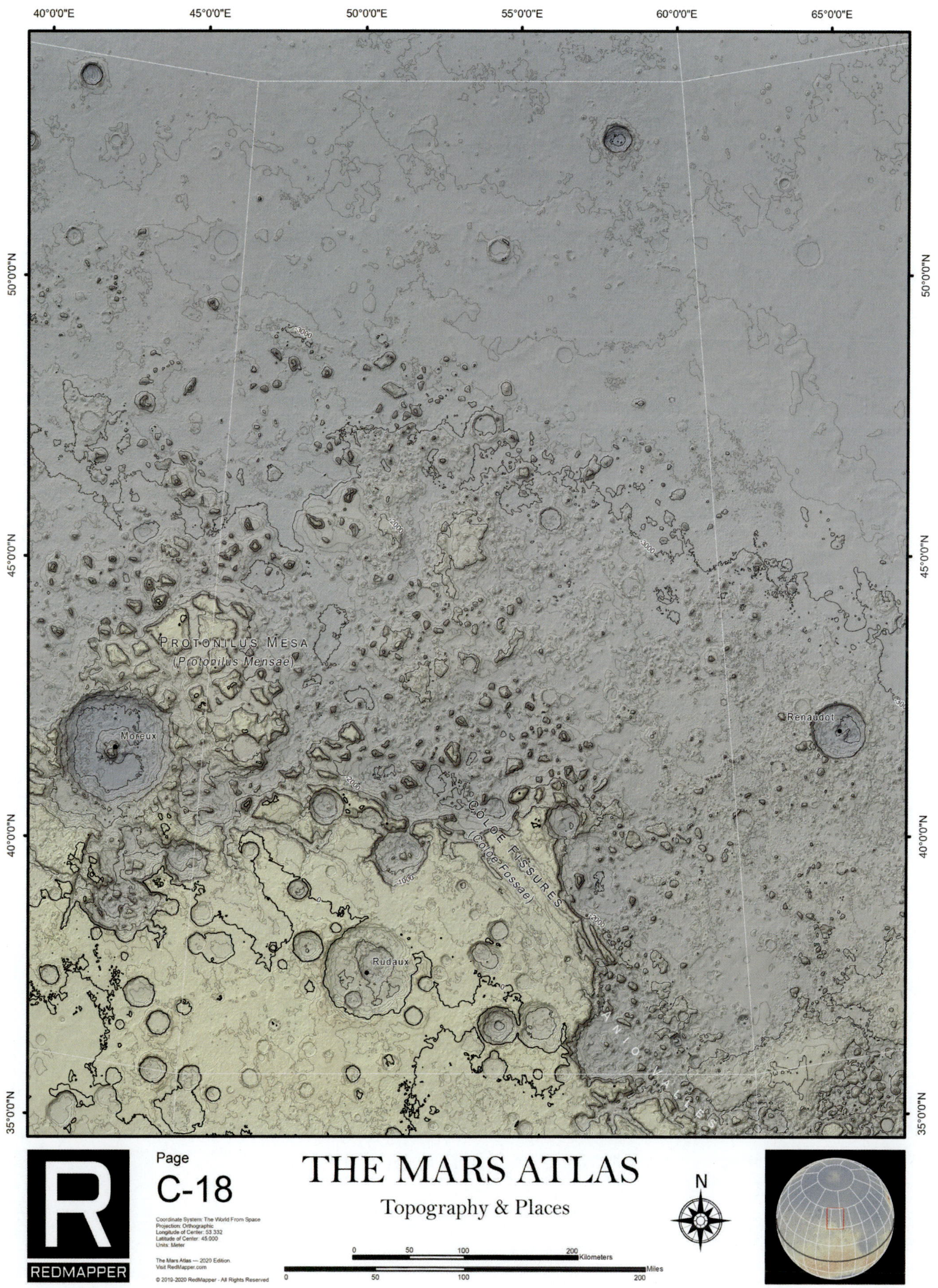
40°0'0"E
45°0'0"E
50°0'0"E
55°0'0"E
60°0'0"E
65°0'0"E
50°0'0"N
45°0'0"N
40°0'0"N
35°0'0"N
-3000
-2000
-1000
PROTONILUS MESA
(Protonilus Mensae)
Moreux
Renaudot
COLOE FISSURES
(Coloe Fossae)
Rudaux
ANIO VALLES
R
REDMAPPER
Page
C-18
Coordinate System: The World From Space
Projection: Orthographic
Longitude of Center: 53.332
Latitude of Center: 45.000
Units: Meter
The Mars Atlas — 2020 Edition.
Visit RedMapper.com
© 2019-2020 RedMapper - All Rights Reserved
THE MARS ATLAS
Topography & Places
N
0 50 100 200 Kilometers
0 50 100 200 Miles

55°0'0"E
60°0'0"E
65°0'0"E
70°0'0"E
75°0'0"E
80°0'0"E
50°0'0"N
45°0'0"N
40°0'0"N
35°0'0"N
COLOE FISSURES
(Coloe Fossae)
-3000
Renaudot
-4000
NILI HILLS
(Colles Nili)
-2000
Page
C-19
Coordinate System: The World From Space
Projection: Orthographic
Longitude of Center: 66.665
Latitude of Center: 45.000
Units: Meter
The Mars Atlas --- 2020 Edition.
Visit RedMapper.com
© 2019-2020 RedMapper - All Rights Reserved
THE MARS ATLAS
Topography & Places
N
0 50 100 200 Kilometers
0 50 100 200 Miles
REDMAPPER

R
REDMAPPER

Page
C-20

Coordinate System: The World From Space
Projection: Orthographic
Longitude of Center: 80.000
Latitude of Center: 45.000
Units: Meter

The Mars Atlas — 2020 Edition.
Visit RedMapper.com

THE MARS ATLAS

Topography & Places

N

0 50 100 200 Kilometers
0 50 100 200 Miles

R
REDMAPPER

Page
C-21

Coordinate System: The World From Space
Projection: Orthographic
Longitude of Center: 93.332
Latitude of Center: 45.000
Units: Meter

The Mars Atlas — 2020 Edition.
Visit RedMapper.com

THE MARS ATLAS

Topography & Places

N

0 50 100 200 Kilometers

0 50 100 200 Miles

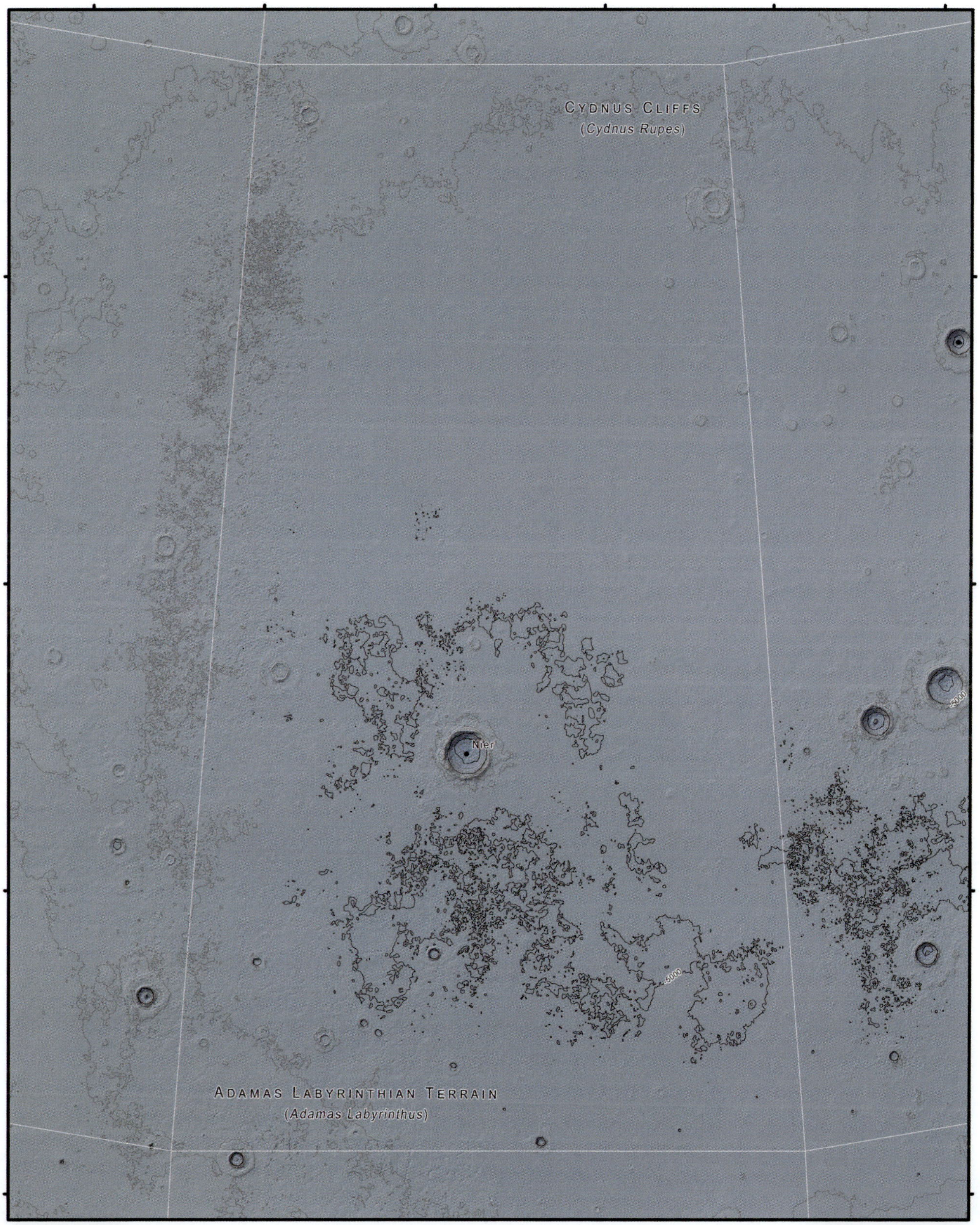

95°0'0"E 100°0'0"E 105°0'0"E 110°0'0"E 115°0'0"E 120°0'0"E

50°0'0"N

45°0'0"N

40°0'0"N

35°0'0"N

CYDNUS CLIFFS
(*Cydnus Rupes*)

Nier

ADAMAS LABYRINTHIAN TERRAIN
(*Adamas Labyrinthus*)

Page
C-22

Coordinate System: The World From Space
Projection: Orthographic
Longitude of Center: 106.665
Latitude of Center: 45.000
Units: Meter

The Mars Atlas — 2020 Edition.
Visit RedMapper.com

THE MARS ATLAS

Topography & Places

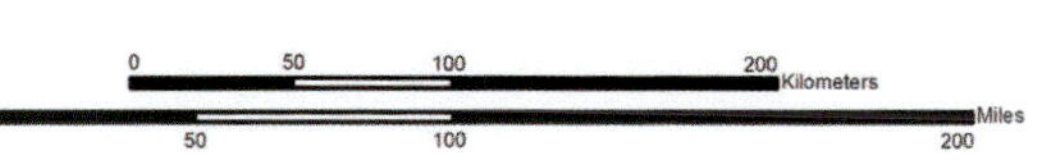

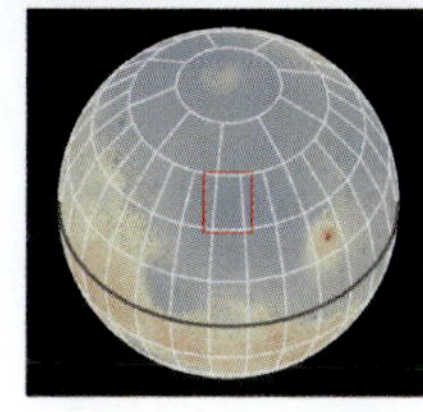

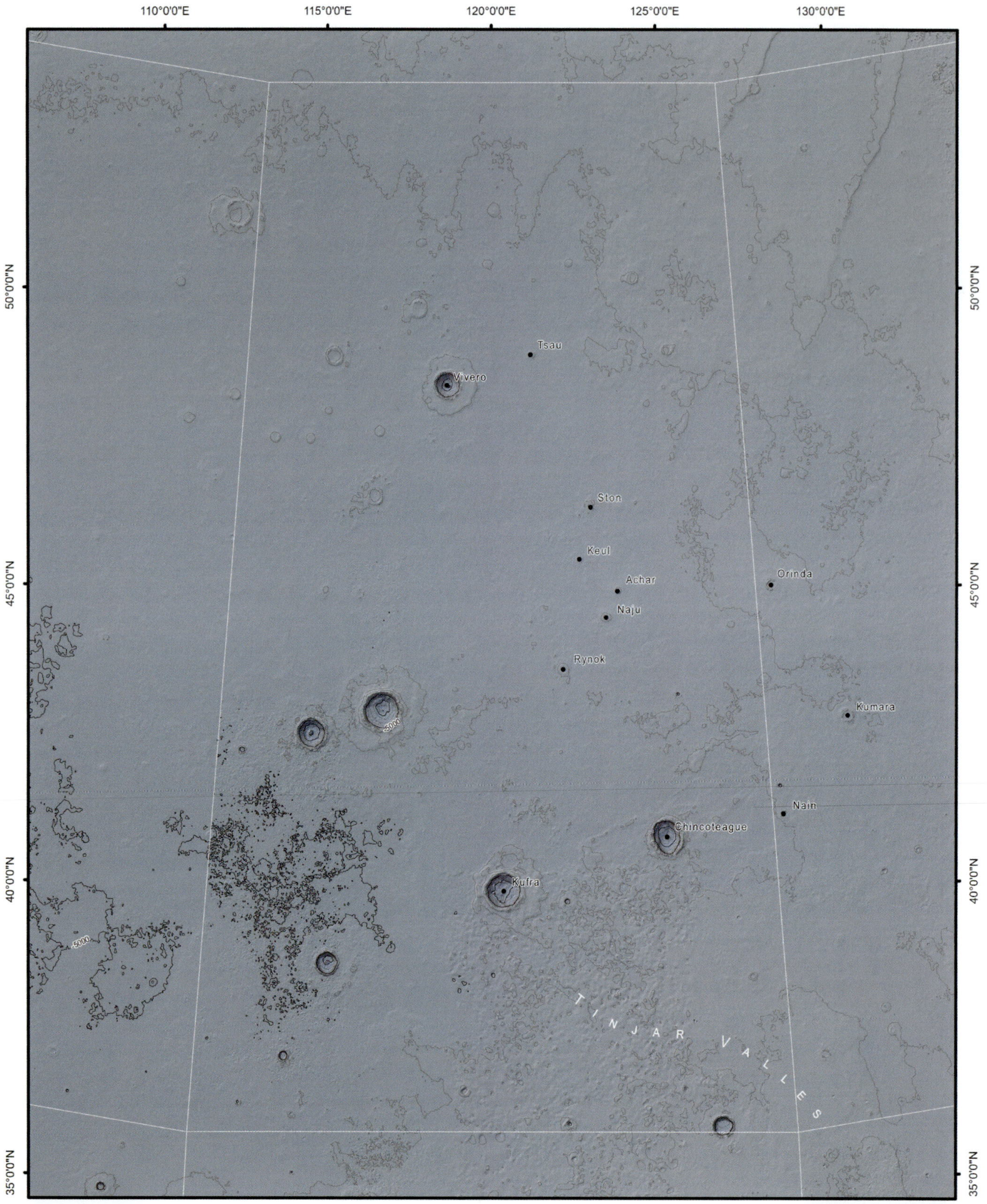

110°0'0"E
115°0'0"E
120°0'0"E
125°0'0"E
130°0'0"E

50°0'0"N
45°0'0"N
40°0'0"N
35°0'0"N

Tsau
Vivero
Ston
Keul
Achar
Orinda
Naju
Rynok
Kumara
-5000
Nain
Chincoteague
Kufra
-5000
T I N J A R V A L L E S

Page

C-23

Coordinate System: The World From Space
Projection: Orthographic
Longitude of Center: 120.000
Latitude of Center: 45.000
Units: Meter

The Mars Atlas — 2020 Edition.
Visit RedMapper.com

THE MARS ATLAS

Topography & Places

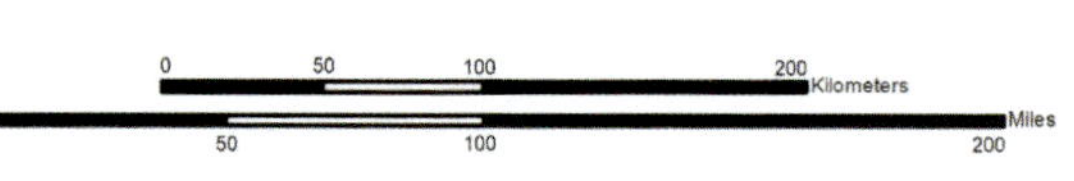

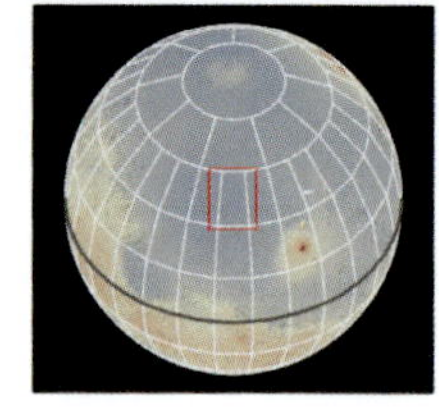

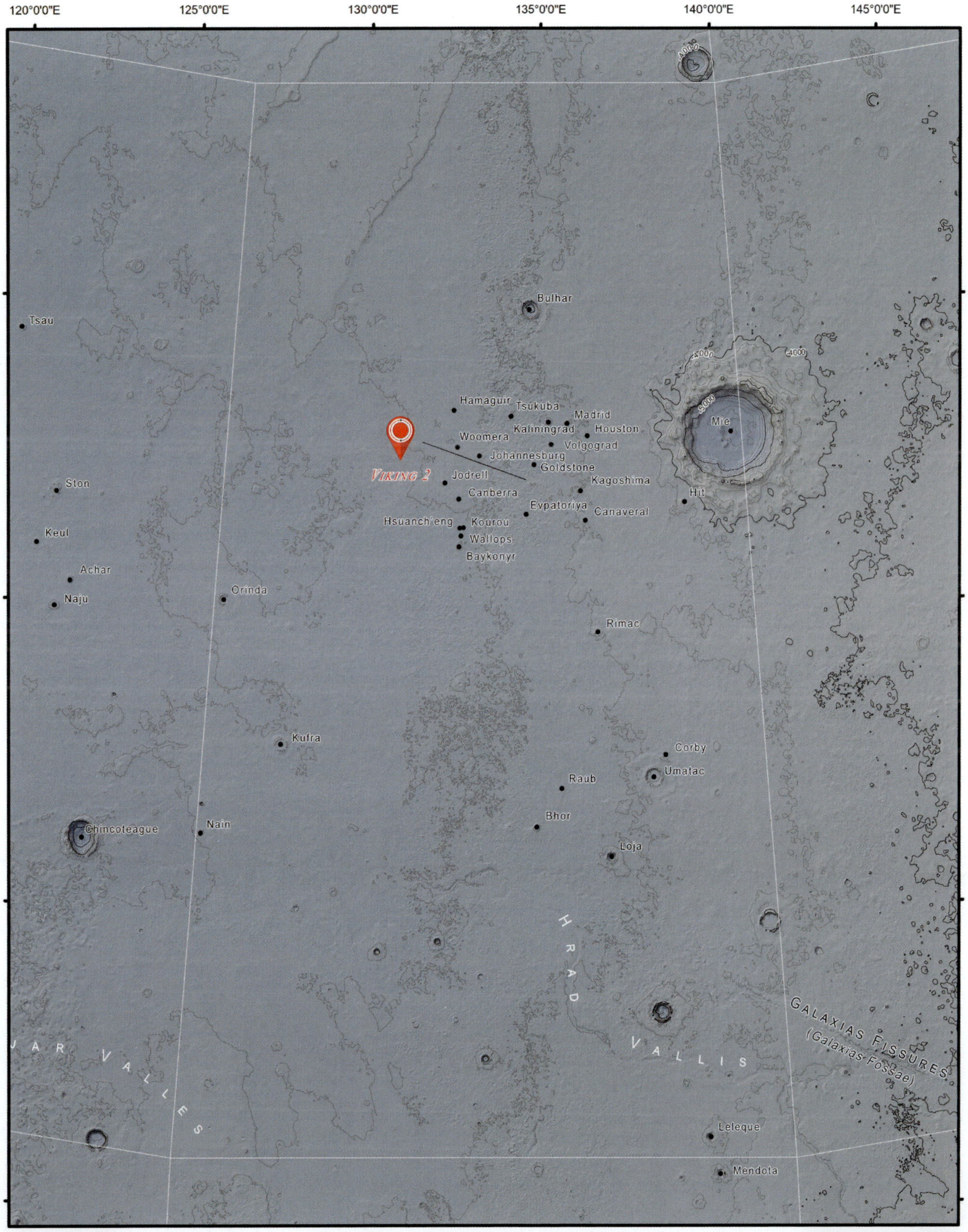

120°0'0"E 125°0'0"E 130°0'0"E 135°0'0"E 140°0'0"E 145°0'0"E

50°0'0"N 45°0'0"N 40°0'0"N 35°0'0"N

Tsau
Bulhar
Hamaguir
Tsukuba
Madrid
Kaliningrad
Houston
Woomera
Volgograd
Johannesburg
Goldstone
VIKING 2
Jodrell
Kagoshima
Mie
Hit
Canberra
Evpatoriya
Canaveral
Hsuanch'eng
Kourou
Wallops
Baykonyr
Ston
Keul
Achar
Naju
Orinda
Rimac
Kufra
Corby
Umatac
Raub
Bhor
Chincoteague
Nain
Loja
HRAD VALLIS
GALAXIAS FISSURES (Galaxias Fossae)
JAR VALLES
Leleque
Mendota

Page
C-24

Coordinate System: The World From Space
Projection: Orthographic
Longitude of Center: 133.332
Latitude of Center: 45.000
Units: Meter

The Mars Atlas — 2020 Edition.
Visit RedMapper.com

THE MARS ATLAS

Topography & Places

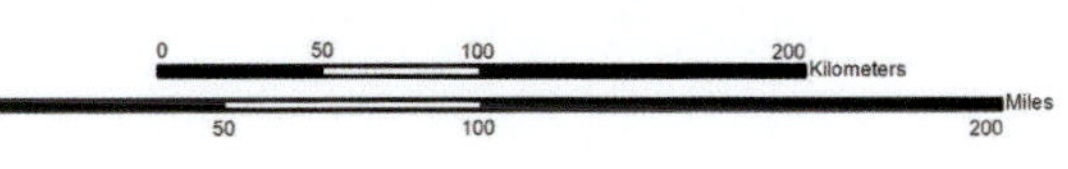

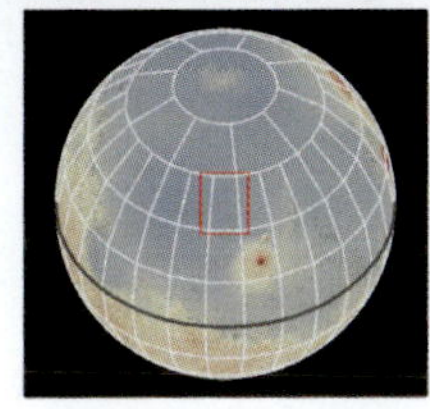

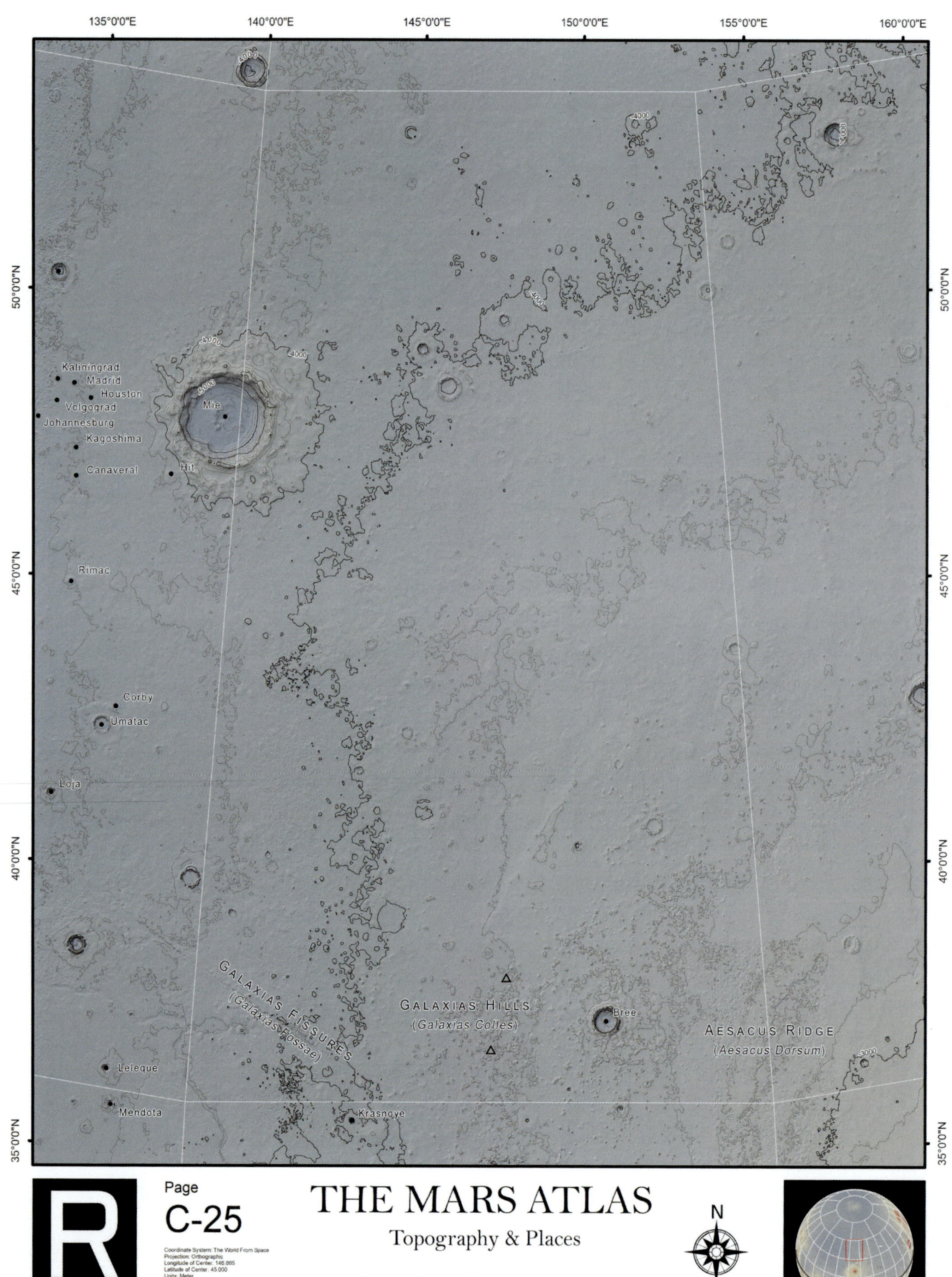

REDMAPPER

Page
C-25

Coordinate System: The World From Space
Projection: Orthographic
Longitude of Center: 146.865
Latitude of Center: 45.000
Units: Meter

The Mars Atlas — 2020 Edition.
Visit RedMapper.com

THE MARS ATLAS
Topography & Places

N

0 50 100 200 Kilometers

0 50 100 200 Miles

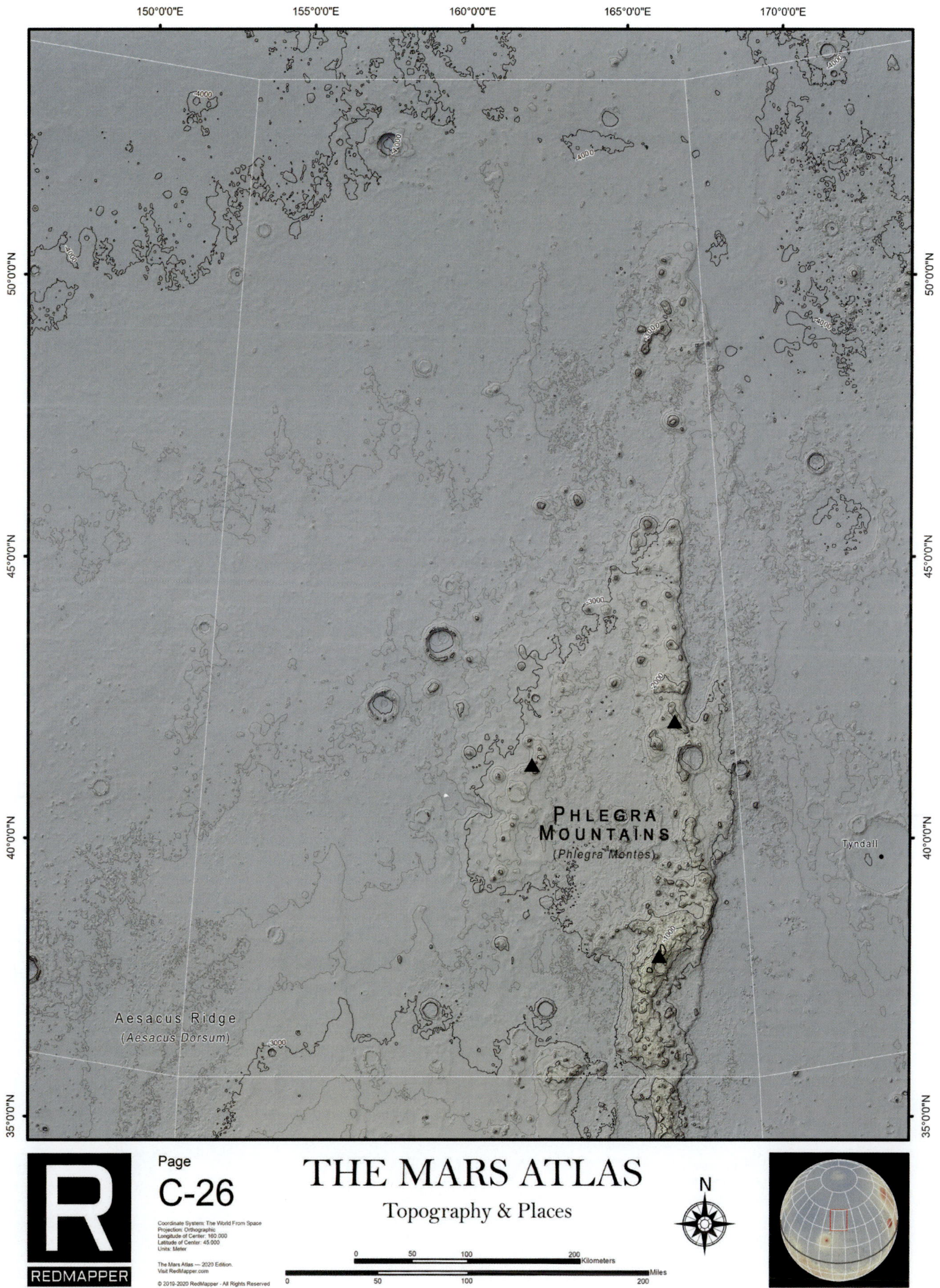
150°0'0"E
155°0'0"E
160°0'0"E
165°0'0"E
170°0'0"E
50°0'0"N
45°0'0"N
40°0'0"N
35°0'0"N
4000
3000
2000
1000
PHLEGRA MOUNTAINS
(Phlegra Montes)
Tyndall
Aesacus Ridge
(Aesacus Dorsum)
R
REDMAPPER
Page
C-26
Coordinate System: The World From Space
Projection: Orthographic
Longitude of Center: 160.000
Latitude of Center: 45.000
Units: Meter
The Mars Atlas — 2020 Edition.
Visit RedMapper.com
© 2019-2020 RedMapper - All Rights Reserved
THE MARS ATLAS
Topography & Places
N
0 50 100 200 Kilometers
0 50 100 200 Miles

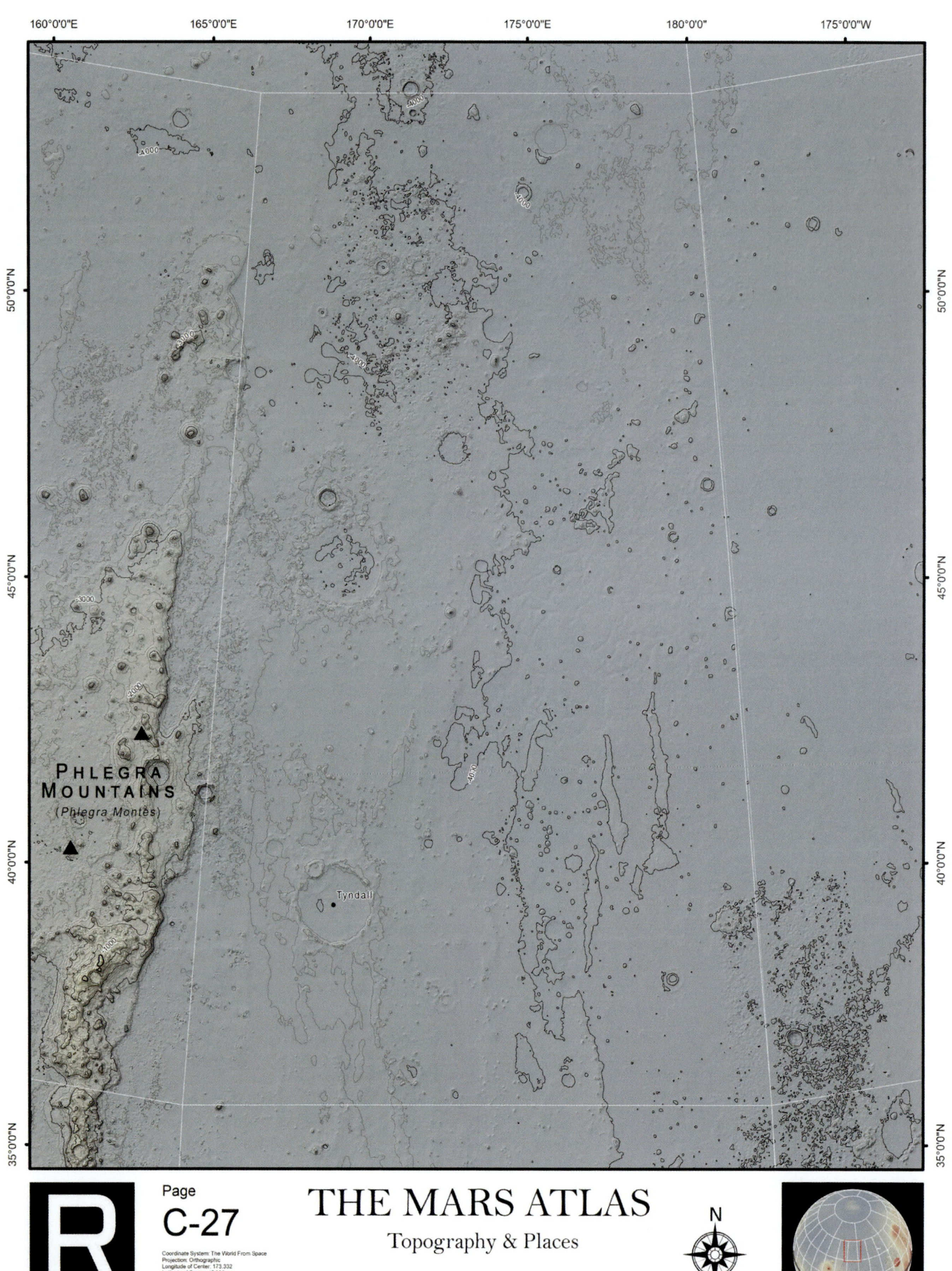

R
REDMAPPER

Page
C-27

Coordinate System: The World From Space
Projection: Orthographic
Longitude of Center: 173.332
Latitude of Center: 45.000
Units: Meter

The Mars Atlas — 2020 Edition.
Visit RedMapper.com

THE MARS ATLAS

Topography & Places

N

0 50 100 200 Kilometers

0 50 100 200 Miles

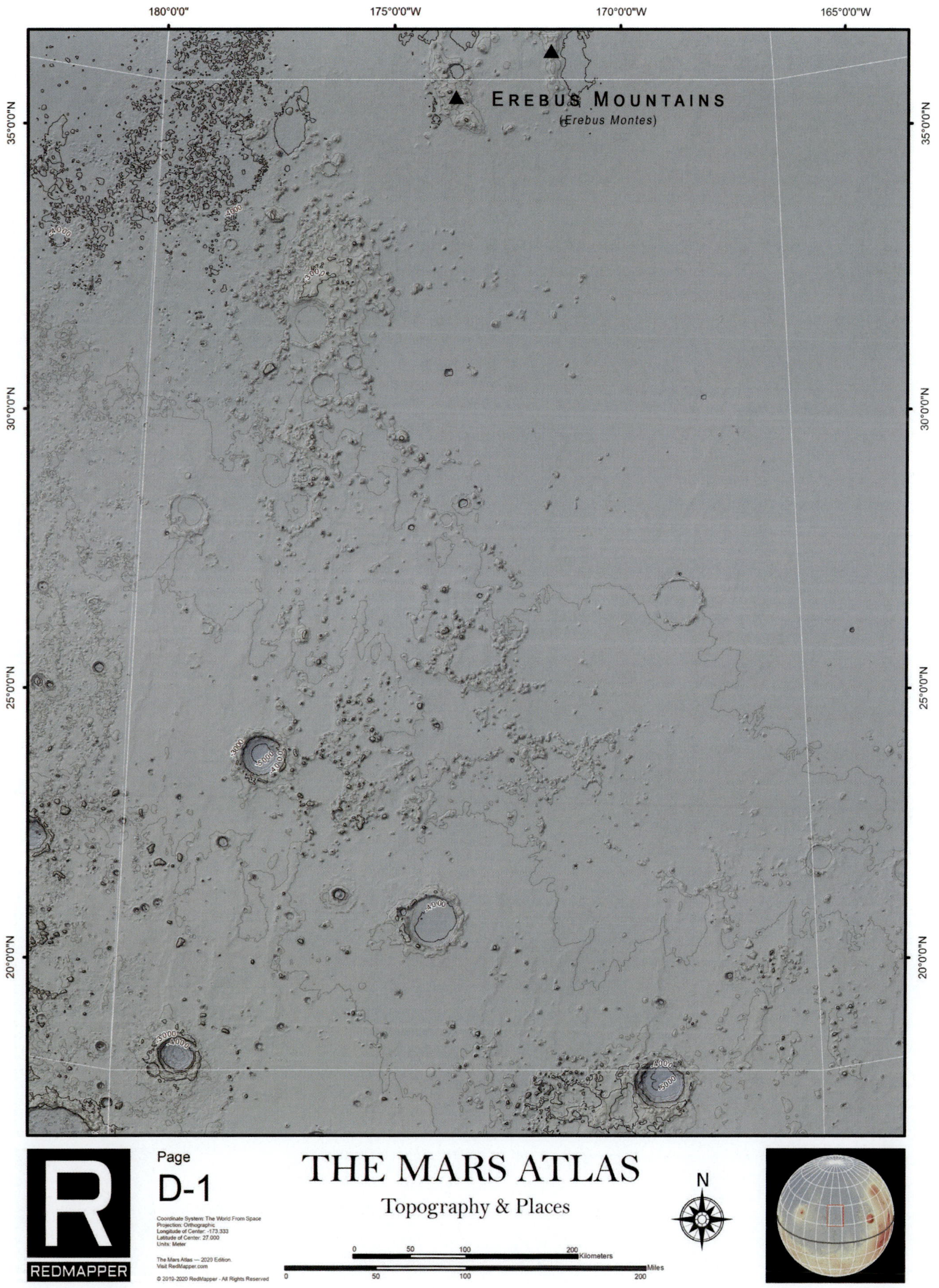
180°0'0"
175°0'0"W
170°0'0"W
165°0'0"W
35°0'0"N
30°0'0"N
25°0'0"N
20°0'0"N
EREBUS MOUNTAINS
(Erebus Montes)
R
REDMAPPER
Page
D-1
Coordinate System: The World From Space
Projection: Orthographic
Longitude of Center: -173.333
Latitude of Center: 27.000
Units: Meter
The Mars Atlas — 2020 Edition.
Visit RedMapper.com
© 2019-2020 RedMapper - All Rights Reserved
THE MARS ATLAS
Topography & Places
0 50 100 200 Kilometers
0 50 100 200 Miles
N

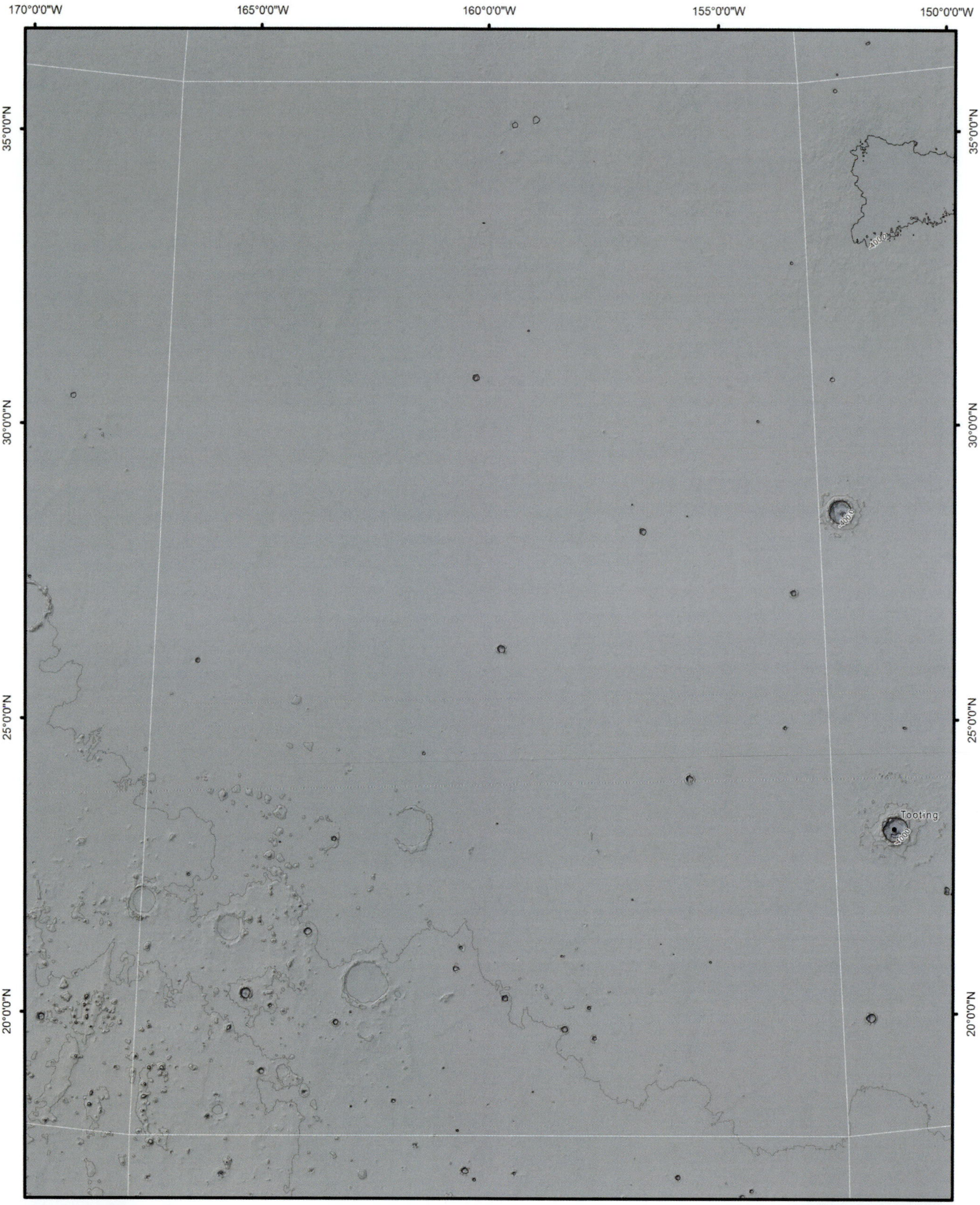

170°0'0"W 165°0'0"W 160°0'0"W 155°0'0"W 150°0'0"W

35°0'0"N

30°0'0"N

25°0'0"N

Tooting

20°0'0"N

Page

D-2

Coordinate System: The World From Space
Projection: Orthographic
Longitude of Center: -160.000
Latitude of Center: 27.000
Units: Meter

The Mars Atlas — 2020 Edition.
Visit RedMapper.com

THE MARS ATLAS

Topography & Places

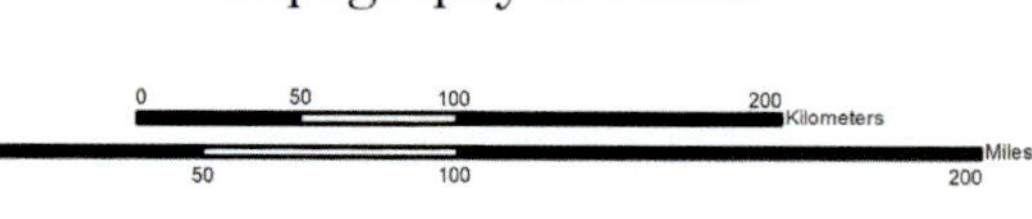

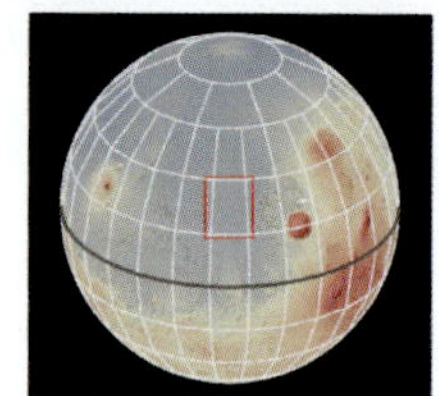

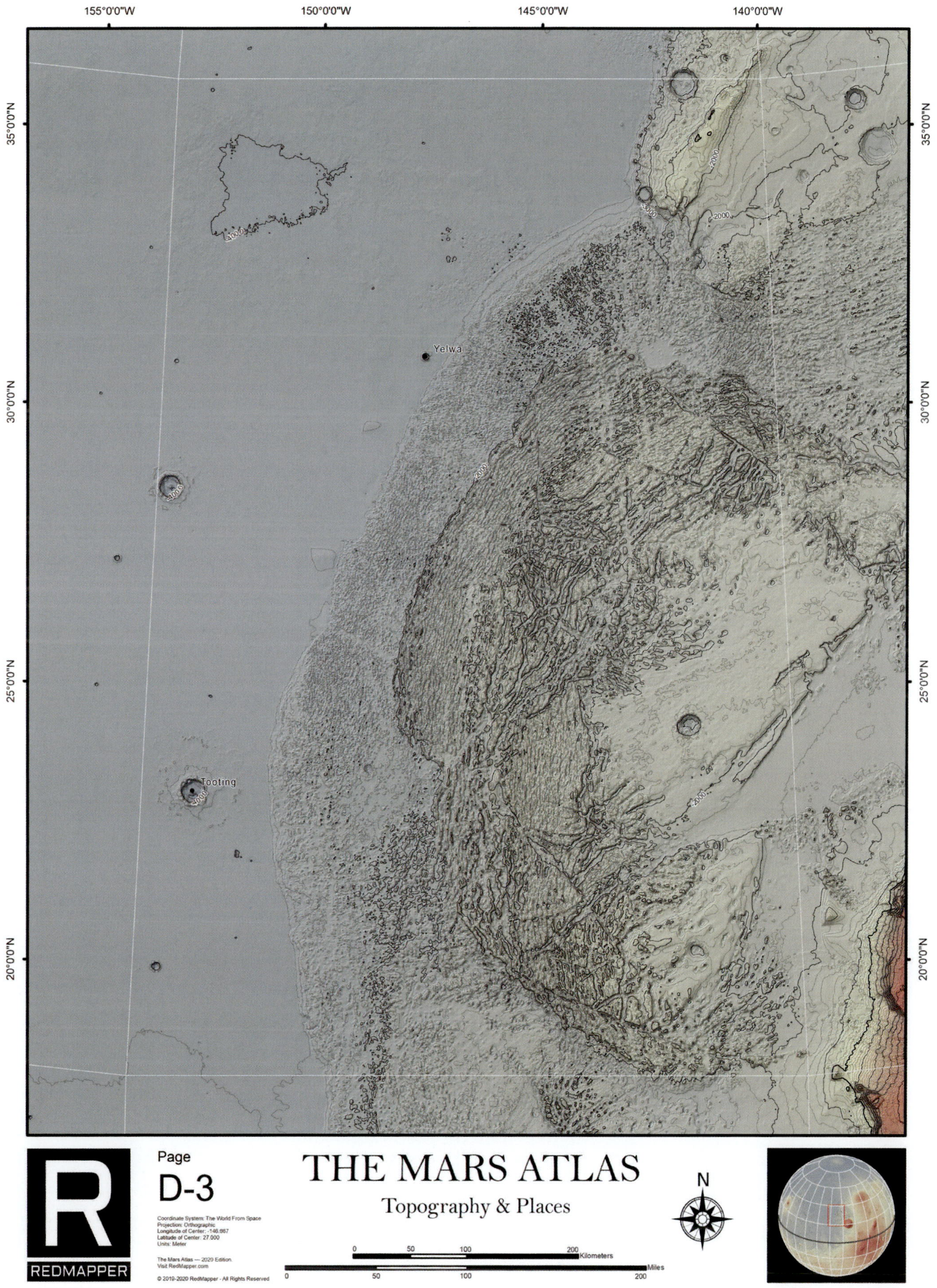
155°0'0"W
150°0'0"W
145°0'0"W
140°0'0"W
35°0'0"N
30°0'0"N
25°0'0"N
20°0'0"N
Yelwa
Tooting
2000
3000
R
REDMAPPER
Page
D-3
Coordinate System: The World From Space
Projection: Orthographic
Longitude of Center: -146.667
Latitude of Center: 27.000
Units: Meter
The Mars Atlas — 2020 Edition.
Visit RedMapper.com
© 2019-2020 RedMapper - All Rights Reserved
THE MARS ATLAS
Topography & Places
N
0 50 100 200 Kilometers
0 50 100 200 Miles

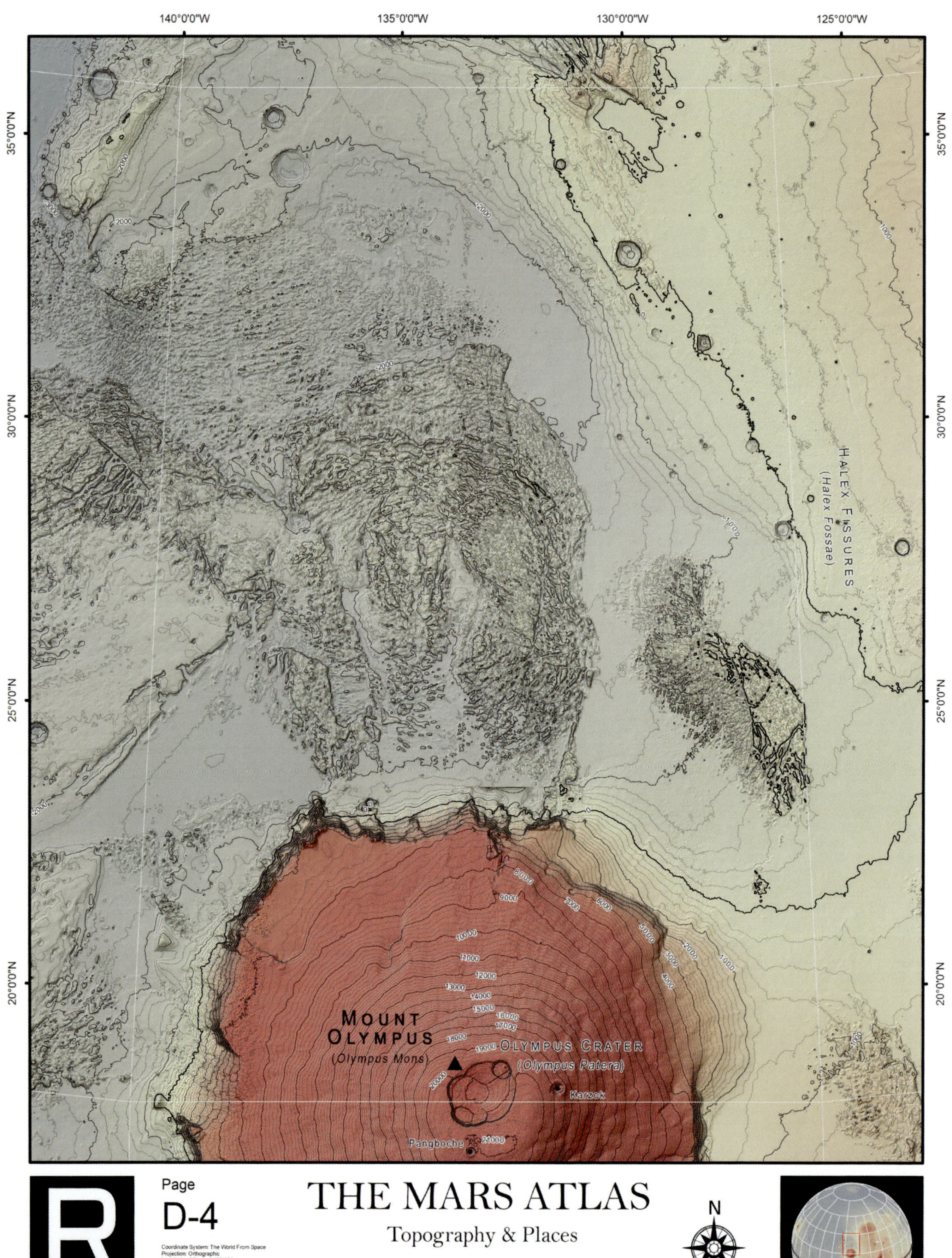

R
REDMAPPER

Page
D-4

Coordinate System: The World From Space
Projection: Orthographic
Longitude of Center: -133.334
Latitude of Center: 27.000
Units: Meter

The Mars Atlas — 2020 Edition.
Visit RedMapper.com

THE MARS ATLAS

Topography & Places

N

0 50 100 200 Kilometers

0 50 100 200 Miles

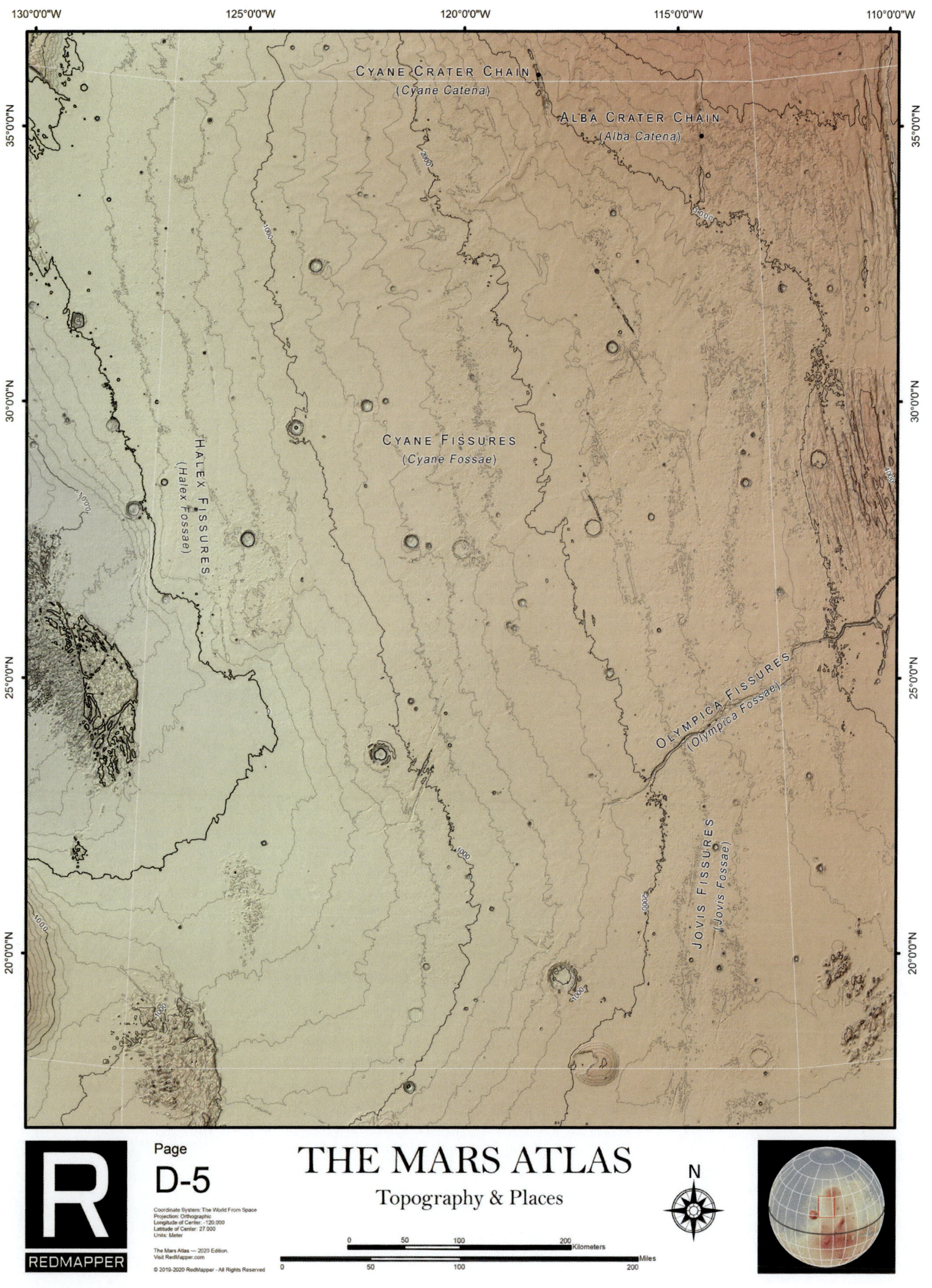

R
REDMAPPER

Page
D-5

Coordinate System: The World From Space
Projection: Orthographic
Longitude of Center: -120.000
Latitude of Center: 27.000
Units: Meter

The Mars Atlas — 2020 Edition.
Visit RedMapper.com

THE MARS ATLAS

Topography & Places

N

0 50 100 200 Kilometers
0 50 100 200 Miles

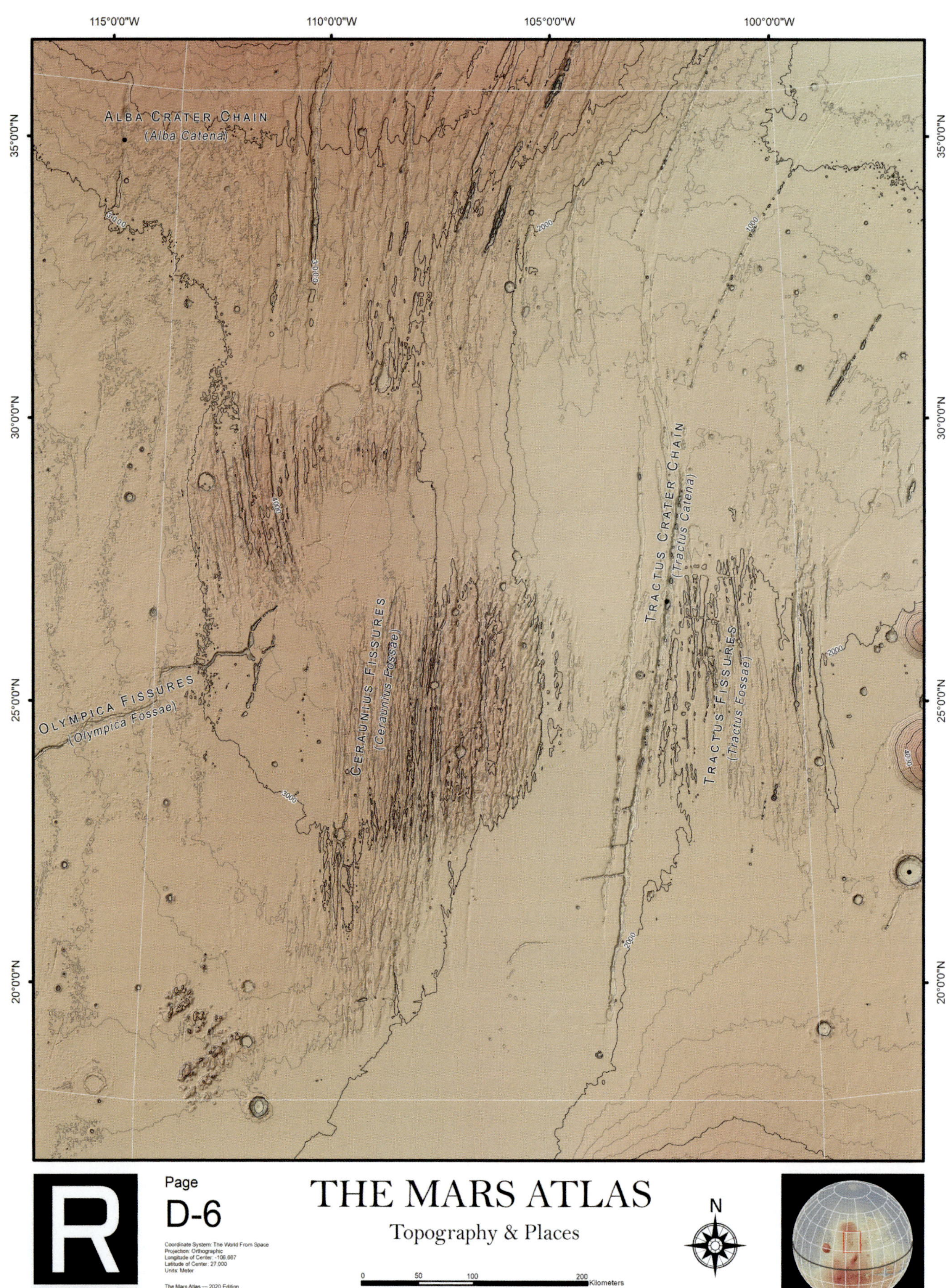

Page
D-6

Coordinate System: The World From Space
Projection: Orthographic
Longitude of Center: -106.667
Latitude of Center: 27.000
Units: Meter

The Mars Atlas — 2020 Edition.
Visit RedMapper.com

THE MARS ATLAS

Topography & Places

R
REDMAPPER

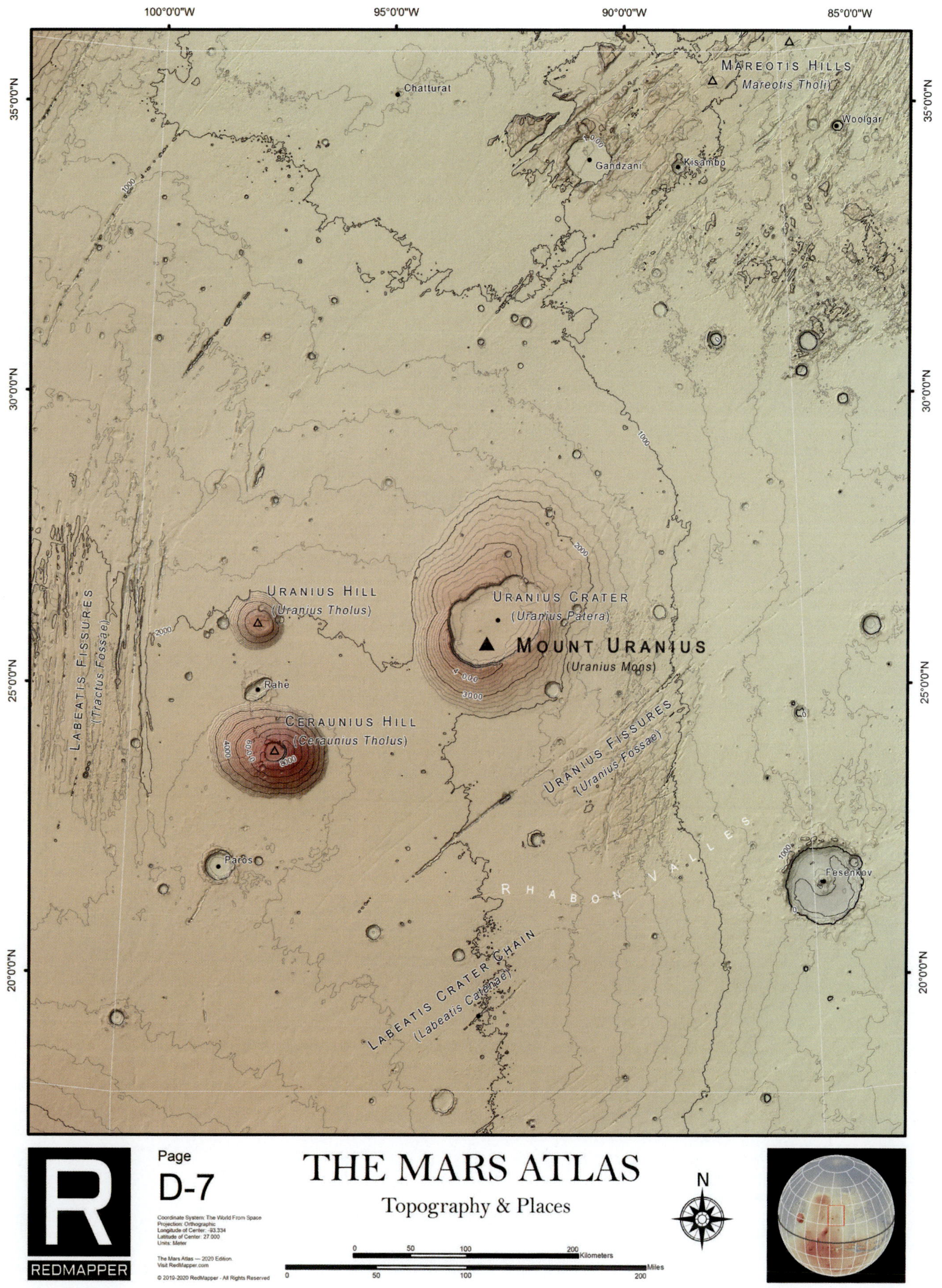

Page
D-7

Coordinate System: The World From Space
Projection: Orthographic
Longitude of Center: -93.334
Latitude of Center: 27.000
Units: Meter

The Mars Atlas — 2020 Edition.
Visit RedMapper.com

THE MARS ATLAS

Topography & Places

0 50 100 200 Kilometers

0 50 100 200 Miles

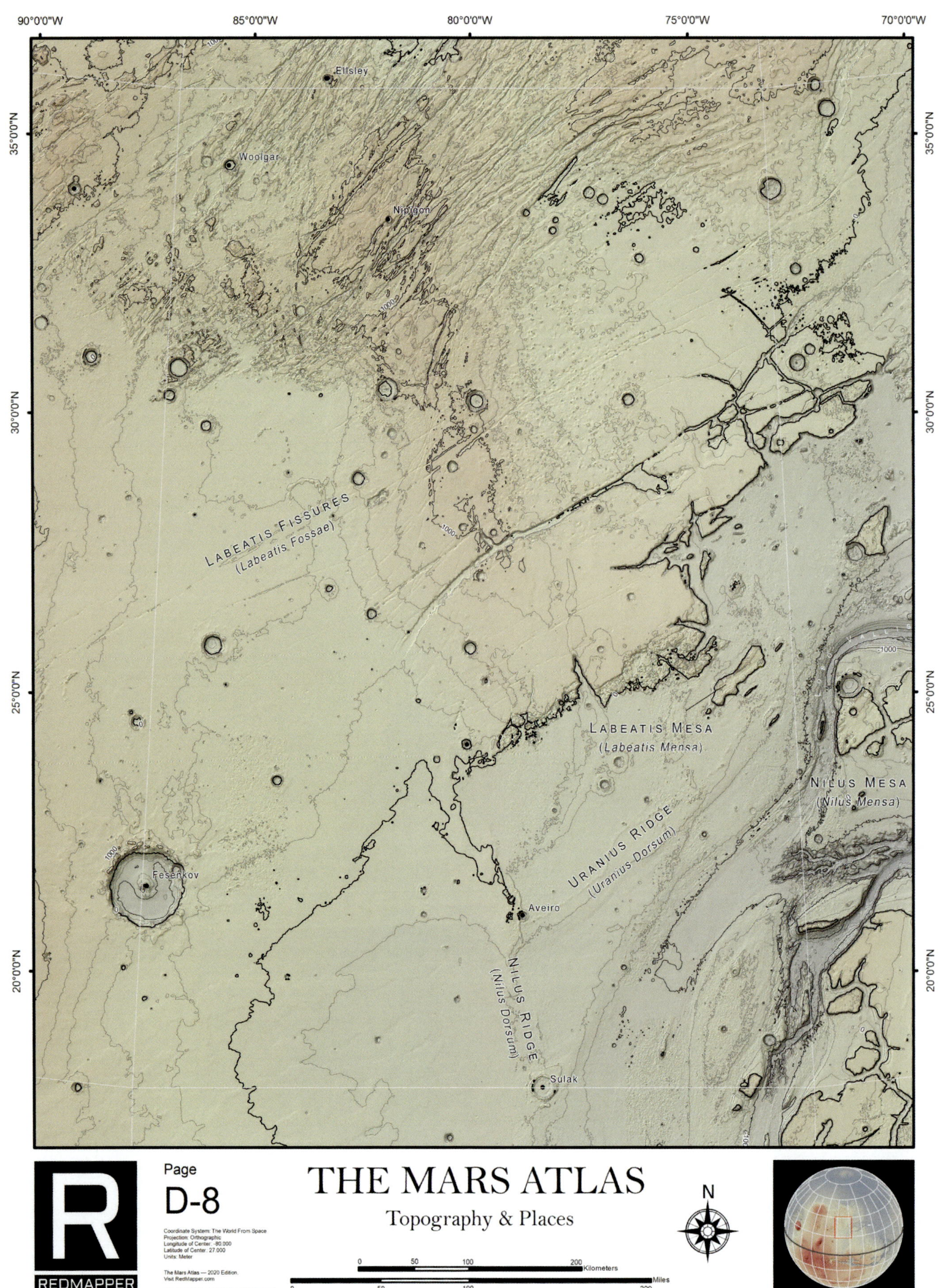
90°0'0"W
85°0'0"W
80°0'0"W
75°0'0"W
70°0'0"W
35°0'0"N
30°0'0"N
25°0'0"N
20°0'0"N
Ellsley
Woolgar
Nipigon
LABEATIS FISSURES
(Labeatis Fossae)
LABEATIS MESA
(Labeatis Mensa)
NILUS MESA
(Nilus Mensa)
URANIUS RIDGE
(Uranius Dorsum)
Fesenkov
Aveiro
NILUS RIDGE
(Nilus Dorsum)
Sulak
1000
R
REDMAPPER
Page
D-8
Coordinate System: The World From Space
Projection: Orthographic
Longitude of Center: -80.000
Latitude of Center: 27.000
Units: Meter
The Mars Atlas ---- 2020 Edition.
Visit RedMapper.com
© 2019-2020 RedMapper - All Rights Reserved
THE MARS ATLAS
Topography & Places
N
0 50 100 200 Kilometers
0 50 100 200 Miles

75°0'0"W
70°0'0"W
65°0'0"W
60°0'0"W
35°0'0"N
30°0'0"N
25°0'0"N
20°0'0"N
TEMPE MESA
(Tempe Mensae)
LABEATIS MESA
(Labeatis Mensae)
SACRA MESA
(Sacra Mensa)
NILUS MESA
(Nilus Mensae)
LUNAE MESA
(Lunae Mensae)
SACRA FISSURES
(Sacra Fossae)
Sharonov
Rauch
Canso
Herculaneum
Pompeii
-1000
-2000
-3000
Page
D-9
THE MARS ATLAS
Topography & Places
Coordinate System: The World From Space
Projection: Orthographic
Longitude of Center: -66.667
Latitude of Center: 27.000
Units: Meter
The Mars Atlas — 2020 Edition.
Visit RedMapper.com
© 2019-2020 RedMapper - All Rights Reserved
REDMAPPER
N
0 50 100 200 Kilometers
0 50 100 200 Miles

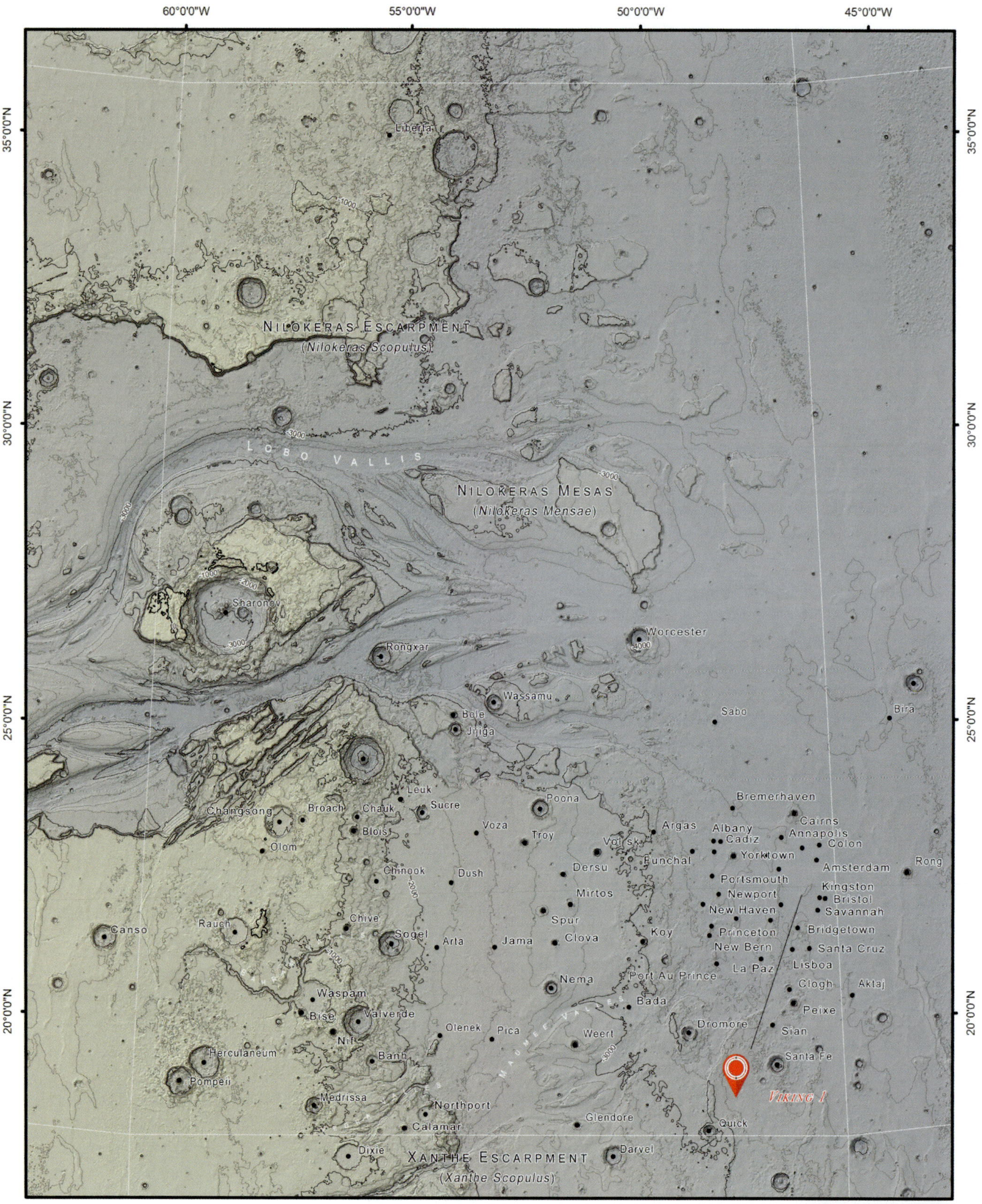

60°0'0"W
55°0'0"W
50°0'0"W
45°0'0"W
35°0'0"N
30°0'0"N
25°0'0"N
20°0'0"N

Liberta
NILOKERAS ESCARPMENT
(Nilokeras Scopulus)
LOBO VALLIS
NILOKERAS MESAS
(Nilokeras Mensae)
Sharonov
Rongxar
Worcester
Wassamu
Bole
Jijiga
Sabo
Bira
Leuk
Sucre
Poona
Bremerhaven
Changsong
Broach
Chauk
Blois
Olom
Voza
Troy
Volsk
Argas
Albany
Cadiz
Cairns
Annapolis
Colon
Funchal
Yorktown
Rong
Chinook
Dush
Dersu
Amsterdam
Portsmouth
Mirtos
Newport
Kingston
Bristol
New Haven
Savannah
Spur
Chive
Canso
Rauch
Sogel
Arta
Jama
Clova
Koy
Princeton
Bridgetown
New Bern
Santa Cruz
La Paz
Lisboa
Port Au Prince
Nema
Clogh
Aktaj
Waspam
Bada
Bise
Valverde
Peixe
Dromore
Olenek
Pica
Sian
Weert
Nif
Herculaneum
Banh
Santa Fe
MAUMEE VALLES
Pompeii
VIKING 1
Medrissa
Northport
Glendore
Calamar
Quick
Darvel
Dixie
XANTHE ESCARPMENT
(Xanthe Scopulus)
1000
2000
3000
4000

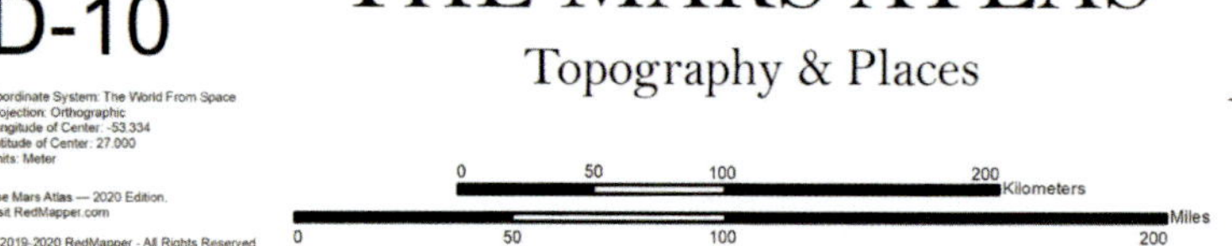

Coordinate System: The World From Space
Projection: Orthographic
Longitude of Center: -53.334
Latitude of Center: 27.000
Units: Meter

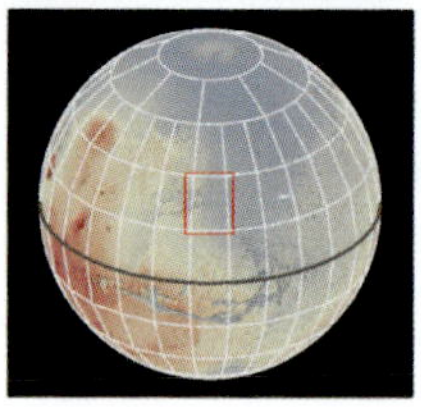

The Mars Atlas — 2020 Edition.
Visit RedMapper.com

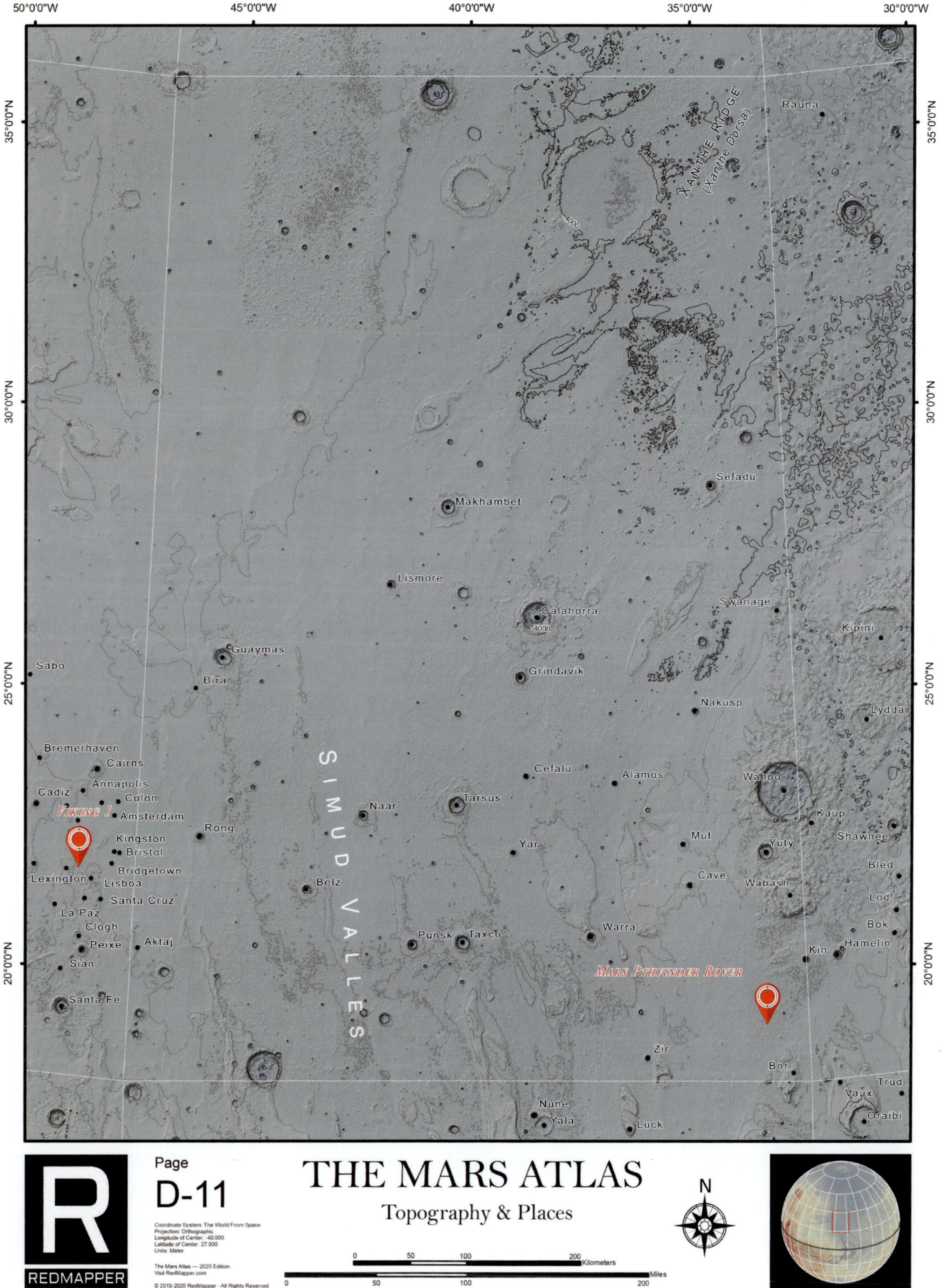

R
REDMAPPER

Page
D-11

Coordinate System: The World From Space
Projection: Orthographic
Longitude of Center: -40.000
Latitude of Center: 27.000
Units: Meter

The Mars Atlas — 2020 Edition.
Visit RedMapper.com

THE MARS ATLAS

Topography & Places

N

0 50 100 200 Kilometers

0 50 100 200 Miles

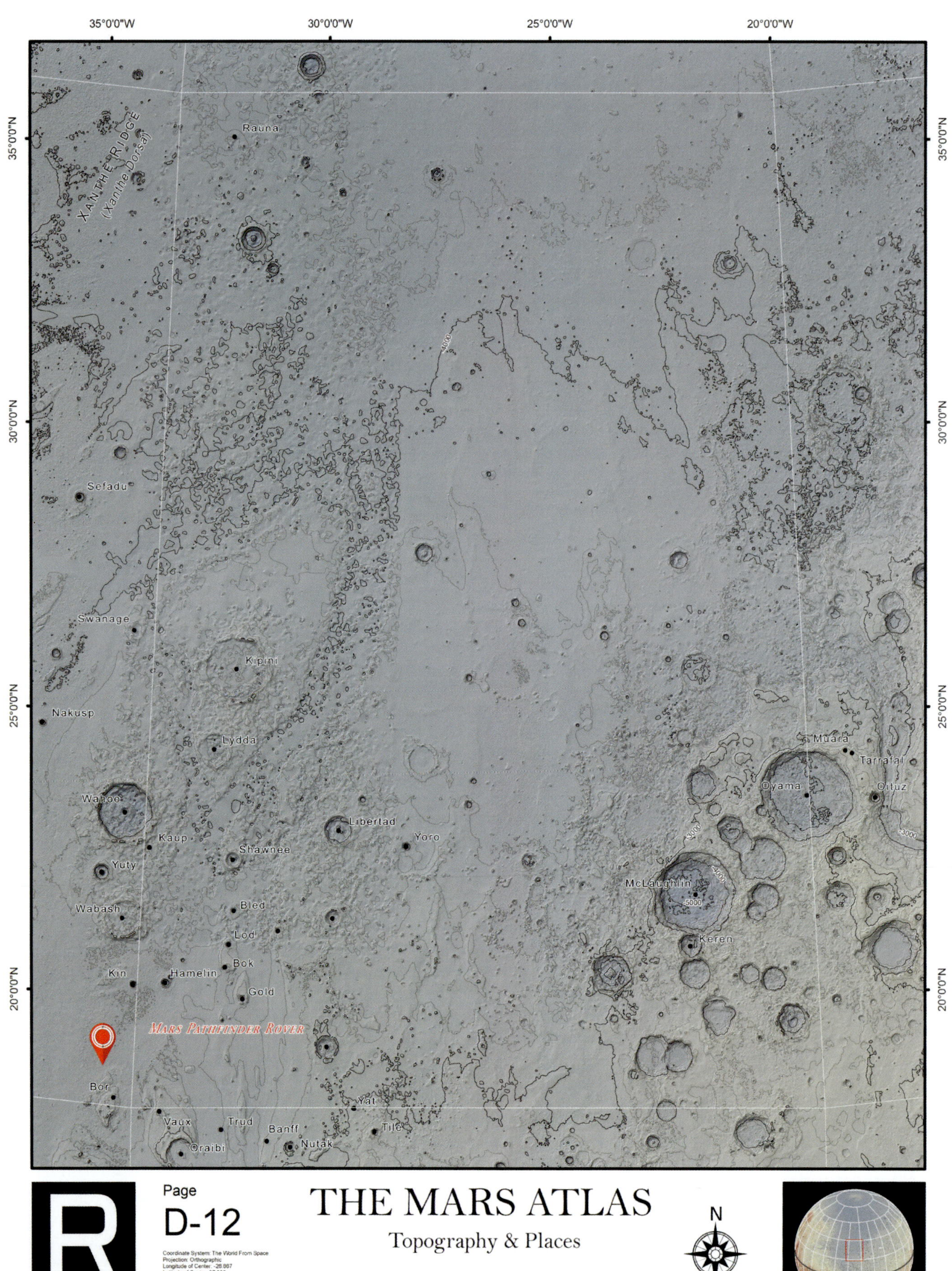

R
REDMAPPER

Page
D-12

Coordinate System: The World From Space
Projection: Orthographic
Longitude of Center: -26.887
Latitude of Center: 27.000
Units: Meter

The Mars Atlas — 2020 Edition.
Visit RedMapper.com

THE MARS ATLAS
Topography & Places

N

0 50 100 200 Kilometers
0 50 100 200 Miles

Page
D-13

THE MARS ATLAS

Topography & Places

Coordinate System: The World From Space
Projection: Orthographic
Longitude of Center: -13.334
Latitude of Center: 27.000
Units: Meter

The Mars Atlas — 2020 Edition
Visit RedMapper.com

REDMAPPER

N

0 50 100 200 Kilometers
0 50 100 200 Miles

R
REDMAPPER

Page
D-14

Coordinate System: The World From Space
Projection: Orthographic
Longitude of Center: 0.000
Latitude of Center: 27.000
Units: Meter

The Mars Atlas — 2020 Edition.
Visit RedMapper.com

THE MARS ATLAS

Topography & Places

N

0 50 100 200 Kilometers

0 50 100 200 Miles

5°0'0"E
10°0'0"E
15°0'0"E
20°0'0"E
35°0'0"N
30°0'0"N
25°0'0"N
20°0'0"N
SILOE CRATER
(Siloe Patera)
ISMENIUS BASIN
(Ismenius Cavus)
Focas
Cerulli
HIDDEKEL BASIN
(Hiddekel Cavus)
Maggini
-2000
-3000
-1000
R
REDMAPPER
Page
D-15
Coordinate System: The World From Space
Projection: Orthographic
Longitude of Center: 13.333
Latitude of Center: 27.000
Units: Meter
The Mars Atlas — 2020 Edition.
Visit RedMapper.com
© 2019-2020 RedMapper - All Rights Reserved
THE MARS ATLAS
Topography & Places
N
0 50 100 200 Kilometers
0 50 100 200 Miles

R
REDMAPPER

Page
D-16

Coordinate System: The World From Space
Projection: Orthographic
Longitude of Center: 26.666
Latitude of Center: 27.000
Units: Meter

The Mars Atlas — 2020 Edition.
Visit RedMapper.com

THE MARS ATLAS

Topography & Places

N

0 50 100 200 Kilometers

0 50 100 200 Miles

30°0'0"E
35°0'0"E
40°0'0"E
45°0'0"E
50°0'0"E
35°0'0"N
30°0'0"N
25°0'0"N
20°0'0"N
Quenisset
Phison Crater
(Phison Patera)
Luzin
Tikhonravov
Cassini
Indus Vallis
-1000
-2000
R
REDMAPPER
Page
D-17
Coordinate System: The World From Space
Projection: Orthographic
Longitude of Center: 40.000
Latitude of Center: 27.000
Units: Meter
The Mars Atlas — 2020 Edition.
Visit RedMapper.com
© 2019-2020 RedMapper - All Rights Reserved
THE MARS ATLAS
Topography & Places
N
0 50 100 200 Kilometers
0 50 100 200 Miles

R
REDMAPPER

Page
D-18

Coordinate System: The World From Space
Projection: Orthographic
Longitude of Center: 53.332
Latitude of Center: 27.000
Units: Meter

The Mars Atlas — 2020 Edition.
Visit RedMapper.com

THE MARS ATLAS

Topography & Places

N

0 50 100 200 Kilometers

0 50 100 200 Miles

60°0'0"E
65°0'0"E
70°0'0"E
75°0'0"E
35°0'0"N
30°0'0"N
25°0'0"N
20°0'0"N
NILOSYRTIS MESAS
(Nilosyrtis Mensae)
1000
2000
Baldet
Antoniadi
Negril
Toro
R
REDMAPPER
Page
D-19
Coordinate System: The World From Space
Projection: Orthographic
Longitude of Center: 66.666
Latitude of Center: 27.000
Units: Meter
The Mars Atlas 2020 Edition
Visit RedMapper.com
© 2019-2020 RedMapper - All Rights Reserved
THE MARS ATLAS
Topography & Places
N
0 50 100 200 Kilometers
0 50 100 200 Miles

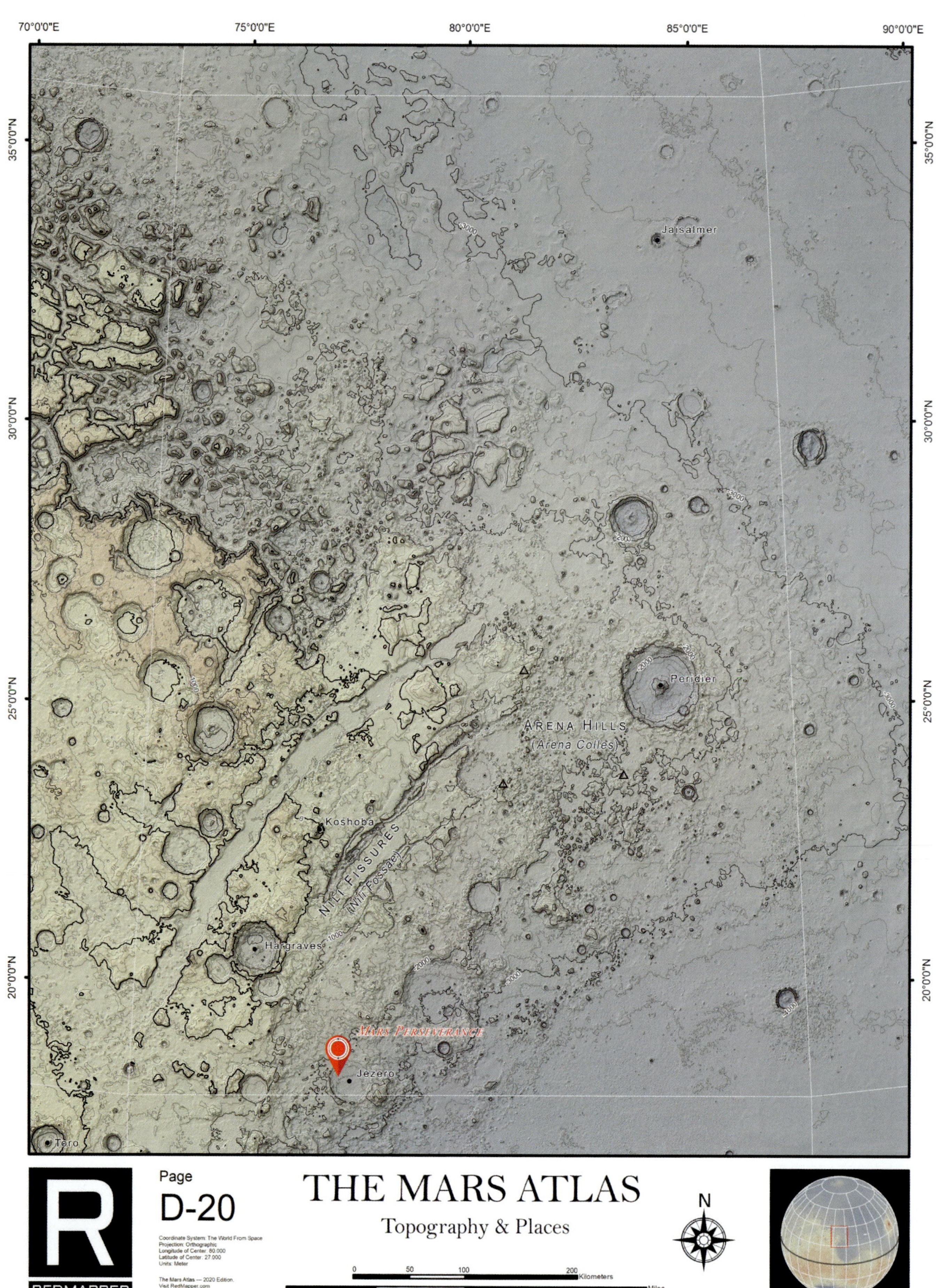

R
REDMAPPER

Page
D-20

Coordinate System: The World From Space
Projection: Orthographic
Longitude of Center: 80.000
Latitude of Center: 27.000
Units: Meter

The Mars Atlas — 2020 Edition.
Visit RedMapper.com

THE MARS ATLAS
Topography & Places

85°0'0"E
90°0'0"E
95°0'0"E
100°0'0"E
35°0'0"N
30°0'0"N
25°0'0"N
20°0'0"N
Astapus Hills
(Astapus Colles)
Jaisalmer
-4000
-5000
-3000
-2000
R
REDMAPPER
Page
D-21
Coordinate System: The World From Space
Projection: Orthographic
Longitude of Center: 93.332
Latitude of Center: 27.000
Units: Meter
The Mars Atlas — 2020 Edition.
Visit RedMapper.com
© 2019-2020 RedMapper - All Rights Reserved
THE MARS ATLAS
Topography & Places
N
0 50 100 200 Kilometers
0 50 100 200 Miles

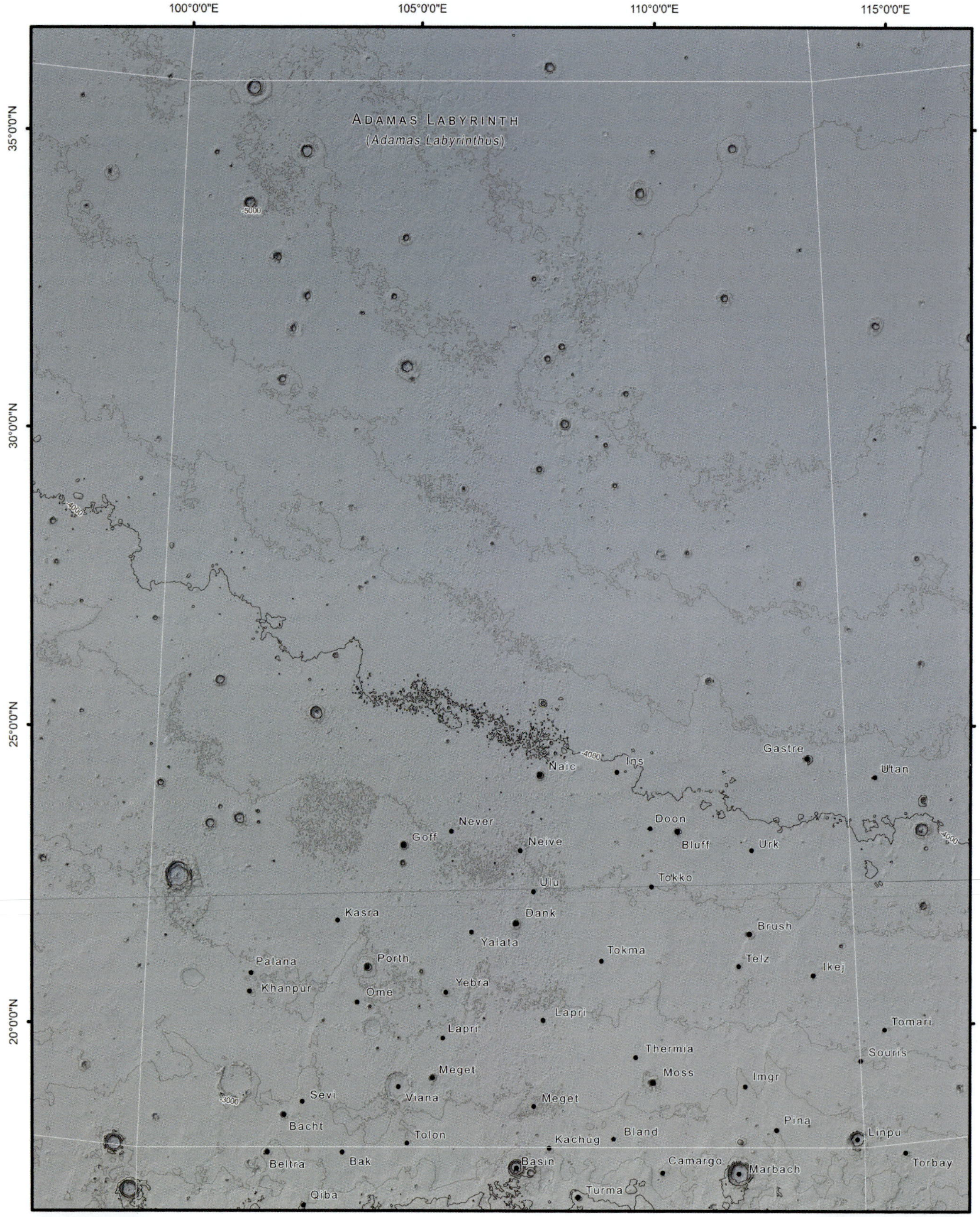

Page
D-22

Coordinate System: The World From Space
Projection: Orthographic
Longitude of Center: 106.865
Latitude of Center: 27.000
Units: Meter

The Mars Atlas — 2020 Edition.
Visit RedMapper.com

THE MARS ATLAS

Topography & Places

0 50 100 200 Kilometers
0 50 100 200 Miles

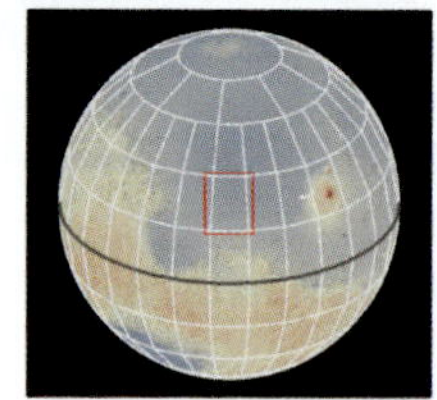

110°0'0"E
115°0'0"E
120°0'0"E
125°0'0"E
130°0'0"E
35°0'0"N
30°0'0"N
25°0'0"N
20°0'0"N
-5000
Bacolor
Utan
Urk
-4000
-4000
-4000
HEPHAESTUS FISSURES
(Hephaestus Fossae)
Brush
Ikej
Tomari
Souris
Pina
Linpu
Tomari
-4000
Page
D-23
Coordinate System: The World From Space
Projection: Orthographic
Longitude of Center: 120.000
Latitude of Center: 27.000
Units: Meter
The Mars Atlas — 2020 Edition.
Visit RedMapper.com
© 2019-2020 RedMapper - All Rights Reserved
REDMAPPER
THE MARS ATLAS
Topography & Places
N
0 50 100 200 Kilometers
0 50 100 200 Miles

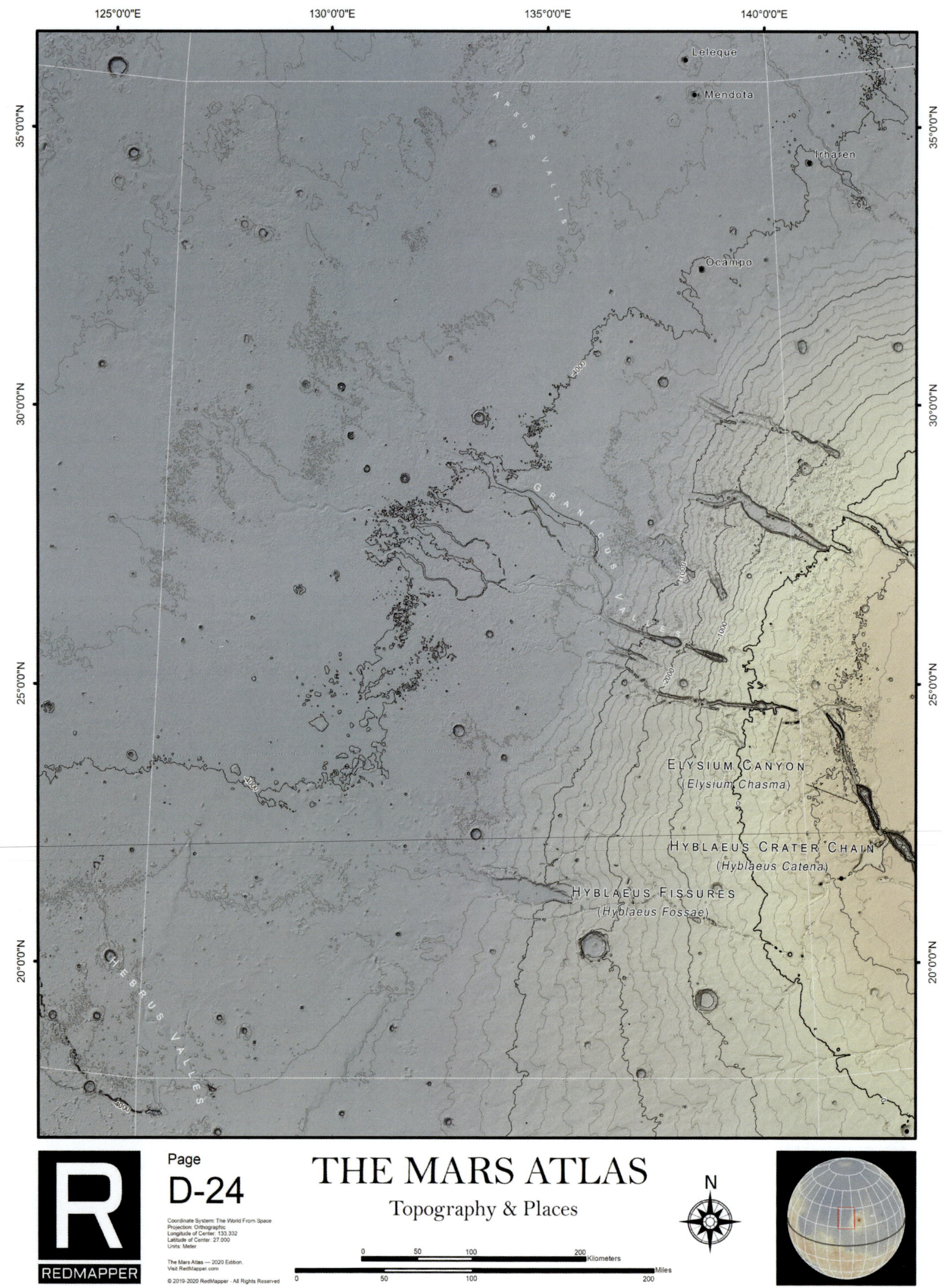
125°0'0"E
130°0'0"E
135°0'0"E
140°0'0"E
35°0'0"N
30°0'0"N
25°0'0"N
20°0'0"N
Leleque
Mendota
Irharen
Ocampo
APSUS VALLIS
GRANICUS VALLES
HEBRUS VALLES
-4000
-3000
-2000
-1000
ELYSIUM CANYON
(Elysium Chasma)
HYBLAEUS CRATER CHAIN
(Hyblaeus Catena)
HYBLAEUS FISSURES
(Hyblaeus Fossae)
R
REDMAPPER
Page
D-24
Coordinate System: The World From Space
Projection: Orthographic
Longitude of Center: 133.332
Latitude of Center: 27.000
Units: Meter
The Mars Atlas — 2020 Edition.
Visit RedMapper.com
© 2019-2020 RedMapper - All Rights Reserved
THE MARS ATLAS
Topography & Places
0
50
100
200
Kilometers
Miles
N

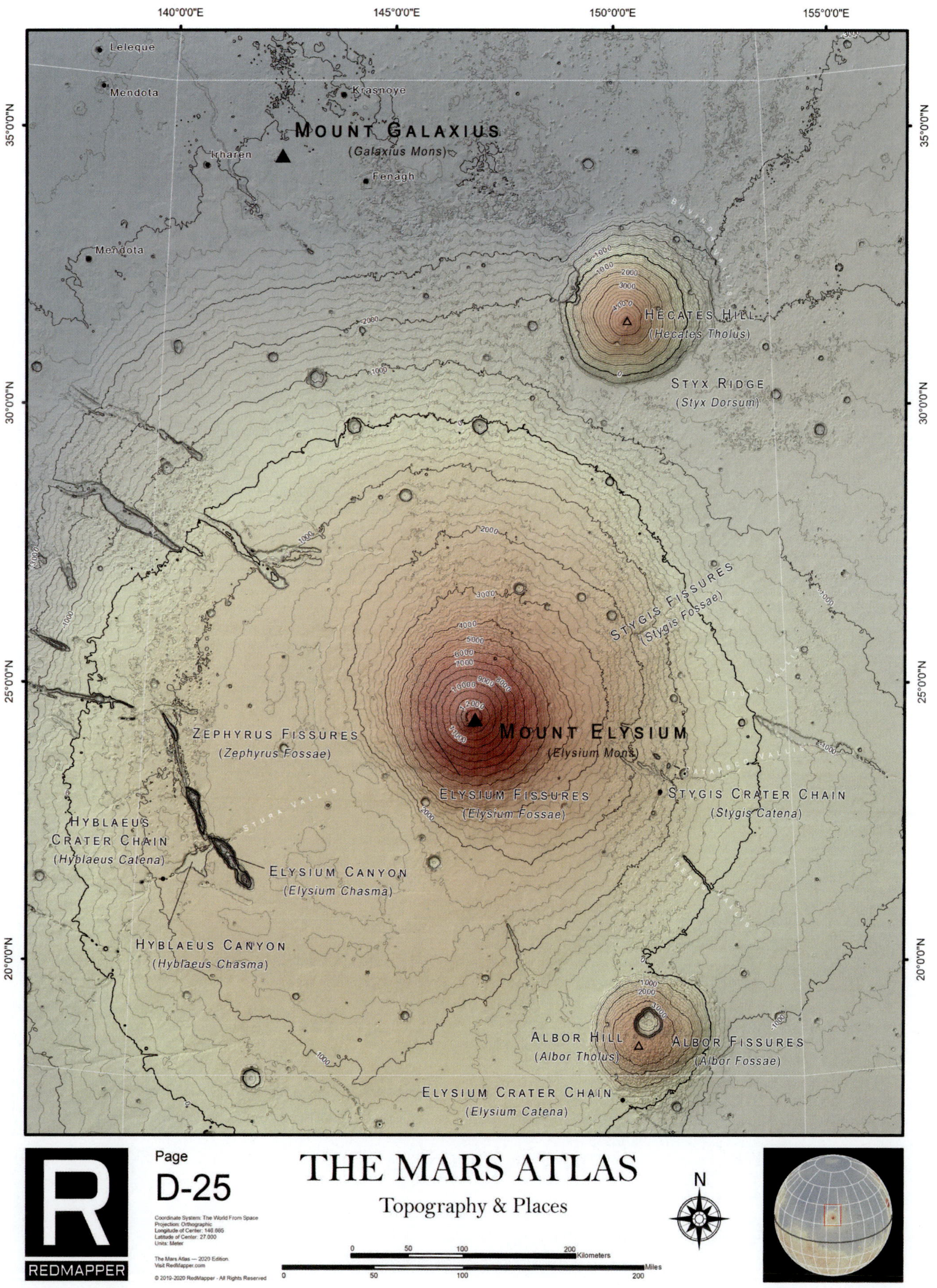
140°0'0"E
145°0'0"E
150°0'0"E
155°0'0"E
35°0'0"N
30°0'0"N
25°0'0"N
20°0'0"N
Leleque
Mendota
Krasnoye
Mount Galaxius
(Galaxius Mons)
Irharen
Fenagh
Mendota
Hecates Hill
(Hecates Tholus)
Styx Ridge
(Styx Dorsum)
Stygis Fissures
(Stygis Fossae)
Mount Elysium
(Elysium Mons)
Zephyrus Fissures
(Zephyrus Fossae)
Elysium Fissures
(Elysium Fossae)
Stygis Crater Chain
(Stygis Catena)
Hyblaeus
Crater Chain
(Hyblaeus Catena)
Stura Vallis
Elysium Canyon
(Elysium Chasma)
Hyblaeus Canyon
(Hyblaeus Chasma)
Albor Hill
(Albor Tholus)
Albor Fissures
(Albor Fossae)
Elysium Crater Chain
(Elysium Catena)
R
REDMAPPER
Page
D-25
Coordinate System: The World From Space
Projection: Orthographic
Longitude of Center: 146.665
Latitude of Center: 27.000
Units: Meter
The Mars Atlas — 2020 Edition.
Visit RedMapper.com
© 2019-2020 RedMapper - All Rights Reserved
THE MARS ATLAS
Topography & Places
N
0 50 100 200 Kilometers
0 50 100 200 Miles

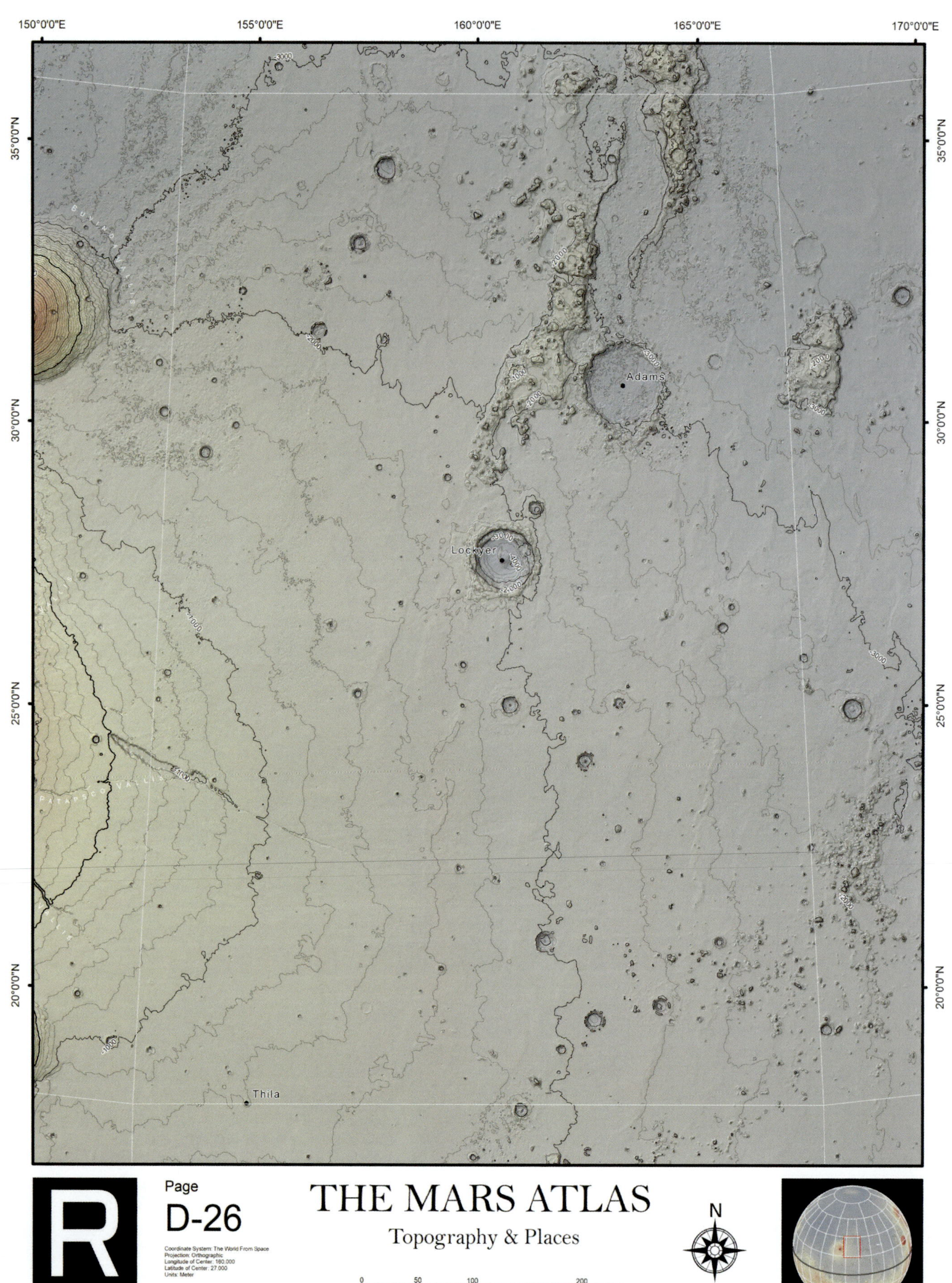

R
REDMAPPER

Page
D-26

Coordinate System: The World From Space
Projection: Orthographic
Longitude of Center: 160.000
Latitude of Center: 27.000
Units: Meter

The Mars Atlas — 2020 Edition.
Visit RedMapper.com

THE MARS ATLAS

Topography & Places

N

0 50 100 200 Kilometers

0 50 100 200 Miles

165°0'0"E
170°0'0"E
175°0'0"E
180°0'0"
35°0'0"N
30°0'0"N
25°0'0"N
20°0'0"N
Phlegra Ridge
(Phlegra Dorsa)
Tartarus Hills
(Tartarus Colles)
Lockyer
R
REDMAPPER
Page
D-27
Coordinate System: The World From Space
Projection: Orthographic
Longitude of Center: 173.332
Latitude of Center: 27.000
Units: Meter
The Mars Atlas — 2020 Edition
Visit RedMapper.com
© 2019-2020 RedMapper - All Rights Reserved
THE MARS ATLAS
Topography & Places
N
0 50 100 200 Kilometers
0 50 100 200 Miles

R
REDMAPPER

Page
E-1

Coordinate System: The World From Space
Projection: Orthographic
Longitude of Center: -173.333
Latitude of Center: 9.000
Units: Meter

The Mars Atlas --- 2020 Edition.
Visit RedMapper.com

THE MARS ATLAS

Topography & Places

N

0 50 100 200 Kilometers

0 50 100 200 Miles

165°0'0"W
160°0'0"W
155°0'0"W
15°0'0"N
10°0'0"N
5°0'0"N
0°0'0"
EUMENIDES RIDGE
(Eumenides Dorsum)
Nicholson
R
REDMAPPER
Page
E-2
Coordinate System: The World From Space
Projection: Orthographic
Longitude of Center: -160.000
Latitude of Center: 9.000
Units: Meter
The Mars Atlas — 2020 Edition.
Visit RedMapper.com
© 2019-2020 RedMapper - All Rights Reserved
THE MARS ATLAS
Topography & Places
N
0 50 100 200 Kilometers
0 50 100 200 Miles

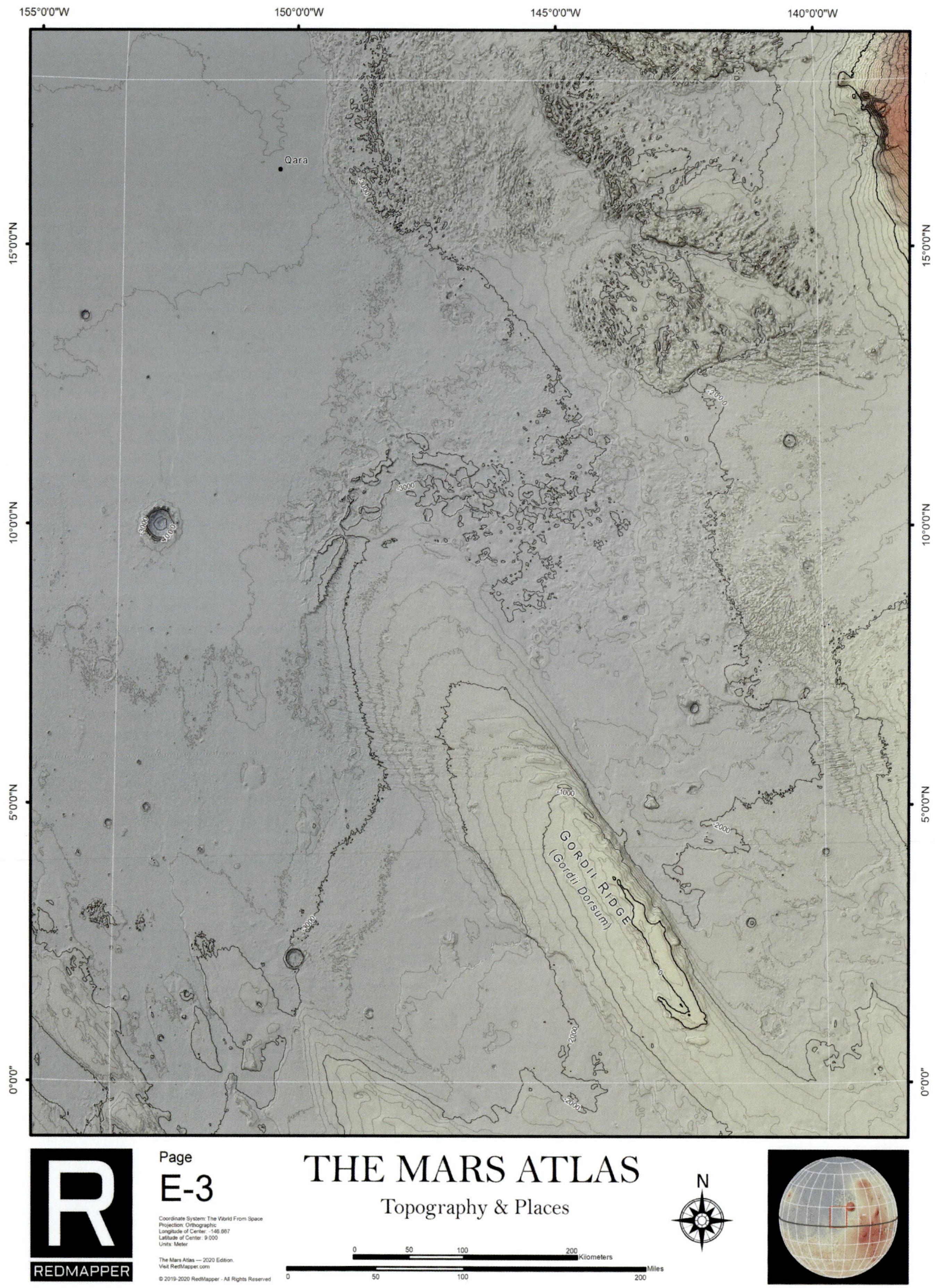

155°0'0"W
150°0'0"W
145°0'0"W
140°0'0"W
15°0'0"N
10°0'0"N
5°0'0"N
0°0'0"
Qara
GORDII RIDGE
(Gordii Dorsum)
-3000
-2000
-1000
0
R
REDMAPPER
Page
E-3
Coordinate System: The World From Space
Projection: Orthographic
Longitude of Center: -146.667
Latitude of Center: 9.000
Units: Meter
The Mars Atlas — 2020 Edition.
Visit RedMapper.com
© 2019-2020 RedMapper - All Rights Reserved
THE MARS ATLAS
Topography & Places
N
0 50 100 200 Kilometers
0 50 100 200 Miles

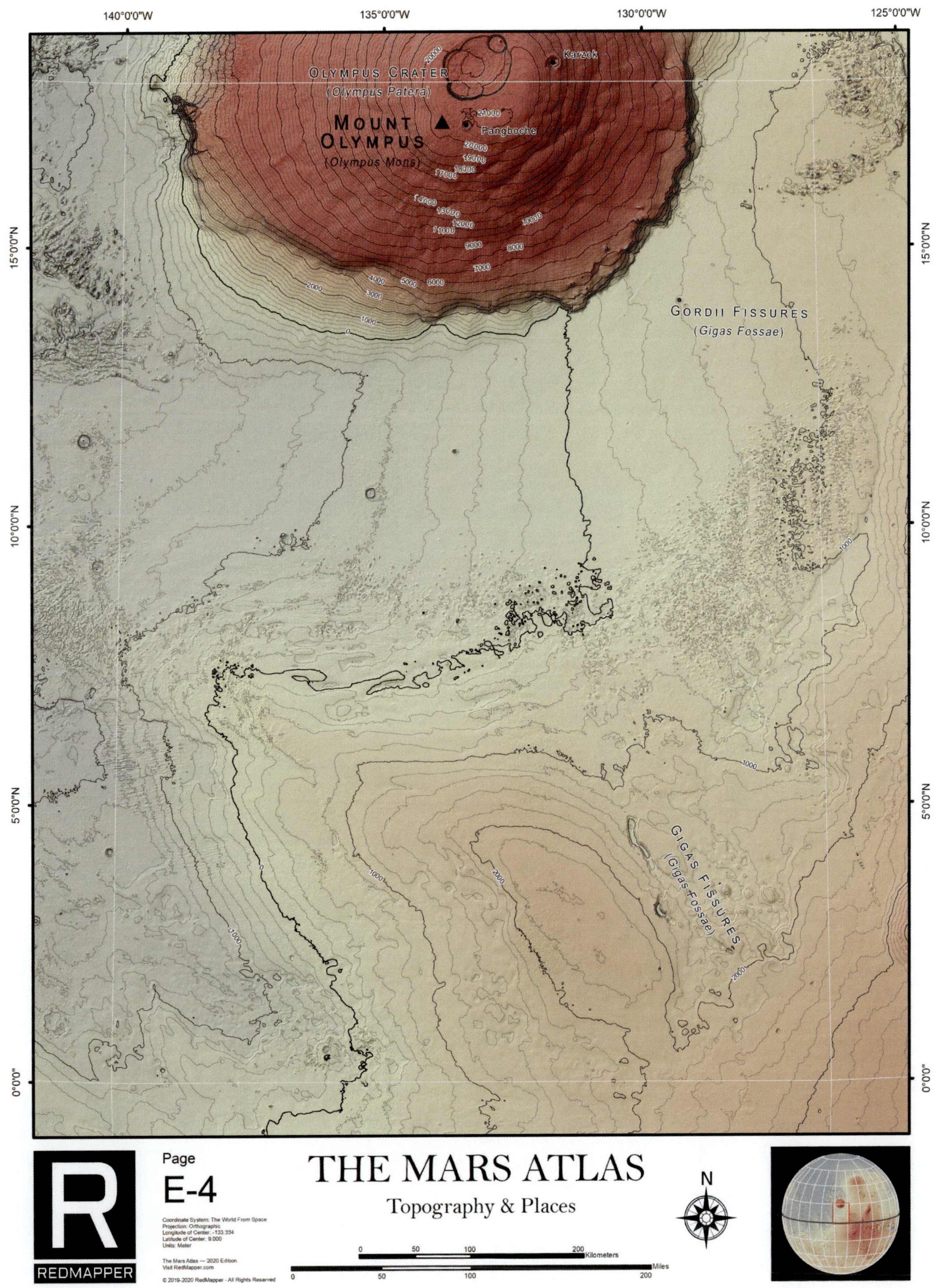
140°0'0"W
135°0'0"W
130°0'0"W
125°0'0"W
15°0'0"N
10°0'0"N
5°0'0"N
0°0'0"
OLYMPUS CRATER
(Olympus Patera)
Karzok
MOUNT OLYMPUS
(Olympus Mons)
Pangboche
GORDII FISSURES
(Gigas Fossae)
GIGAS FISSURES
(Gigas Fossae)
R
REDMAPPER
Page
E-4
Coordinate System: The World From Space
Projection: Orthographic
Longitude of Center: -133.334
Latitude of Center: 9.000
Units: Meter
The Mars Atlas — 2020 Edition.
Visit RedMapper.com
© 2019-2020 RedMapper - All Rights Reserved
THE MARS ATLAS
Topography & Places
N
0 50 100 200 Kilometers
0 50 100 200 Miles

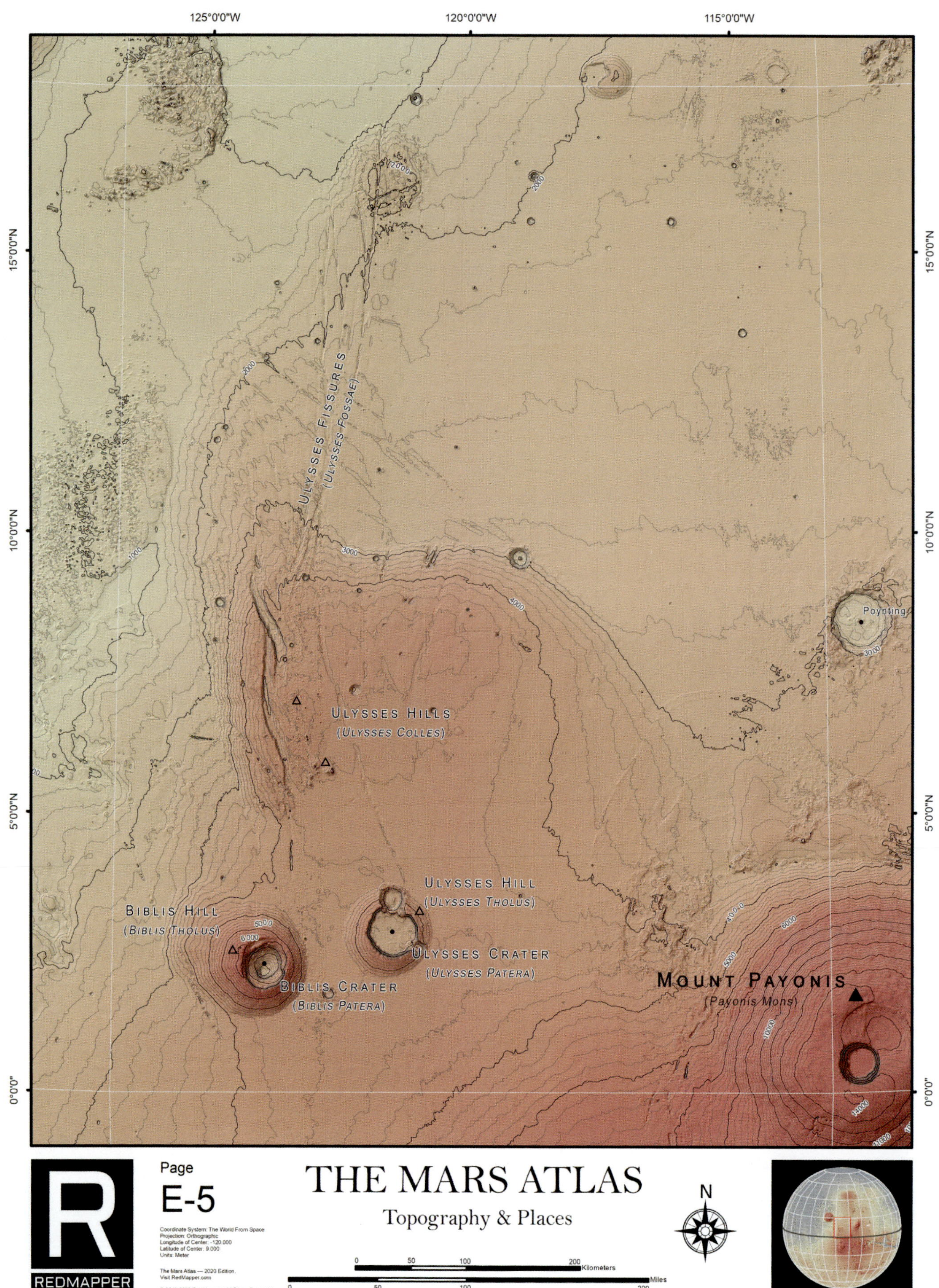

125°0'0"W
120°0'0"W
115°0'0"W
15°0'0"N
10°0'0"N
5°0'0"N
0°0'0"
Ulysses Fissures
(Ulysses Fossae)
Ulysses Hills
(Ulysses Colles)
Ulysses Hill
(Ulysses Tholus)
Ulysses Crater
(Ulysses Patera)
Biblis Hill
(Biblis Tholus)
Biblis Crater
(Biblis Patera)
Mount Payonis
(Payonis Mons)
Poynting
Page
E-5
THE MARS ATLAS
Topography & Places
REDMAPPER
Coordinate System: The World From Space
Projection: Orthographic
Longitude of Center: -120.000
Latitude of Center: 9.000
Units: Meter
The Mars Atlas — 2020 Edition.
Visit RedMapper.com
© 2019-2020 RedMapper · All Rights Reserved
Kilometers
Miles
N

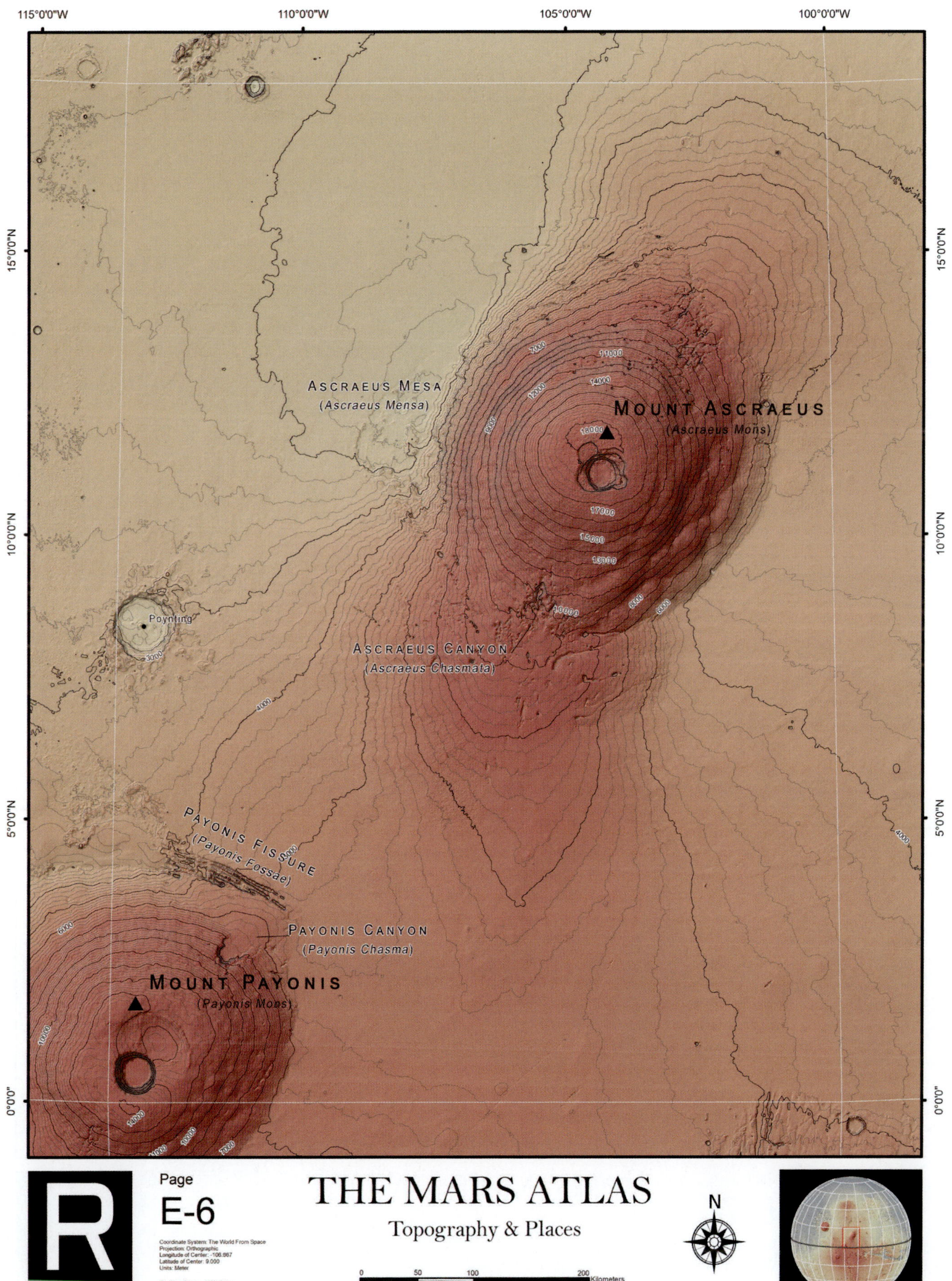

REDMAPPER

Page
E-6

Coordinate System: The World From Space
Projection: Orthographic
Longitude of Center: -106.567
Latitude of Center: 9.000
Units: Meter

Visit RedMapper.com

THE MARS ATLAS

Topography & Places

N

0 50 100 200 Kilometers

0 50 100 200 Miles

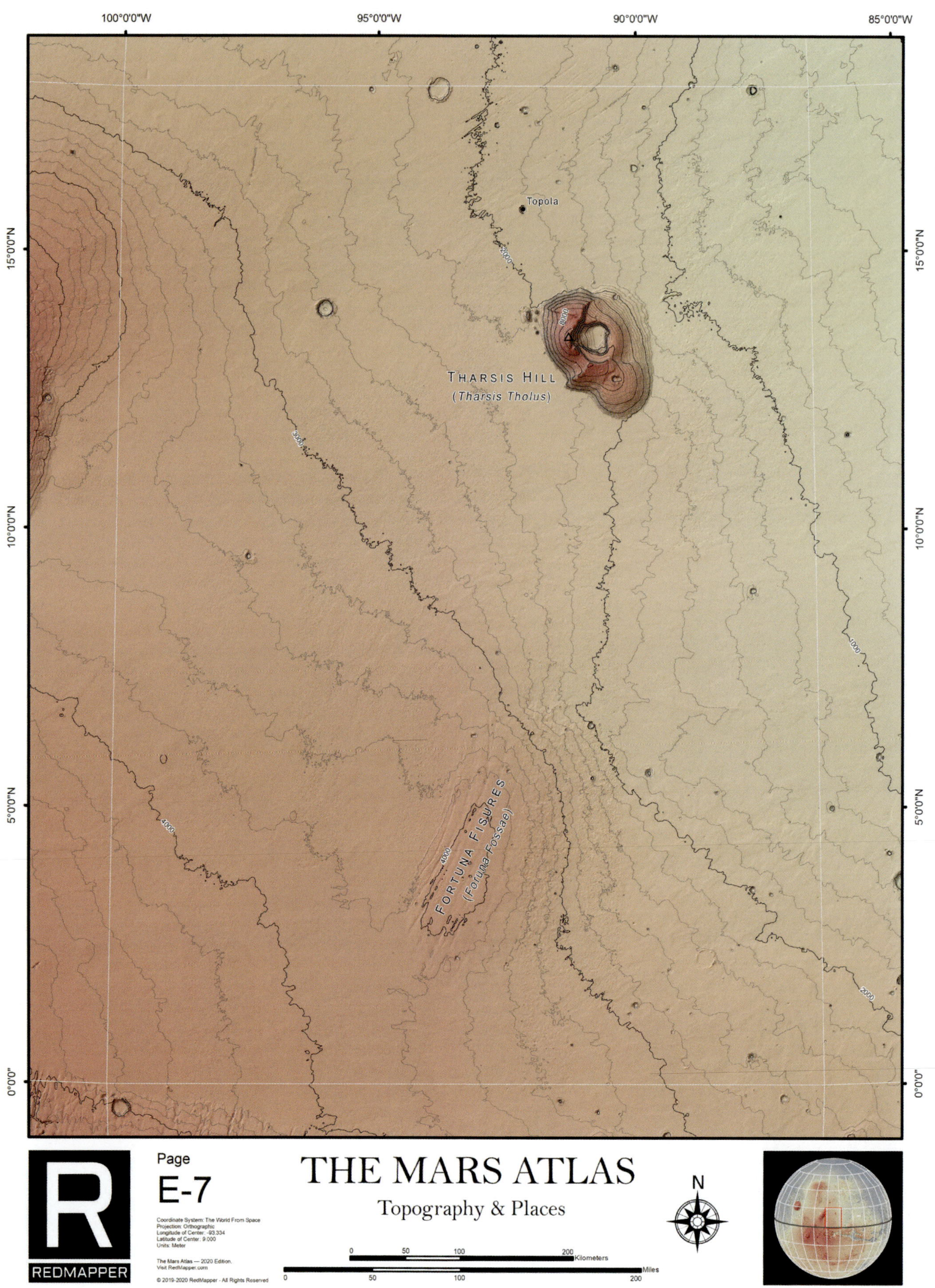

100°0'0"W
95°0'0"W
90°0'0"W
85°0'0"W
15°0'0"N
10°0'0"N
5°0'0"N
0°0'0"
Topola
2000
THARSIS HILL
(Tharsis Tholus)
3000
1000
FORTUNA FISURES
(Foruna Fossae)
4000
2000
R
REDMAPPER
Page
E-7
THE MARS ATLAS
Topography & Places
Coordinate System: The World From Space
Projection: Orthographic
Longitude of Center: -93.334
Latitude of Center: 9.000
Units: Meter
The Mars Atlas — 2020 Edition.
Visit RedMapper.com
0
50
100
200
Kilometers
Miles
N

85°0'0"W
80°0'0"W
75°0'0"W
15°0'0"N
10°0'0"N
5°0'0"N
0°0'0"
Sulak
ECHUS MOUNTAINS
(Echus Montes)
ECHUS CANYON
(Echus Chasma)
ECHUS FISSURES
(Echus Fossae)
1000
2000
3000
4000
0
Page
E-8
THE MARS ATLAS
Topography & Places
Coordinate System: The World From Space
Projection: Orthographic
Longitude of Center: -80.000
Latitude of Center: 9.000
Units: Meter
The Mars Atlas — 2020 Edition.
Visit RedMapper.com
© 2019-2020 RedMapper - All Rights Reserved
0
50
100
200
Kilometers
Miles
N
R
REDMAPPER

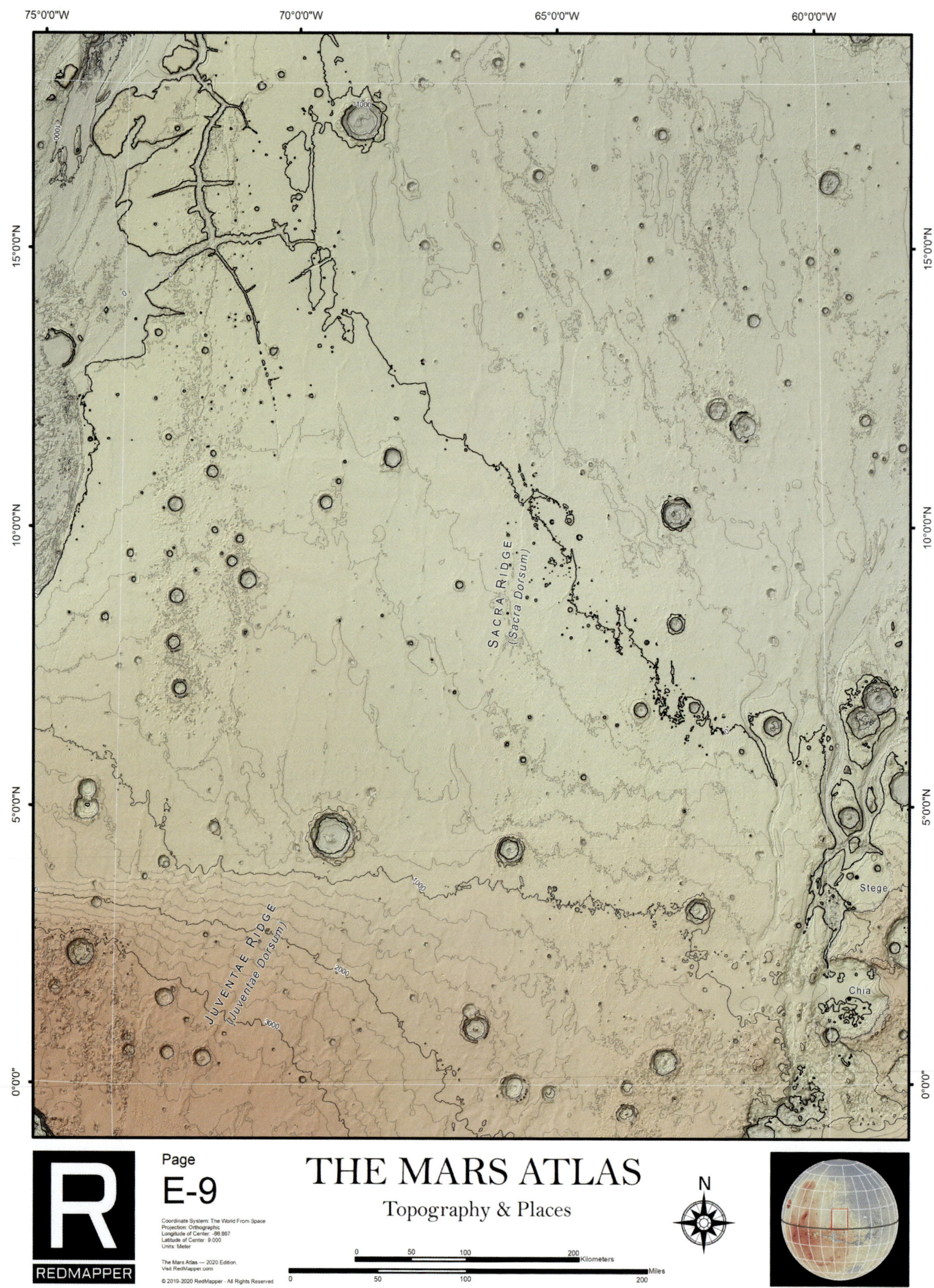
75°0'0"W
70°0'0"W
65°0'0"W
60°0'0"W
15°0'0"N
10°0'0"N
5°0'0"N
0°0'0"
SACRA RIDGE
(Sacra Dorsum)
JUVENTAE RIDGE
(Juventae Dorsum)
Stege
Chia
1000
2000
3000
R
REDMAPPER
Page
E-9
Coordinate System: The World From Space
Projection: Orthographic
Longitude of Center: -66.667
Latitude of Center: 9.000
Units: Meter
The Mars Atlas — 2020 Edition.
Visit RedMapper.com
© 2019-2020 RedMapper - All Rights Reserved
THE MARS ATLAS
Topography & Places
N
0 50 100 200 Kilometers
0 50 100 200 Miles

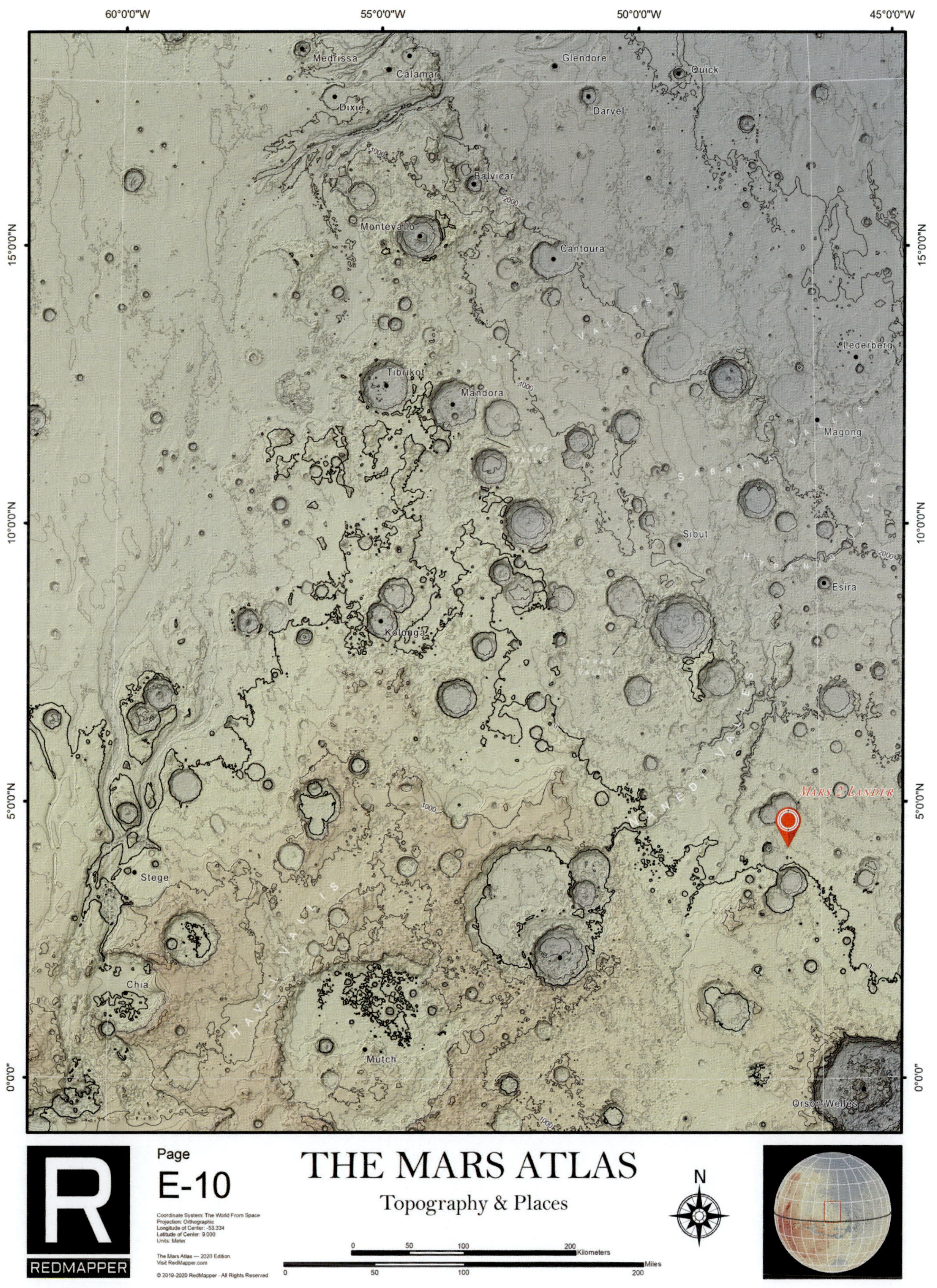
60°0'0"W
55°0'0"W
50°0'0"W
45°0'0"W
15°0'0"N
10°0'0"N
5°0'0"N
0°0'0"
Medrissa
Calamar
Glendore
Quick
Dixie
Darvel
Balvicar
Montevallo
Cantoura
Vistula Valles
Lederberg
Tibrikot
Mandora
Magong
Subur Vallis
Sabrina Vallis
Sibut
Hypanis Valles
Esira
Kolonga
Tiber Vallis
Nanedi Valles
Mars 2 Lander
Stege
Havel Vallis
Chia
Mutch
Orson Welles
1000
2000
Page
E-10
Coordinate System: The World From Space
Projection: Orthographic
Longitude of Center: -53.334
Latitude of Center: 9.000
Units: Meter
The Mars Atlas — 2020 Edition
Visit RedMapper.com
© 2019-2020 RedMapper - All Rights Reserved
R
REDMAPPER
THE MARS ATLAS
Topography & Places
0 50 100 200 Kilometers
0 50 100 200 Miles
N

45°0'0"W
40°0'0"W
35°0'0"W
15°0'0"N
10°0'0"N
5°0'0"N
0°0'0"
SIMUD VALLES
TIU VALLES
OCHUS VALLES
Zir
Bor
Vaux
Nune
Yala
Luck
Oraibi
Ori
Concord
Suf
Lederberg
Magong
Masursky
Esira
Dukhan
Mojave
Mars 2 Lander
Da Vinci
Dia-Cau
Camiling
-3000
-2000
-1000
Page
E-11
Coordinate System: The World From Space
Projection: Orthographic
Longitude of Center: -40.000
Latitude of Center: 9.000
Units: Meter
The Mars Atlas — 2020 Edition.
Visit RedMapper.com
© 2019-2020 RedMapper - All Rights Reserved
THE MARS ATLAS
Topography & Places
N
0
50
100
200
Kilometers
Miles
REDMAPPER

Page
E-12

Coordinate System: The World From Space
Projection: Orthographic
Longitude of Center: -26.667
Latitude of Center: 9.000
Units: Meter

The Mars Atlas — 2020 Edition.
Visit RedMapper.com

THE MARS ATLAS

Topography & Places

N

0 50 100 200 Kilometers
0 50 100 200 Miles

R
REDMAPPER

Page

E-13

Coordinate System: The World From Space
Projection: Orthographic
Longitude of Center: -13.334
Latitude of Center: 9.000
Units: Meter

The Mars Atlas — 2020 Edition.
Visit RedMapper.com

R
REDMAPPER

THE MARS ATLAS

Topography & Places

0 50 100 200 Kilometers
0 50 100 200 Miles

N

R
REDMAPPER

Page
E-14

THE MARS ATLAS
Topography & Places

Coordinate System: The World From Space
Projection: Orthographic
Longitude of Center: 0.000
Latitude of Center: 9.000
Units: Meter

The Mars Atlas — 2020 Edition.
Visit RedMapper.com

0 50 100 200 Kilometers
0 50 100 200 Miles

N

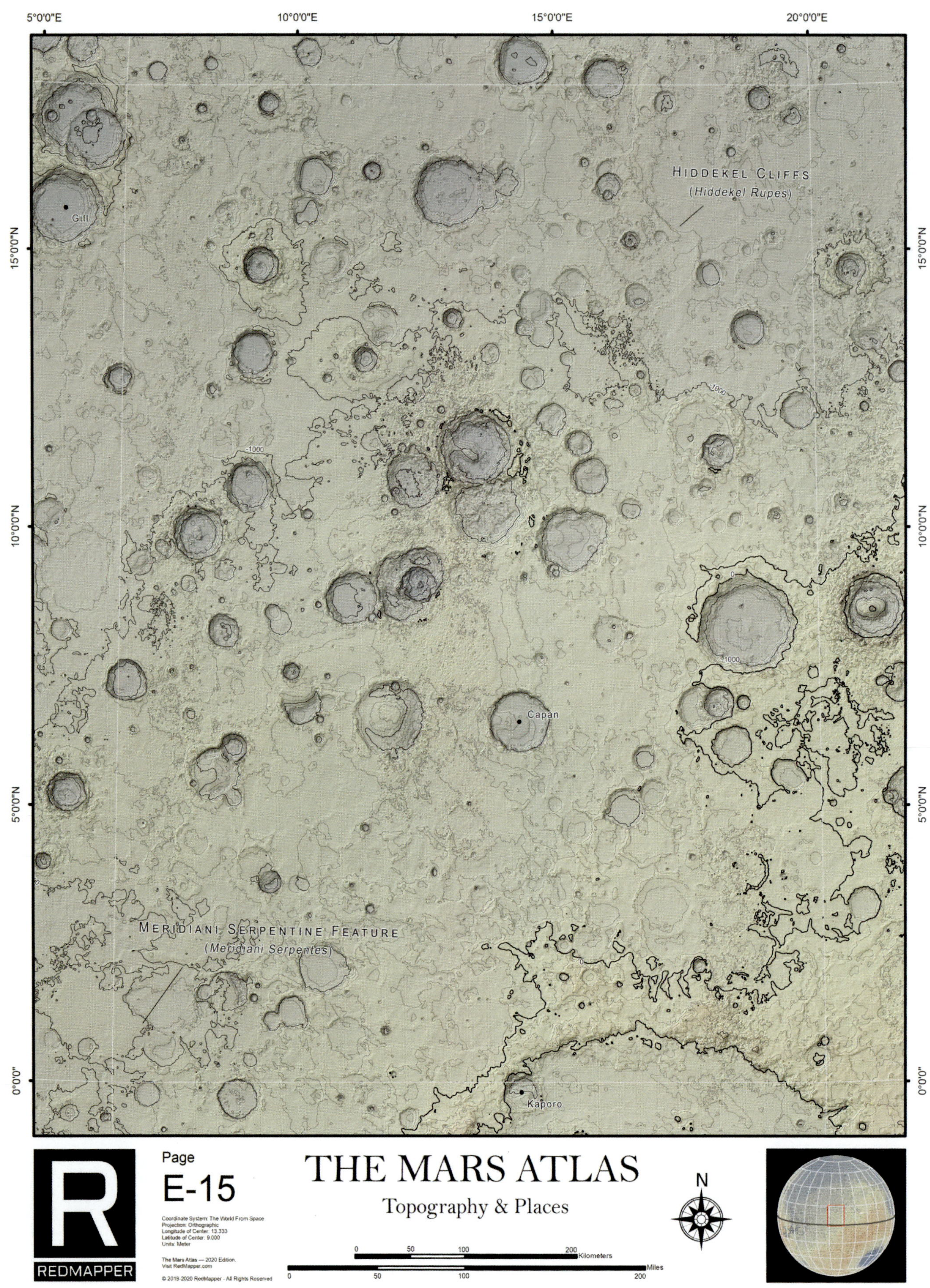
5°0'0"E
10°0'0"E
15°0'0"E
20°0'0"E
15°0'0"N
10°0'0"N
5°0'0"N
0°0'0"
Gill
HIDDEKEL CLIFFS
(Hiddekel Rupes)
-1000
Capan
MERIDIANI SERPENTINE FEATURE
(Meridiani Serpentes)
Kaporo
REDMAPPER
Page
E-15
Coordinate System: The World From Space
Projection: Orthographic
Longitude of Center: 13.333
Latitude of Center: 9.000
Units: Meter
The Mars Atlas — 2020 Edition.
Visit RedMapper.com
© 2019-2020 RedMapper - All Rights Reserved
THE MARS ATLAS
Topography & Places
N
0 50 100 200 Kilometers
0 50 100 200 Miles

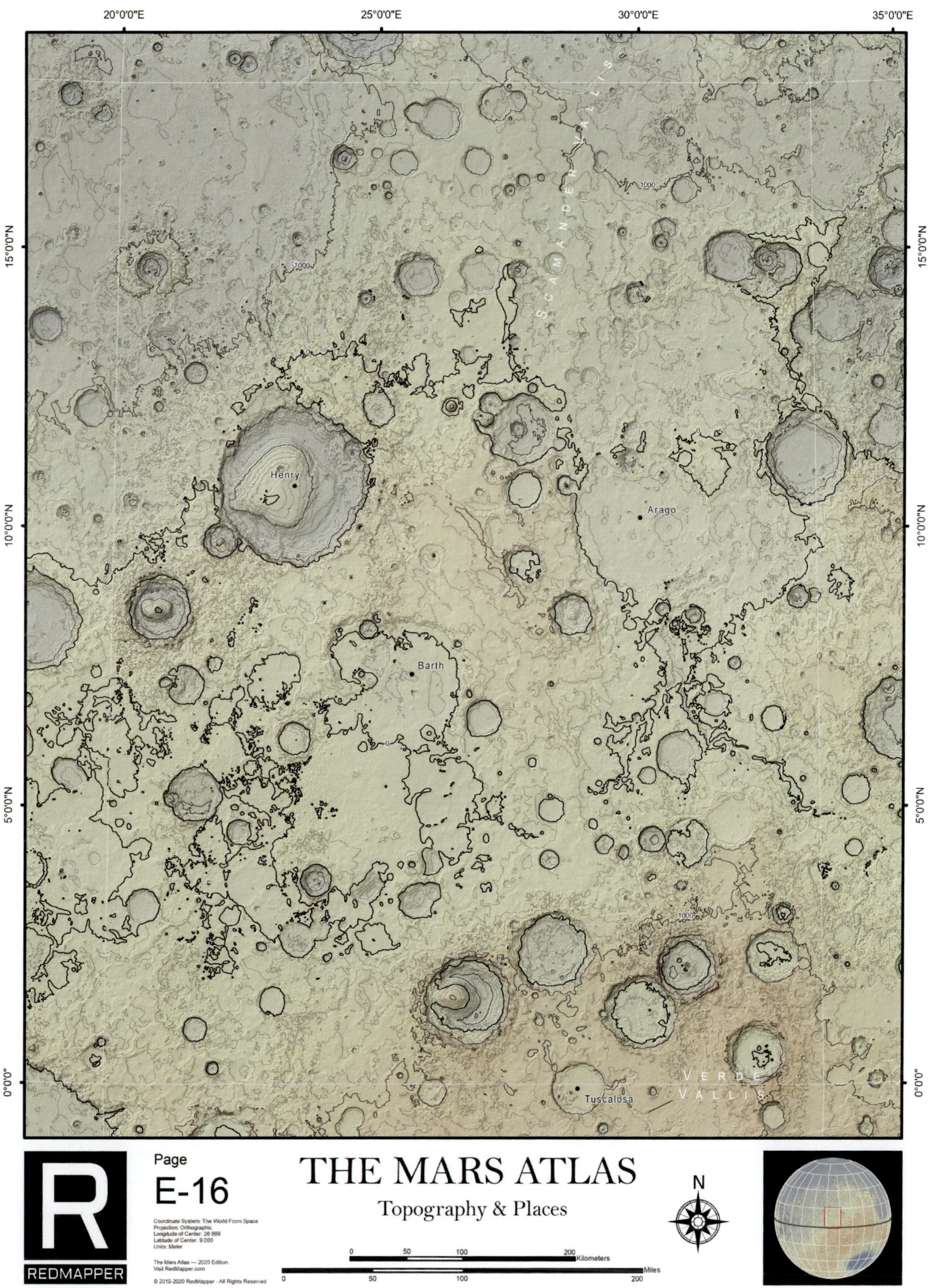
20°0'0"E
25°0'0"E
30°0'0"E
35°0'0"E
15°0'0"N
10°0'0"N
5°0'0"N
0°0'0"
SCAMANDER VALLIS
1000
-1000
Henry
Arago
Barth
0
1000
Tuscalosa
VERDE VALLIS
R
REDMAPPER
Page
E-16
THE MARS ATLAS
Topography & Places
Coordinate System: The World From Space
Projection: Orthographic
Longitude of Center: 26.666
Latitude of Center: 9.000
Units: Meter
The Mars Atlas — 2020 Edition.
Visit RedMapper.com
© 2019-2020 RedMapper - All Rights Reserved
0 50 100 200 Kilometers
0 50 100 200 Miles
N

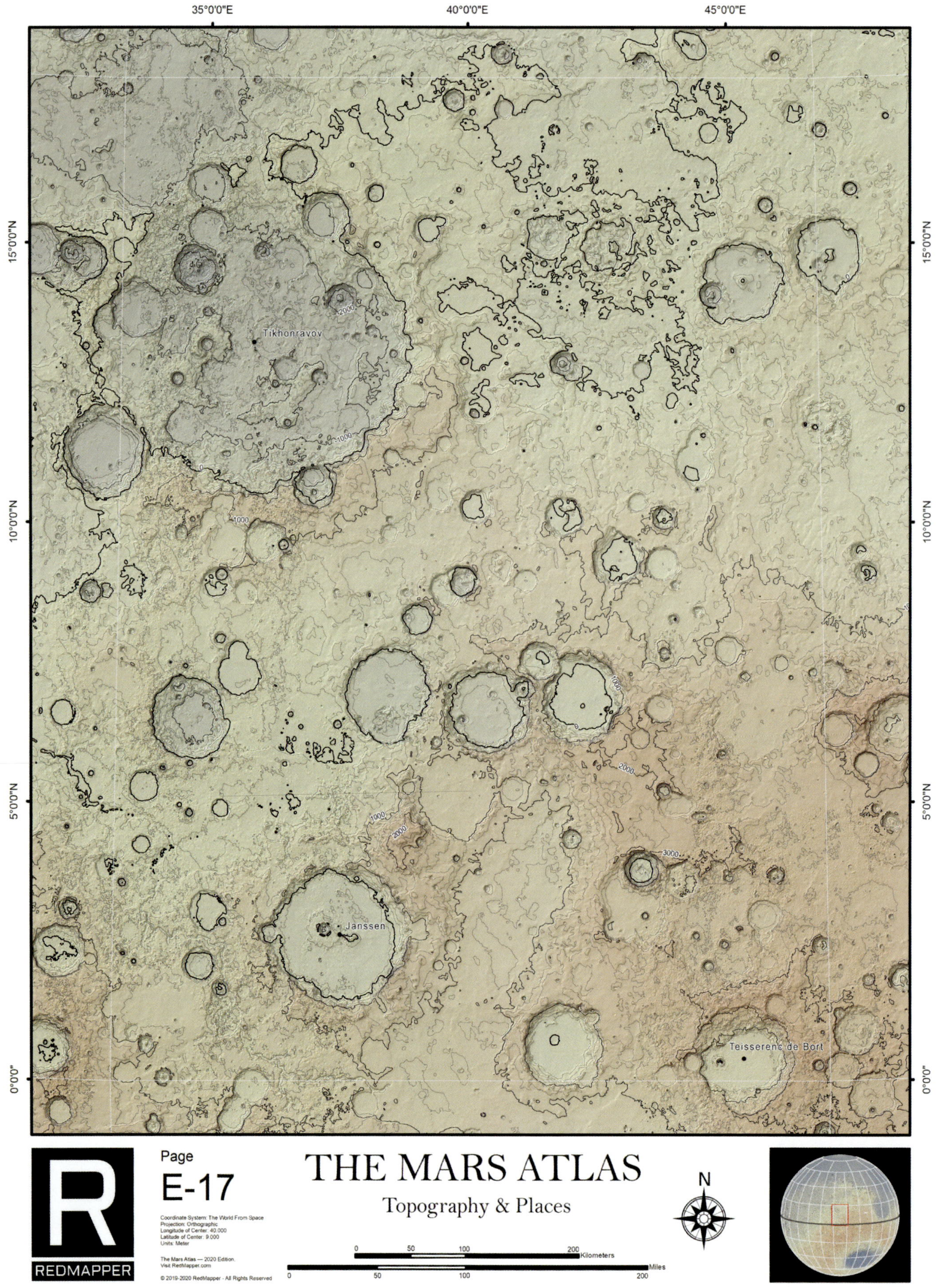
35°0'0"E
40°0'0"E
45°0'0"E
15°0'0"N
10°0'0"N
5°0'0"N
0°0'0"
Tikhonravov
Janssen
Teisserenc de Bort
1000
2000
3000
R
REDMAPPER
Page
E-17
Coordinate System: The World From Space
Projection: Orthographic
Longitude of Center: 40.000
Latitude of Center: 9.000
Units: Meter
The Mars Atlas — 2020 Edition.
Visit RedMapper.com
© 2019-2020 RedMapper - All Rights Reserved
THE MARS ATLAS
Topography & Places
0 50 100 200 Kilometers
0 50 100 200 Miles
N

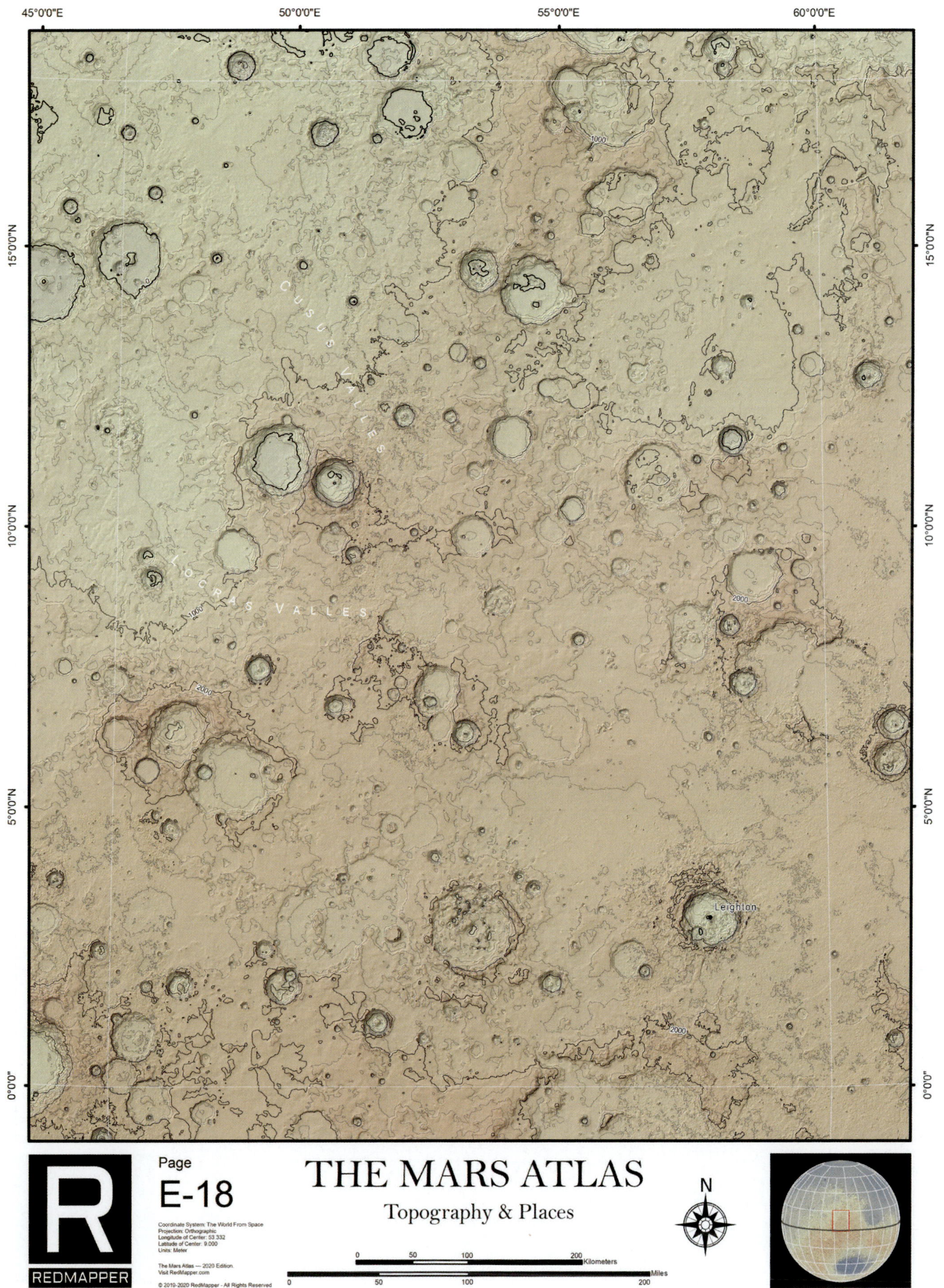

R
REDMAPPER

Page
E-18

Coordinate System: The World From Space
Projection: Orthographic
Longitude of Center: 53.332
Latitude of Center: 9.000
Units: Meter

The Mars Atlas — 2020 Edition.
Visit RedMapper.com

THE MARS ATLAS

Topography & Places

N

0 50 100 200 Kilometers

0 50 100 200 Miles

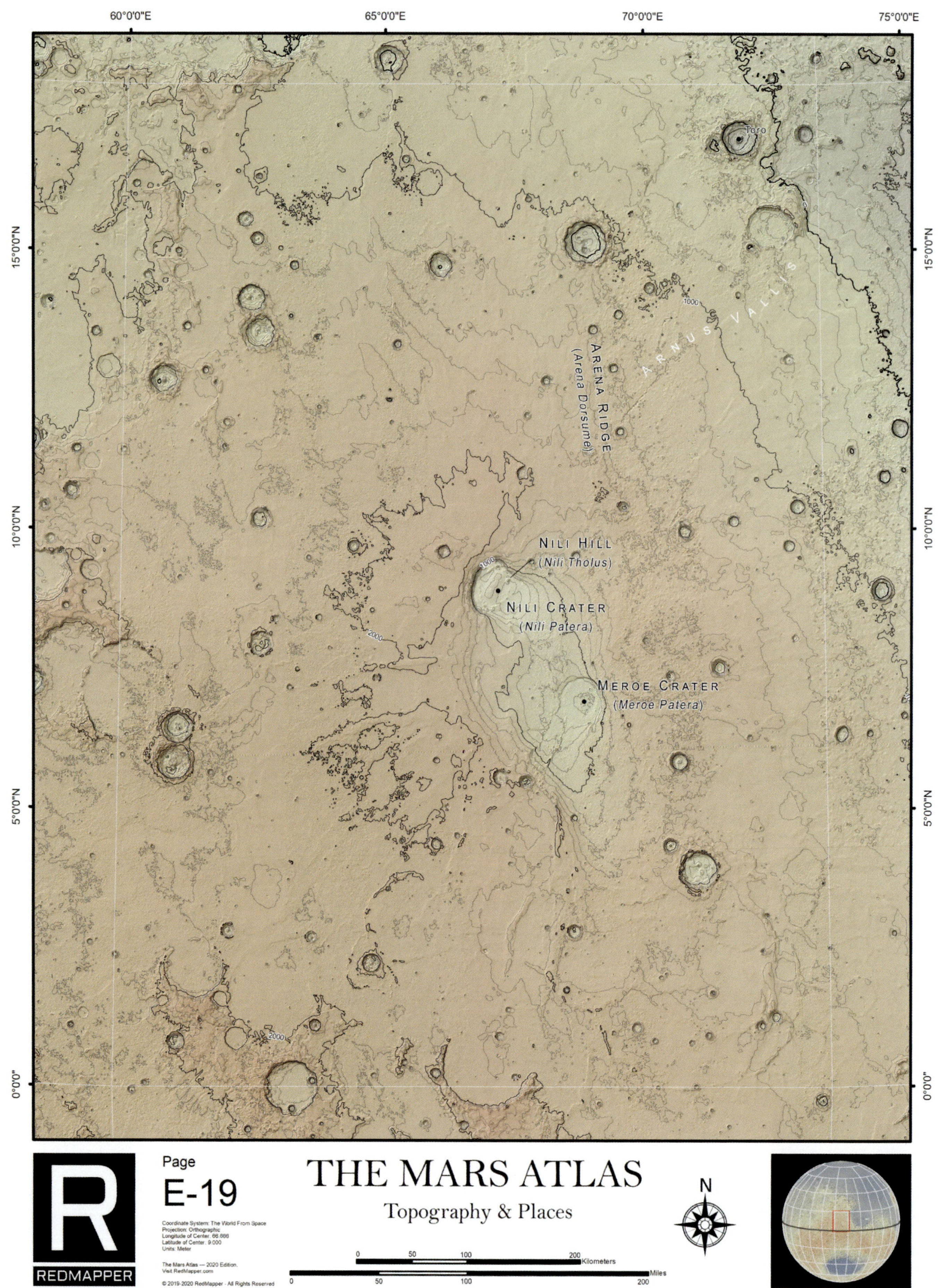
60°0'0"E
65°0'0"E
70°0'0"E
75°0'0"E
15°0'0"N
10°0'0"N
5°0'0"N
0°0'0"
Toro
ARNUS VALLIS
1000
ARENA RIDGE
(Arena Dorsum)
NILI HILL
(Nili Tholus)
NILI CRATER
(Nili Patera)
MEROE CRATER
(Meroe Patera)
2000
REDMAPPER
Page
E-19
Coordinate System: The World From Space
Projection: Orthographic
Longitude of Center: 66.666
Latitude of Center: 9.000
Units: Meter
The Mars Atlas — 2020 Edition.
Visit RedMapper.com
© 2019-2020 RedMapper - All Rights Reserved
THE MARS ATLAS
Topography & Places
N
0 50 100 200 Kilometers
0 50 100 200 Miles

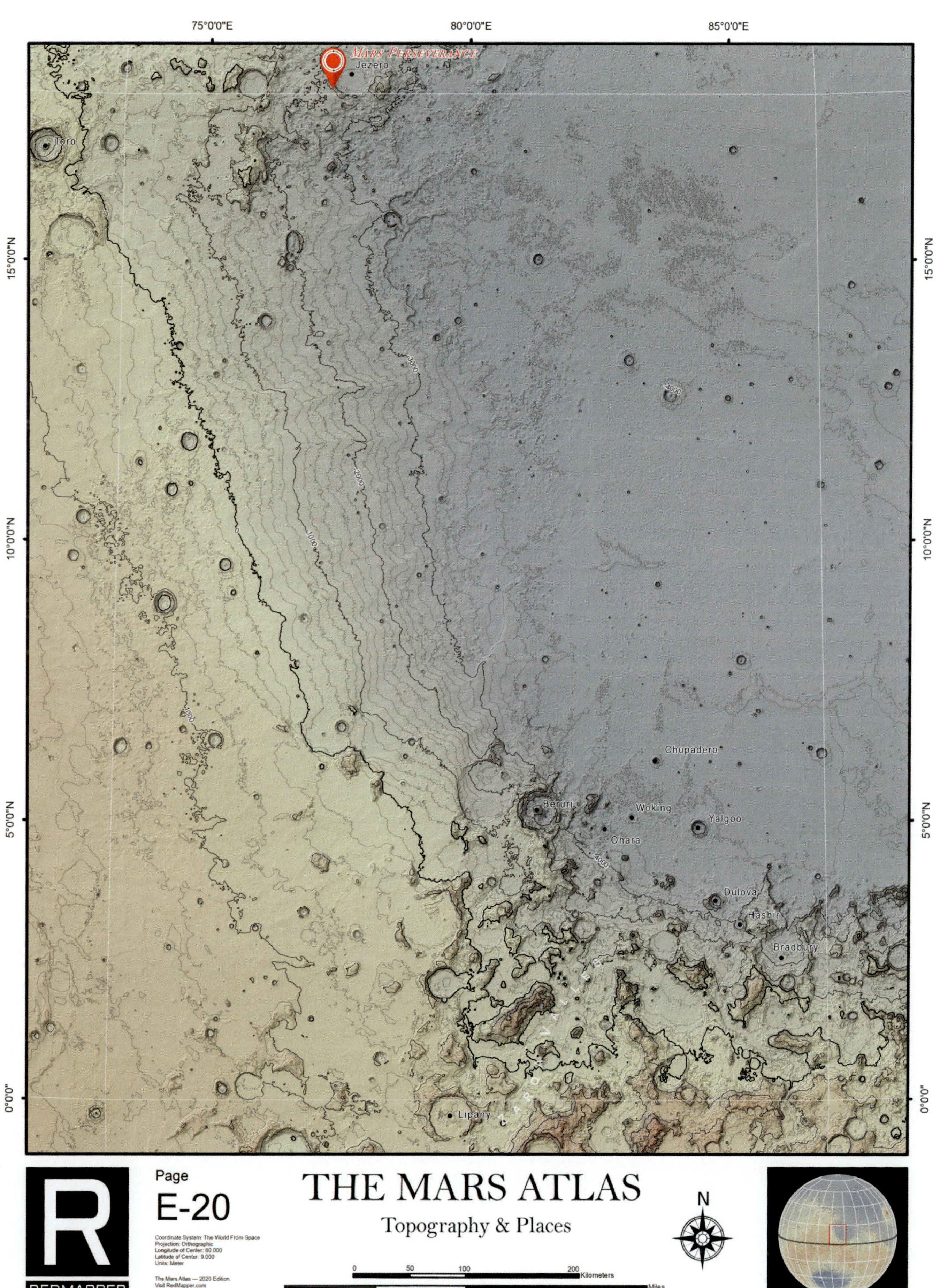

R
REDMAPPER

Page
E-20

Coordinate System: The World From Space
Projection: Orthographic
Longitude of Center: 80.000
Latitude of Center: 9.000
Units: Meter

The Mars Atlas — 2020 Edition.
Visit RedMapper.com

THE MARS ATLAS

Topography & Places

N

0 50 100 200 Kilometers

0 50 100 200 Miles

85°0'0"E
90°0'0"E
95°0'0"E
100°0'0"E
15°0'0"N
10°0'0"N
5°0'0"N
0°0'0"
Isidis Ridge
(Isidis Dorsa)
Beagle 2 Lander
Du Martheray
-4000
-3000
-2000
-1000
1000
2000
Page
E-21
Coordinate System: The World From Space
Projection: Orthographic
Longitude of Center: 93.332
Latitude of Center: 9.000
Units: Meter
The Mars Atlas — 2020 Edition.
Visit RedMapper.com
© 2019-2020 RedMapper - All Rights Reserved
THE MARS ATLAS
Topography & Places
N
0
50
100
200
Kilometers
Miles
REDMAPPER
R

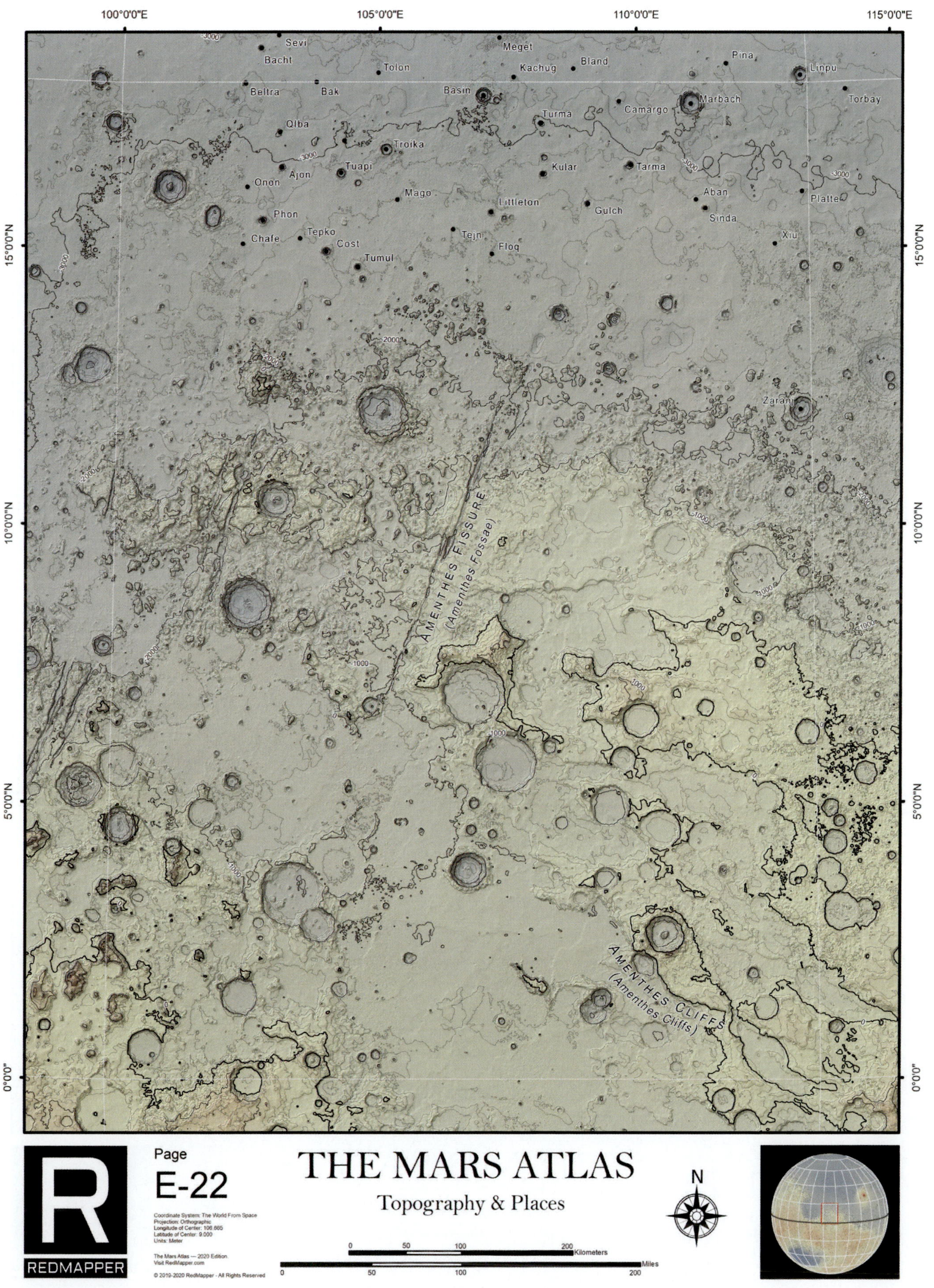
100°0'0"E
105°0'0"E
110°0'0"E
115°0'0"E
15°0'0"N
10°0'0"N
5°0'0"N
0°0'0"
Sevi
Bacht
Meget
Kachug
Bland
Pina
Linpu
Torbay
Tolon
Beltra
Bak
Basin
Camargo
Marbach
Turma
Qlba
Troika
Tuapi
Ajon
Onon
Kular
Tarma
Mago
Aban
Platte
Littleton
Gulch
Sinda
Phon
Tepko
Chafe
Cost
Tejn
Floq
Xiu
Tumul
Zaran
AMENTHES FISSURE
(Amenthes Fossae)
AMENTHES CLIFFS
(Amenthes Cliffs)
-3000
-2000
-1000
1000
0
R
REDMAPPER
Page
E-22
Coordinate System: The World From Space
Projection: Orthographic
Longitude of Center: 106.665
Latitude of Center: 9.000
Units: Meter
The Mars Atlas — 2020 Edition.
Visit RedMapper.com
© 2019-2020 RedMapper - All Rights Reserved
THE MARS ATLAS
Topography & Places
N
0
50
100
200
Kilometers
Miles

115°0'0"E
120°0'0"E
125°0'0"E
15°0'0"N
10°0'0"N
5°0'0"N
0°0'0"
Pina
Linpu
Torbay
Platte
Xiu
Tavua
Tomini
Naryn
Phedra
Zarani
Doba
Canillo
Nepenthese Mesa
(Nepenthese Mensae)
Eilden
Escalante
-3000
-4000
-2000
-1000
0
1000
2000
Page
E-23
THE MARS ATLAS
Topography & Places
Coordinate System: The World From Space
Projection: Orthographic
Longitude of Center: 120.000
Latitude of Center: 9.000
Units: Meter
The Mars Atlas — 2020 Edition.
Visit RedMapper.com
© 2019-2020 RedMapper - All Rights Reserved
REDMAPPER
N
0
50
100
200
Kilometers
Miles

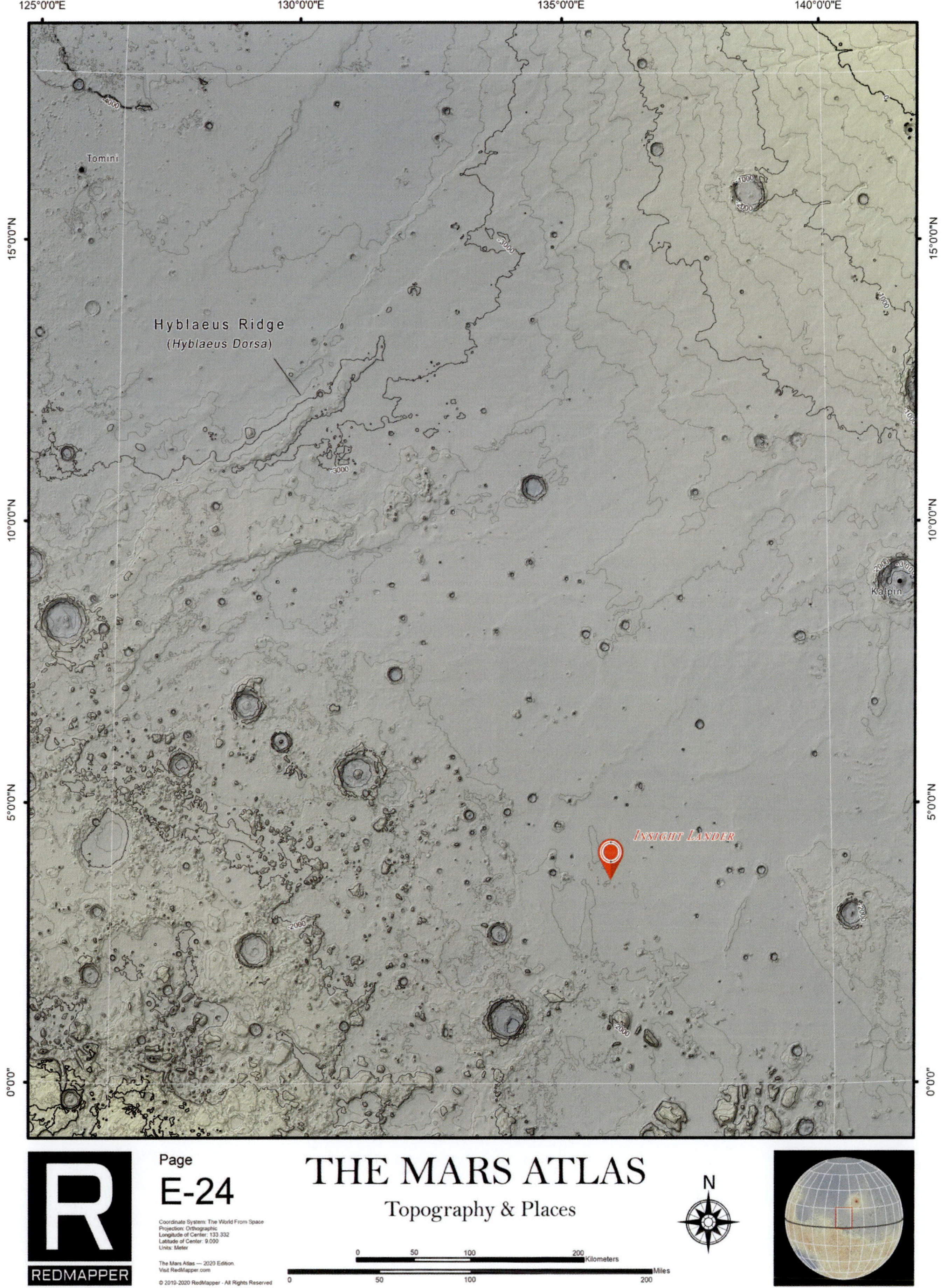
125°0'0"E
130°0'0"E
135°0'0"E
140°0'0"E
15°0'0"N
10°0'0"N
5°0'0"N
0°0'0"
Tomini
Hyblaeus Ridge
(Hyblaeus Dorsa)
Kalpin
Insight Lander
-3000
-1000
-2000
N
R
REDMAPPER
Page
E-24
THE MARS ATLAS
Topography & Places
Coordinate System: The World From Space
Projection: Orthographic
Longitude of Center: 133.332
Latitude of Center: 9.000
Units: Meter
The Mars Atlas — 2020 Edition.
Visit RedMapper.com
0 50 100 200 Kilometers
0 50 100 200 Miles

140°0'0"E
145°0'0"E
150°0'0"E
155°0'0"E
15°0'0"N
10°0'0"N
5°0'0"N
0°0'0"
ALBOR HILL
(Albor Tholus)
ALBOR FISSURE
(Albor Fossae)
ELYSIUM CRATER CHAIN
(Elysium Catena)
Corinto
Eddie
Kalpin
Wafra
Quthing
Gunjur
1000
0
-1000
-2000
-3000
-4000
R
REDMAPPER
Page
E-25
Coordinate System: The World From Space
Projection: Orthographic
Longitude of Center: 146.885
Latitude of Center: 9.000
Units: Meter
The Mars Atlas — 2020 Edition.
Visit RedMapper.com
© 2019-2020 RedMapper - All Rights Reserved
THE MARS ATLAS
Topography & Places
N
0
50
100
200
Kilometers
Miles

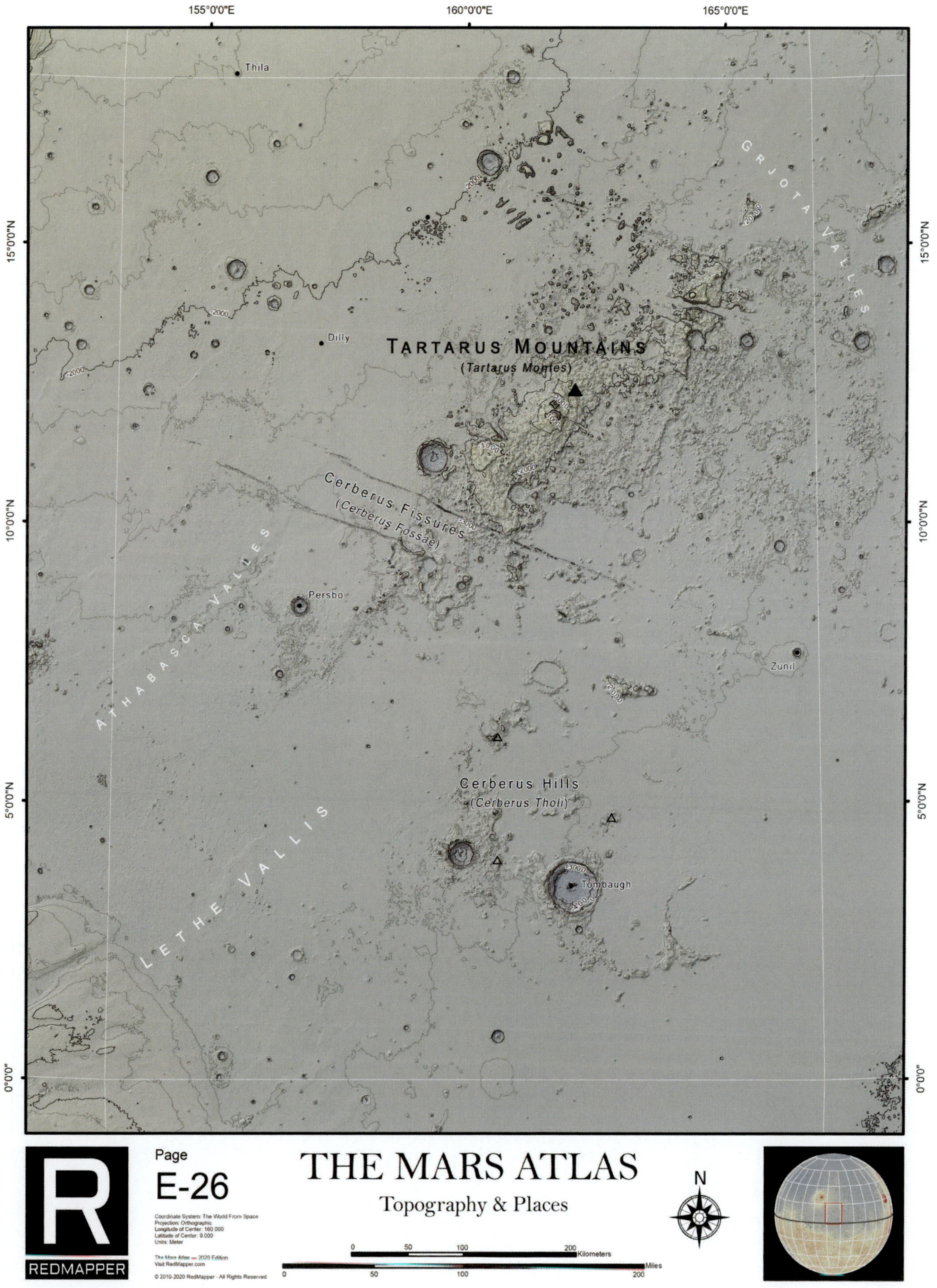

Page

E-26

Coordinate System: The World From Space
Projection: Orthographic
Longitude of Center: 160.000
Latitude of Center: 9.000
Units: Meter

The Mars Atlas — 2020 Edition
Visit RedMapper.com

THE MARS ATLAS

Topography & Places

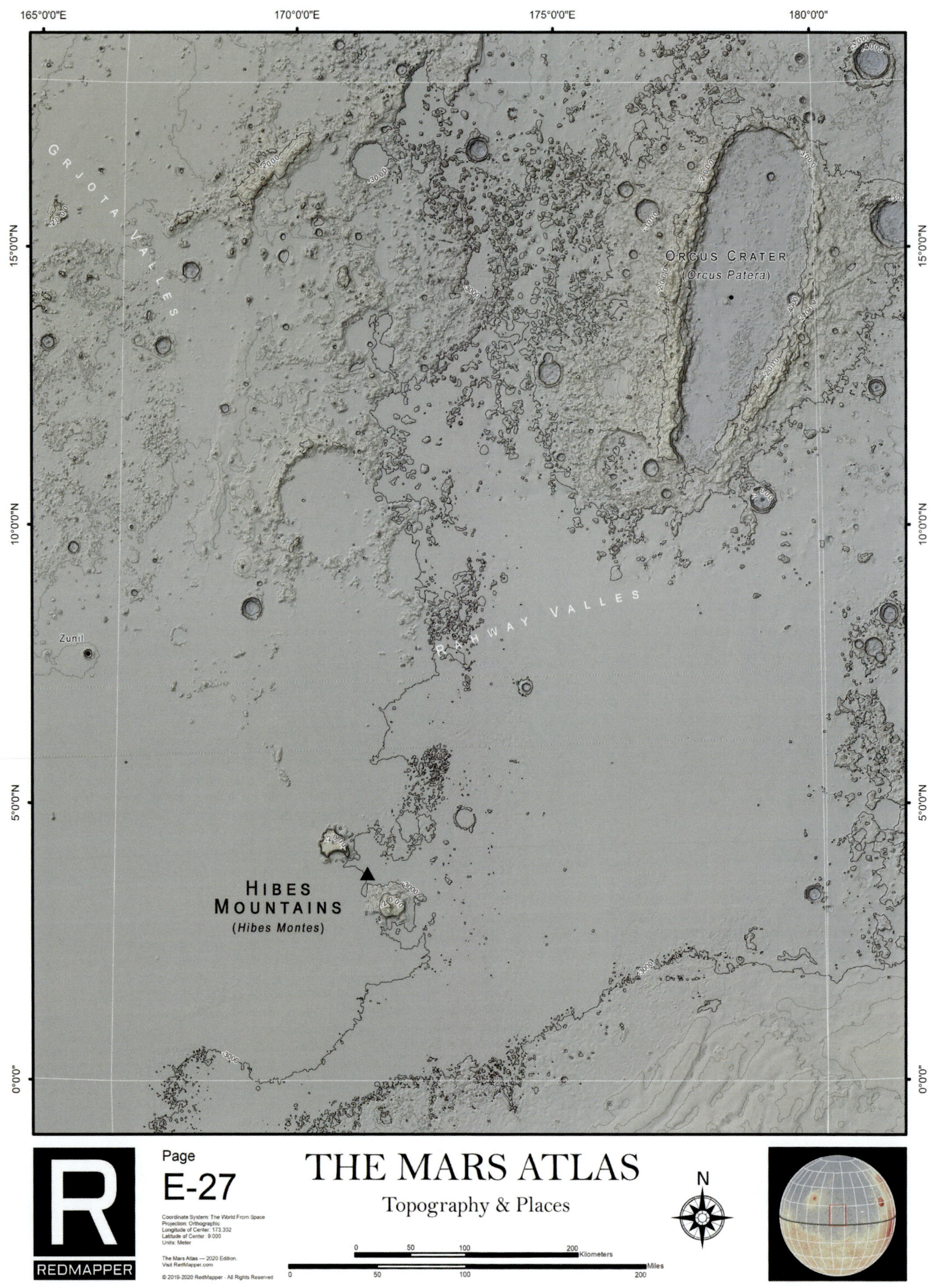

165°0'0"E
170°0'0"E
175°0'0"E
180°0'0"
15°0'0"N
10°0'0"N
5°0'0"N
0°0'0"
GRJOTA VALLES
ORCUS CRATER
(Orcus Patera)
RAHWAY VALLES
Zunil
HIBES MOUNTAINS
(Hibes Montes)
R
REDMAPPER
Page
E-27
Coordinate System: The World From Space
Projection: Orthographic
Longitude of Center: 173.332
Latitude of Center: 9.000
Units: Meter
The Mars Atlas — 2020 Edition.
Visit RedMapper.com
© 2019-2020 RedMapper - All Rights Reserved
THE MARS ATLAS
Topography & Places
N
0 50 100 200 Kilometers
0 50 100 200 Miles

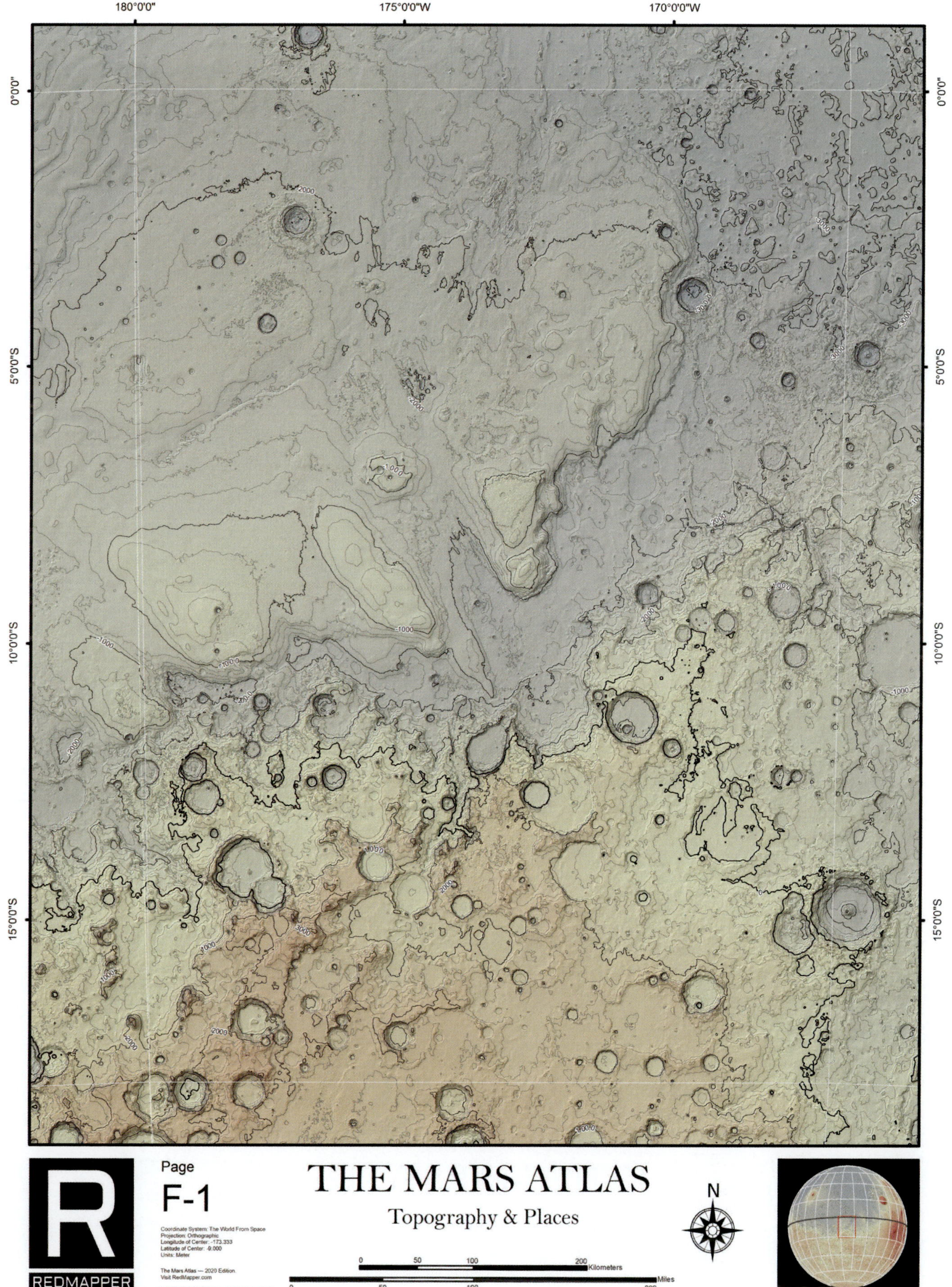

Page

F-1

THE MARS ATLAS

Topography & Places

Coordinate System: The World From Space
Projection: Orthographic
Longitude of Center: -173.333
Latitude of Center: -9.000
Units: Meter

The Mars Atlas — 2020 Edition.
Visit RedMapper.com

0 50 100 200 Kilometers

0 50 100 200 Miles

REDMAPPER

Page
F-2

Coordinate System: The World From Space
Projection: Orthographic
Longitude of Center: -160.000
Latitude of Center: -9.000
Units: Meter

The Mars Atlas — 2020 Edition.
Visit RedMapper.com

THE MARS ATLAS

Topography & Places

N

0 50 100 200 Kilometers

0 50 100 200 Miles

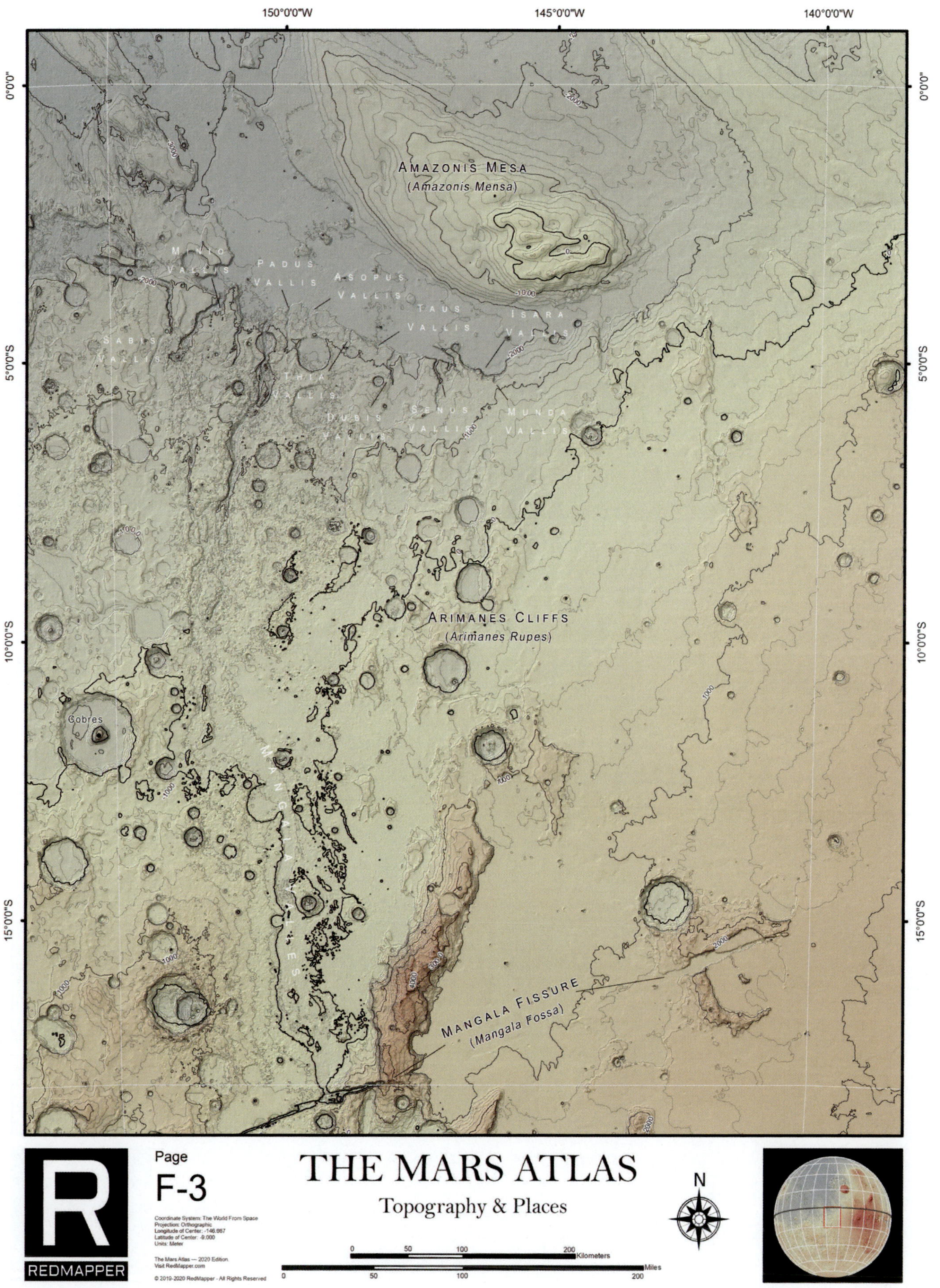
150°0'0"W
145°0'0"W
140°0'0"W
0°0'0"
5°0'0"S
10°0'0"S
15°0'0"S
AMAZONIS MESA
(Amazonis Mensa)
MINIO VALLIS
PADUS VALLIS
ASOPUS VALLIS
TAUS VALLIS
ISARA VALLIS
SABIS VALLIS
THIA VALLIS
DUBIS VALLIS
SENUS VALLIS
MUNDA VALLIS
ARIMANES CLIFFS
(Arimanes Rupes)
Cobres
MANGALA VALLES
MANGALA FISSURE
(Mangala Fossa)
R
REDMAPPER
Page
F-3
Coordinate System: The World From Space
Projection: Orthographic
Longitude of Center: -146.667
Latitude of Center: -9.000
Units: Meter
The Mars Atlas — 2020 Edition.
Visit RedMapper.com
© 2019-2020 RedMapper - All Rights Reserved
THE MARS ATLAS
Topography & Places
N
0 50 100 200 Kilometers
0 50 100 200 Miles

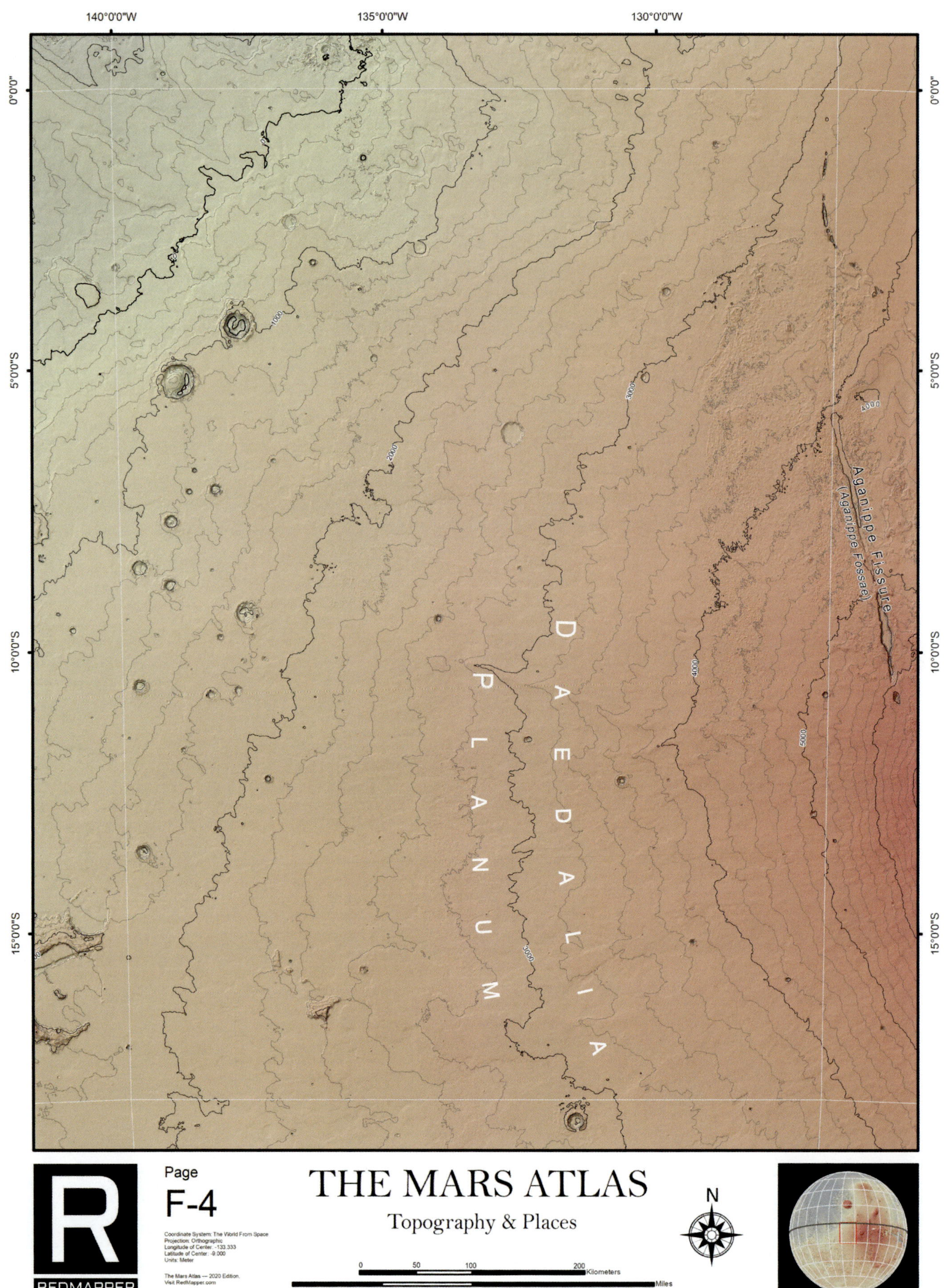

Page
F-4

THE MARS ATLAS

Topography & Places

Coordinate System: The World From Space
Projection: Orthographic
Longitude of Center: -133.333
Latitude of Center: -9.000
Units: Meter

The Mars Atlas — 2020 Edition.
Visit RedMapper.com

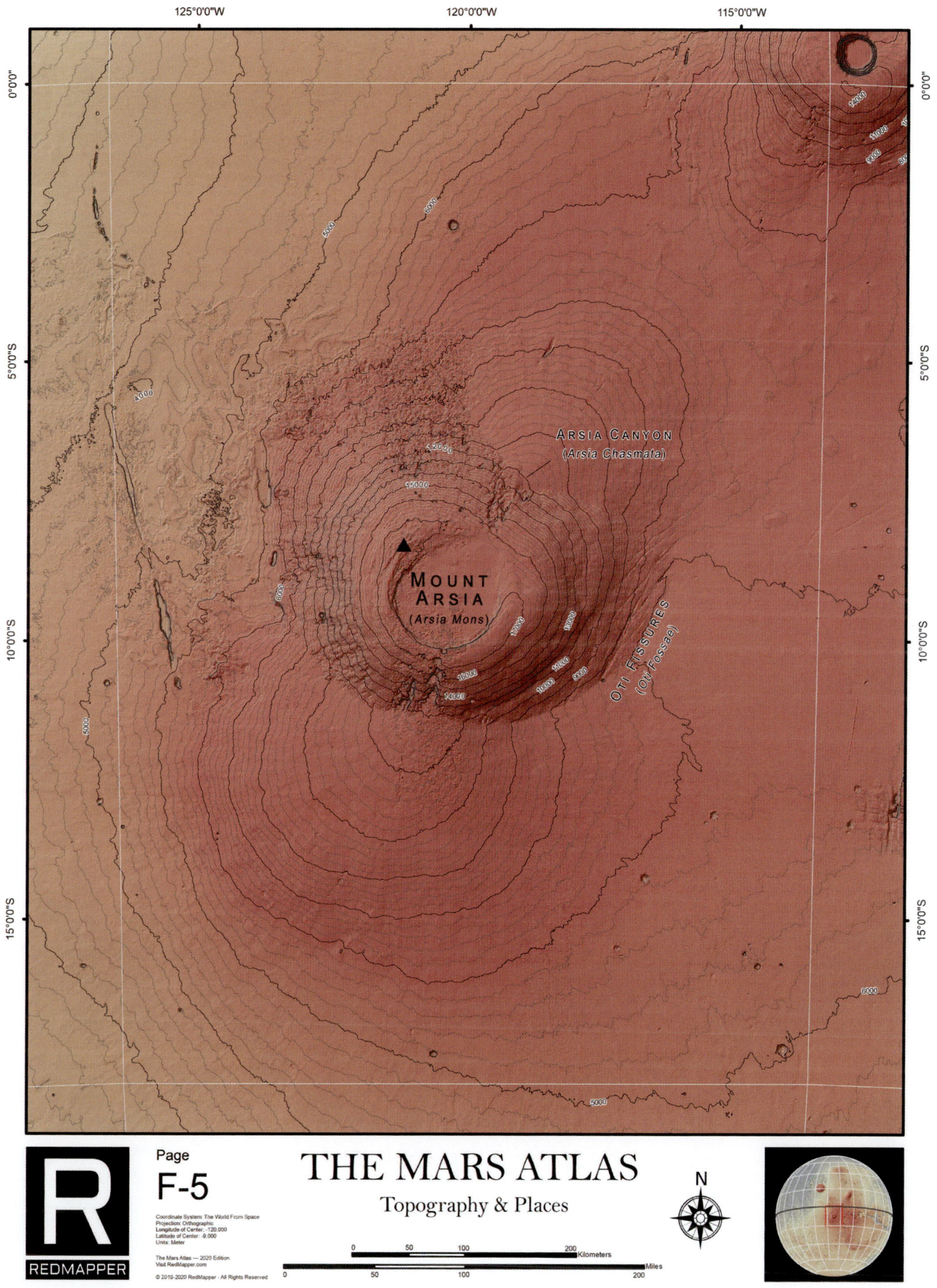
125°0'0"W
120°0'0"W
115°0'0"W
0°0'0"
5°0'0"S
10°0'0"S
15°0'0"S
Arsia Canyon
(Arsia Chasmata)
Mount
Arsia
(Arsia Mons)
Oti Fissures
(Oti Fossae)
12000
15000
14000
11000
10000
9000
6000
5000
4000
R
REDMAPPER
Page
F-5
Coordinate System: The World From Space
Projection: Orthographic
Longitude of Center: -120.000
Latitude of Center: -9.000
Units: Meter
The Mars Atlas — 2020 Edition.
Visit RedMapper.com
© 2019-2020 RedMapper - All Rights Reserved
THE MARS ATLAS
Topography & Places
N
0
50
100
200
Kilometers
Miles

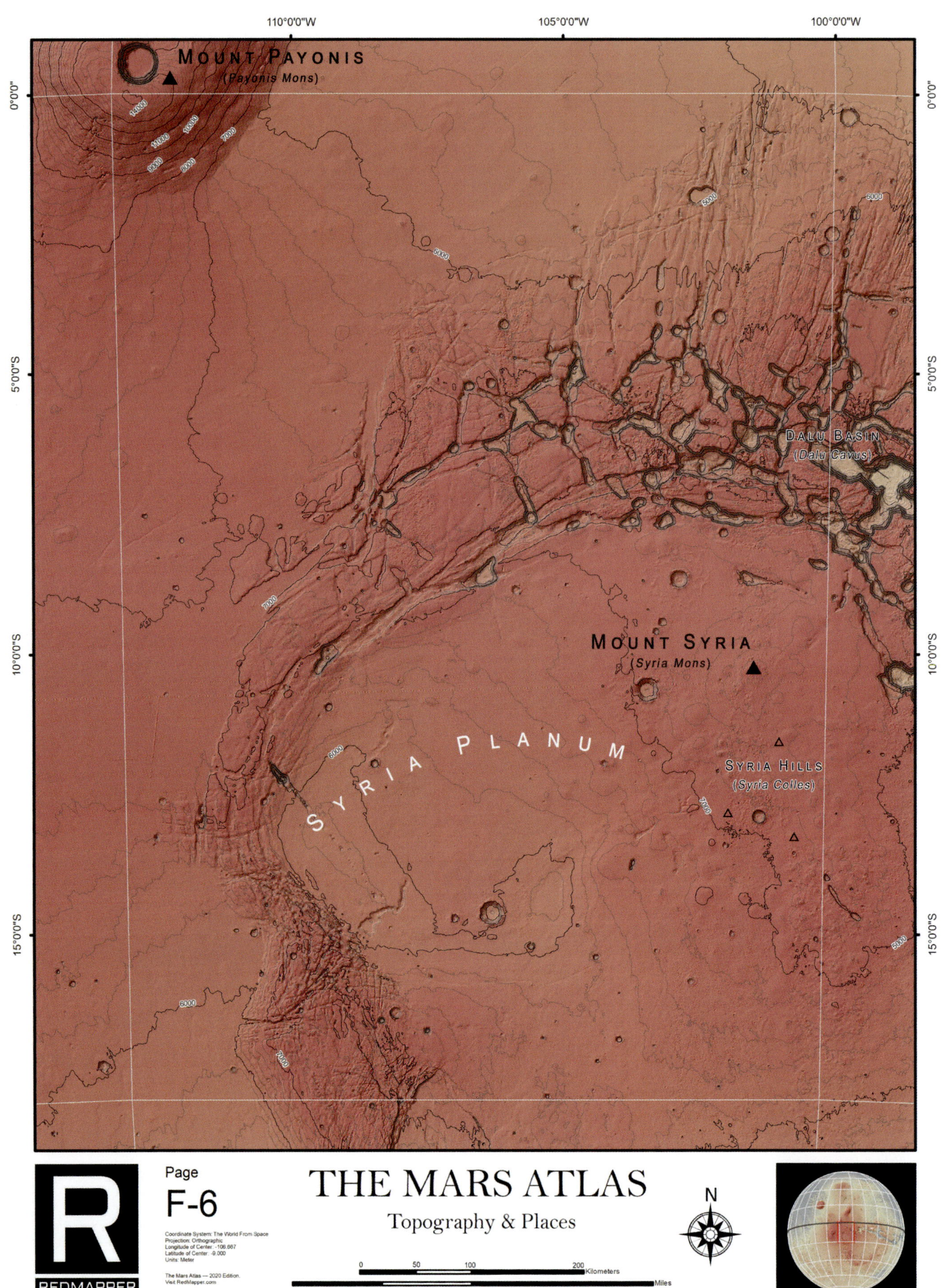
110°0'0"W
105°0'0"W
100°0'0"W
0°0'0"
5°0'0"S
10°0'0"S
15°0'0"S
MOUNT PAYONIS
(Payonis Mons)
DALU BASIN
(Dalu Cavus)
MOUNT SYRIA
(Syria Mons)
SYRIA PLANUM
SYRIA HILLS
(Syria Colles)
14000
11000
10000
9000
8000
7000
6000
5000
R
REDMAPPER
Page
F-6
Coordinate System: The World From Space
Projection: Orthographic
Longitude of Center: -106.667
Latitude of Center: -9.000
Units: Meter
The Mars Atlas — 2020 Edition.
Visit RedMapper.com
© 2019-2020 RedMapper - All Rights Reserved
THE MARS ATLAS
Topography & Places
N
0 50 100 200 Kilometers
0 50 100 200 Miles

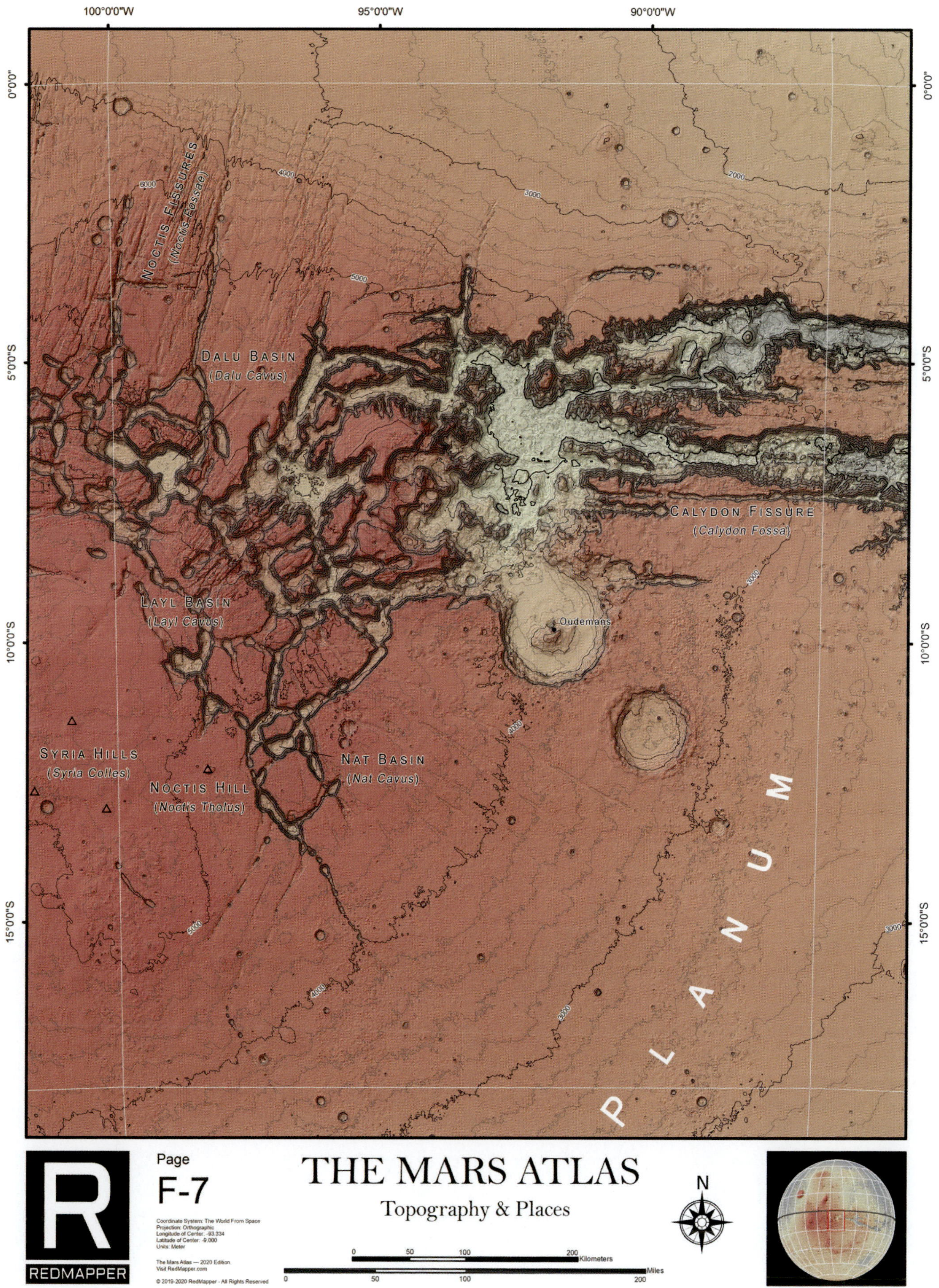

100°0'0"W
95°0'0"W
90°0'0"W
0°0'0"
5°0'0"S
10°0'0"S
15°0'0"S
NOCTIS FISSURES
(Noctis Fossae)
DALU BASIN
(Dalu Cavus)
CALYDON FISSURE
(Calydon Fossa)
LAYL BASIN
(Layl Cavus)
Oudemans
SYRIA HILLS
(Syria Colles)
NOCTIS HILL
(Noctis Tholus)
NAT BASIN
(Nat Cavus)
PLANUM
6000
5000
4000
3000
2000
R
REDMAPPER
Page
F-7
Coordinate System: The World From Space
Projection: Orthographic
Longitude of Center: -93.334
Latitude of Center: -9.000
Units: Meter
The Mars Atlas — 2020 Edition.
Visit RedMapper.com
© 2019-2020 RedMapper - All Rights Reserved
THE MARS ATLAS
Topography & Places
N
0 50 100 200 Kilometers
0 50 100 200 Miles

85°0'0"W
80°0'0"W
75°0'0"W
0°0'0"
5°0'0"S
10°0'0"S
15°0'0"S
Hebes Mesa
(Hebes Mensa)
Perrotin
Tithonium Canyon
(Tithonium Chasma)
Ophir Mesa
(Ophir Mensa)
Tithoniae Fissures
(Tithoniae Fossae)
Ius Canyon
(Ius Chasma)
Tithoniae Crater Chain
(Tithoniae Catenae)
Ceti Mesa
(Ceti Mensa)
Candor Mesa
(Ceti Mensa)
Nia Hill
(Nia Tholus)
Calydon Fissure
(Calydon Fossa)
Ius Mesa
(Ius Mensa)
Louros Valles
Geryon
Mountains
(Geryon Montes)
Melas Mesa
(Melas Mensa)
Sinai Ridge
(Sinae Dorsa)
Sinai Fissure
(Sinae Fossae)
Planum
2000
3000
4000
R
REDMAPPER
Page
F-8
Coordinate System: The World From Space
Projection: Orthographic
Longitude of Center: -80.000
Latitude of Center: -9.000
Units: Meter
The Mars Atlas — 2020 Edition.
Visit RedMapper.com
© 2019-2020 RedMapper - All Rights Reserved
THE MARS ATLAS
Topography & Places
N
0 50 100 200 Kilometers
0 50 100 200 Miles

70°0'0"W
65°0'0"W
60°0'0"W
0°0'0"
5°0'0"S
10°0'0"S
15°0'0"S
DITTAINO VALLIS
SABRIK VALLIS
GANGES CRATER CHAIN
(Ganges Catena)
OPHIR MESA
(Ophir Mensa)
BAETIS CANYONS
(Baetis Chasmata)
CETI CANYONS
(Ceti Chasmata)
JUVENTAE CANYON
(Juventae Chasma)
CANDOR MESA
(Candor Mensa)
CANDOR CANYONS
(Candor Chasmata)
NIA MESA
(Nia Mensa)
JUVENTAE MESA
(Juventae Mensa)
CHRYSAS MESA
(Chrysas Mensa)
HYDRAE BASIN
(Hydrae Cavus)
Pital
MENSA CANYON
(Mensa Chasma)
MELAS MESA
(Melas Mensa)
Avan
Garni
Sibiti
COPRATES MESA
(Coprates Mensa)
COPRATES CANYON
(Coprates Chasma)
NIA FISSURES
(Nia Fossae)
COPRATES CRATER CHAIN
(Coprates Catena)
MOUNT COPRATES
(Coprates Mons)
Arima
3000
4000
5000
6000
Page
F-9
Coordinate System: The World From Space
Projection: Orthographic
Longitude of Center: -66.667
Latitude of Center: -9.000
Units: Meter
The Mars Atlas — 2020 Edition.
Visit RedMapper.com
© 2019-2020 RedMapper - All Rights Reserved
REDMAPPER
THE MARS ATLAS
Topography & Places
N
0 50 100 200 Kilometers
0 50 100 200 Miles

Page
F-10

Coordinate System: The World From Space
Projection: Orthographic
Longitude of Center: -53.334
Latitude of Center: -9.000
Units: Meter

The Mars Atlas — 2020 Edition.
Visit RedMapper.com

THE MARS ATLAS

Topography & Places

45°0'0"W
40°0'0"W
35°0'0"W
0°0'0"
5°0'0"S
10°0'0"S
15°0'0"S
Orson Welles
Dia Cau
Camiling
Rypin
Chimbote
Mega
Windfall
Wicklow
Zulanka
West
East
Oglala
Pinglo
Tuskegee
Bamba
Locana
Bahn
Huancayo
Tarata
Manti
Spry
Balboa
Vals
Groves
Twin
Thalweg
Kaid
Sitka
Berseba
Conches
Rakke
Chinju
Lachute
Manah
Butte
Stobs
Byske
Kong
Azusa
Quorn
Timbuktu
Creel
Innsbruck
Wink
Kholm
Dael
Batoka
Sfax
Paks
Rincon
Glide
Capri Canyon
(Capri Chasma)
Ganges Canyon
(Ganges Chasma)
Eos Mesa
(Eos Mensa)
Columbia Valles
Marineris
Eos Canyon
(Eos Chasma)
Capri Mesa
(Capri Mensa)
Luba
1000
R
REDMAPPER
Page
F-11
Coordinate System: The World From Space
Projection: Orthographic
Longitude of Center: -40.000
Latitude of Center: -9.000
Units: Meter
The Mars Atlas — 2020 Edition.
Visit RedMapper.com
© 2019-2020 RedMapper - All Rights Reserved
THE MARS ATLAS
Topography & Places
0
50
100
200
Kilometers
Miles
N

Page

F-12

Coordinate System: The World From Space
Projection: Orthographic
Longitude of Center: 26.667
Latitude of Center: -9.000
Units: Meter

The Mars Atlas — 2020 Edition.
Visit RedMapper.com

THE MARS ATLAS

Topography & Places

N

0 50 100 200 Kilometers

0 50 100 200 Miles

R
REDMAPPER

R
REDMAPPER

Page
F-13

Coordinate System: The World From Space
Projection: Orthographic
Longitude of Center: -13.334
Latitude of Center: -9.000
Units: Meter

The Mars Atlas — 2020 Edition.
Visit RedMapper.com

THE MARS ATLAS
Topography & Places

N

0 50 100 200 Kilometers
0 50 100 200 Miles

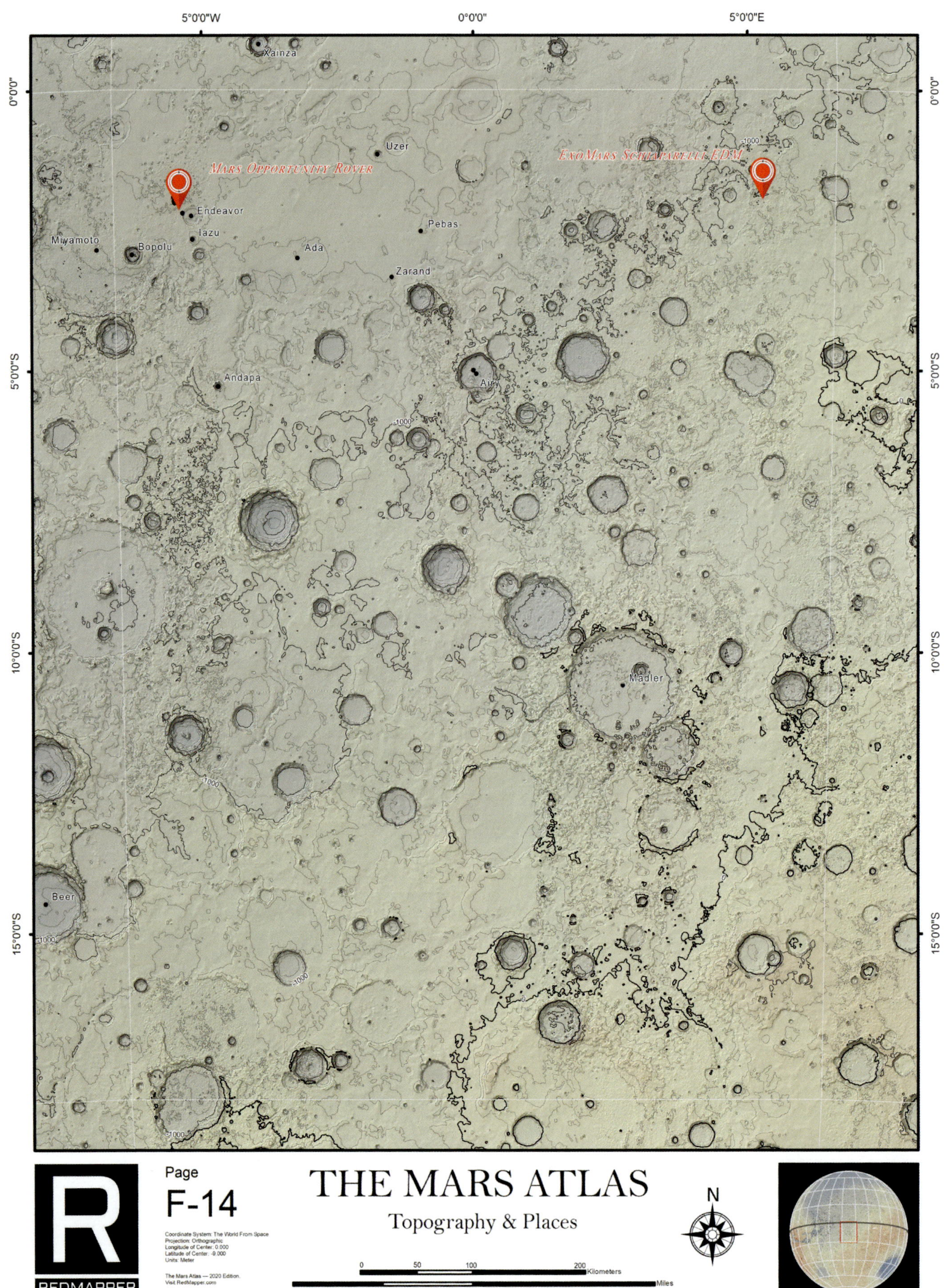

R
REDMAPPER

Page

F-14

Coordinate System: The World From Space
Projection: Orthographic
Longitude of Center: 0.000
Latitude of Center: -9.000
Units: Meter

The Mars Atlas — 2020 Edition.
Visit RedMapper.com

THE MARS ATLAS

Topography & Places

N

0 50 100 200 Kilometers

0 50 100 200 Miles

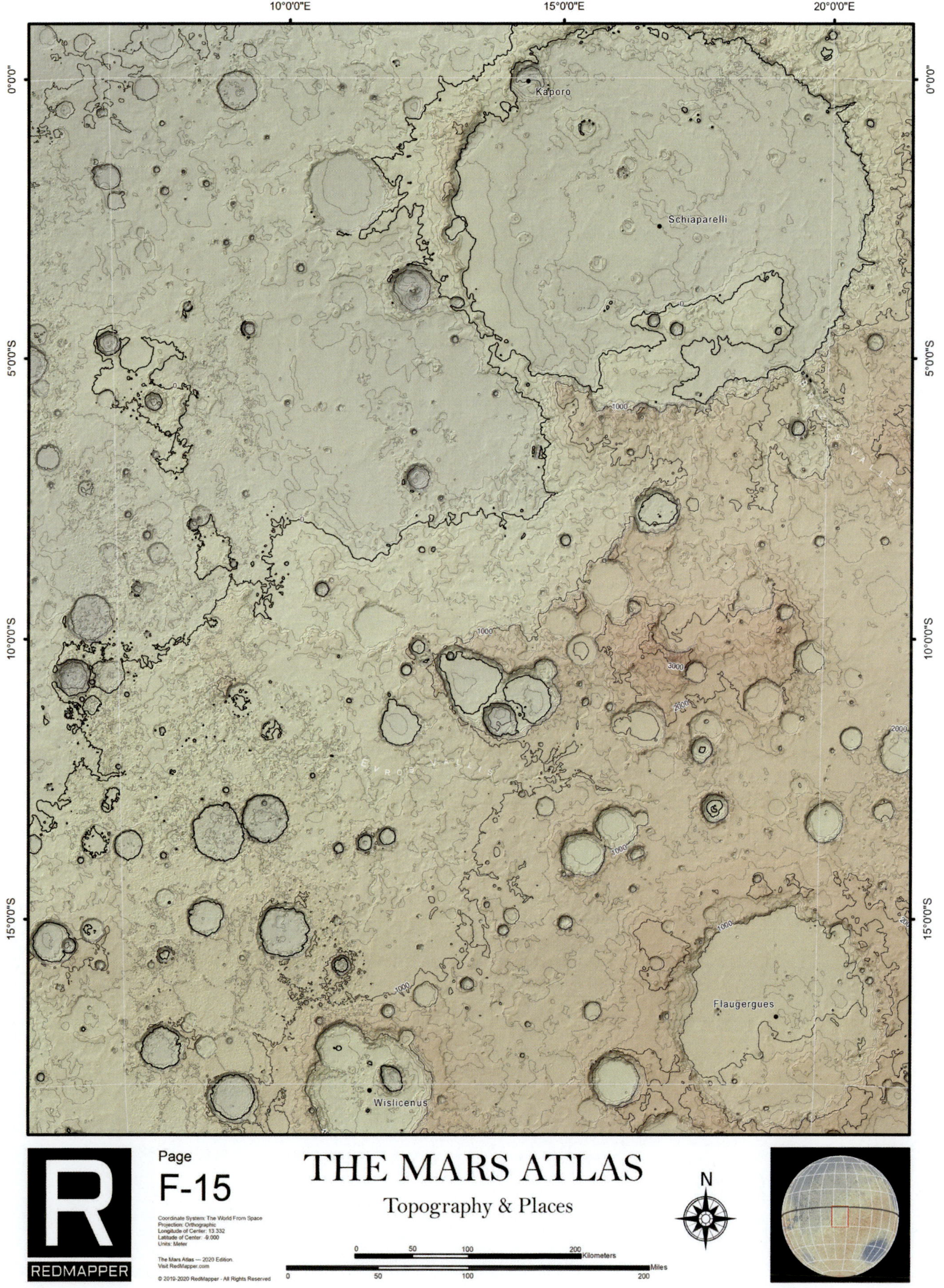

R
REDMAPPER

Page
F-15

THE MARS ATLAS
Topography & Places

Coordinate System: The World From Space
Projection: Orthographic
Longitude of Center: 13.332
Latitude of Center: -9.000
Units: Meter

The Mars Atlas — 2020 Edition.
Visit RedMapper.com

N

0 50 100 200 Kilometers
0 50 100 200 Miles

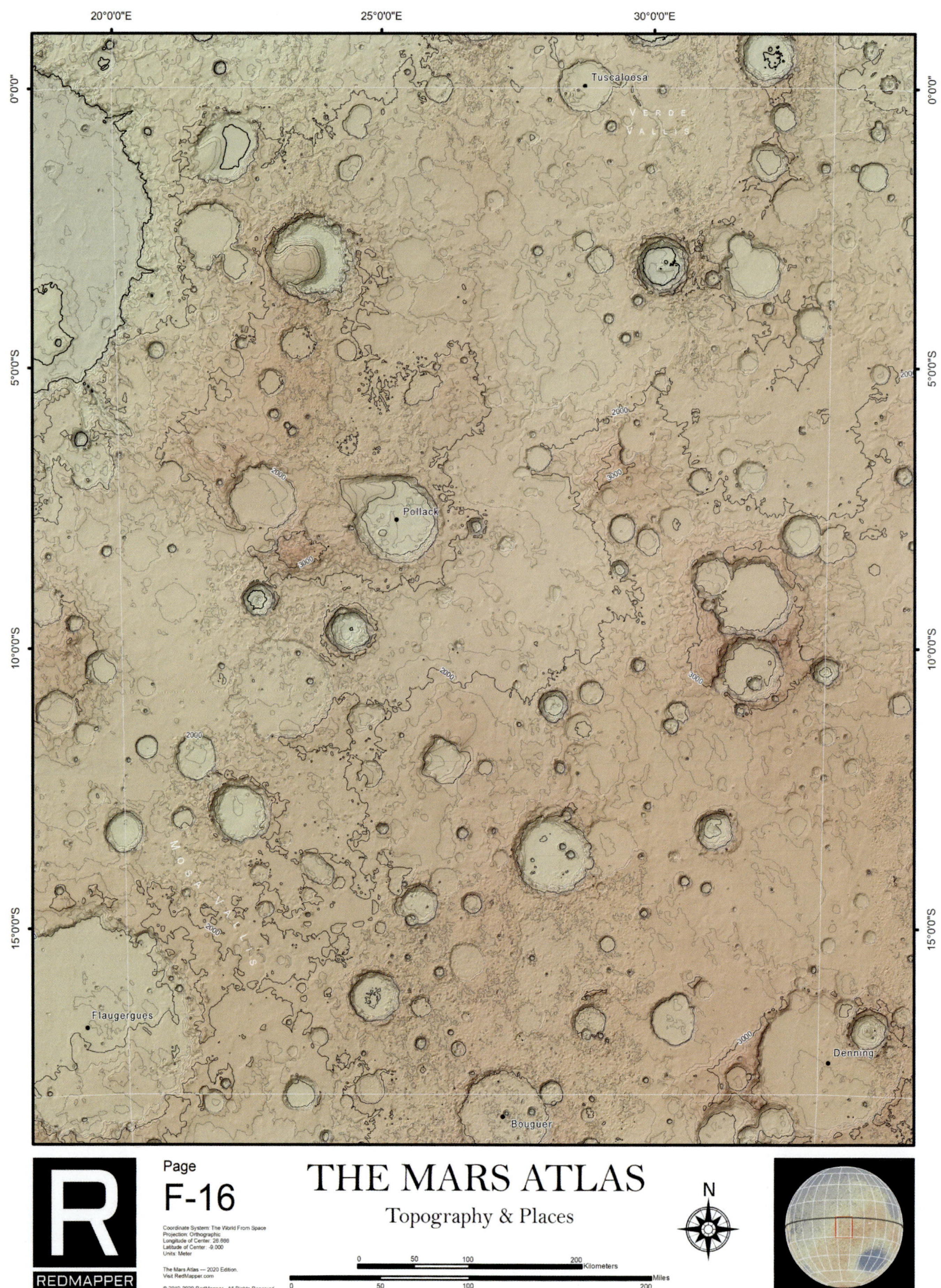
20°0'0"E
25°0'0"E
30°0'0"E
0°0'0"
5°0'0"S
10°0'0"S
15°0'0"S
Tuscaloosa
VERDE VALLIS
Pollack
2000
3000
MOSA VALLIS
Flaugergues
Denning
Bouguer
REDMAPPER
Page
F-16
Coordinate System: The World From Space
Projection: Orthographic
Longitude of Center: 26.666
Latitude of Center: -9.000
Units: Meter
The Mars Atlas — 2020 Edition.
Visit RedMapper.com
© 2019-2020 RedMapper - All Rights Reserved
THE MARS ATLAS
Topography & Places
N
0 50 100 200 Kilometers
0 50 100 200 Miles

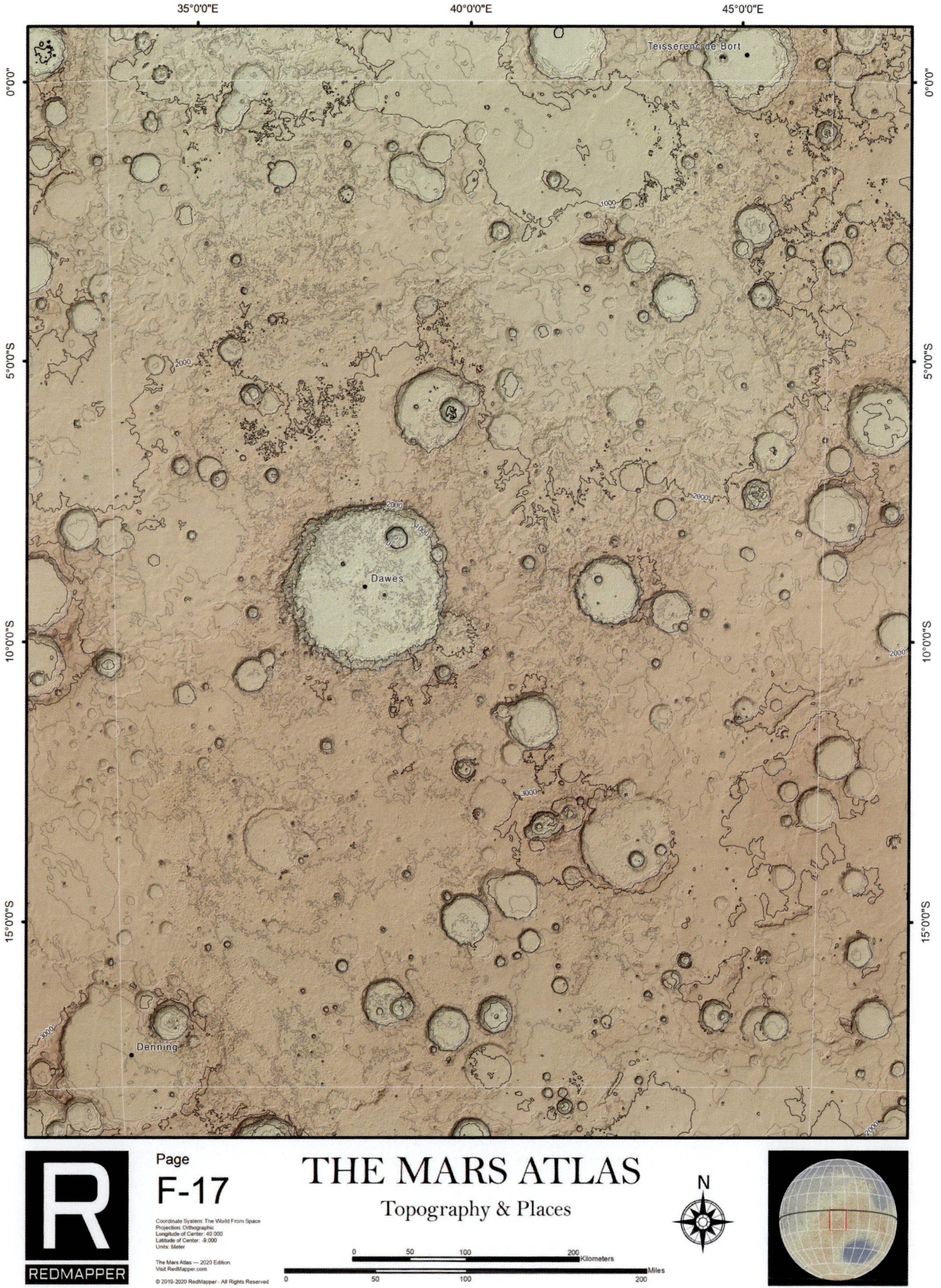
35°0'0"E
40°0'0"E
45°0'0"E
0°0'0"
5°0'0"S
10°0'0"S
15°0'0"S
Teisserenc de Bort
Dawes
Denning
1000
2000
3000
R
REDMAPPER
Page
F-17
Coordinate System: The World From Space
Projection: Orthographic
Longitude of Center: 40.000
Latitude of Center: -9.000
Units: Meter
The Mars Atlas — 2020 Edition.
Visit RedMapper.com
THE MARS ATLAS
Topography & Places
N
0 50 100 200 Kilometers
0 50 100 200 Miles

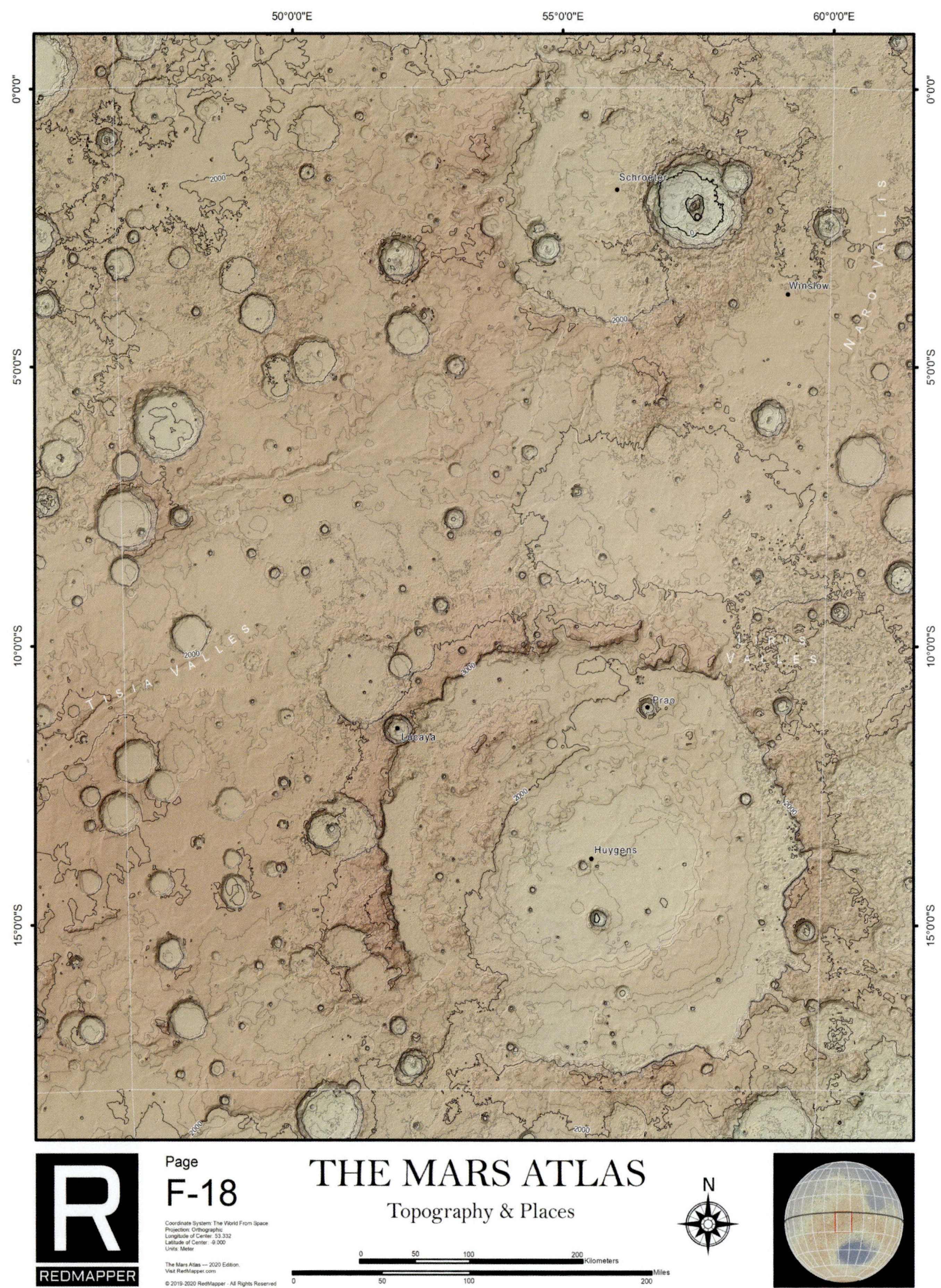

R
REDMAPPER

Page
F-18

THE MARS ATLAS

Topography & Places

Coordinate System: The World From Space
Projection: Orthographic
Longitude of Center: 53.332
Latitude of Center: -9.000
Units: Meter

The Mars Atlas --- 2020 Edition.
Visit RedMapper.com

0 50 100 200 Kilometers

0 50 100 200 Miles

N

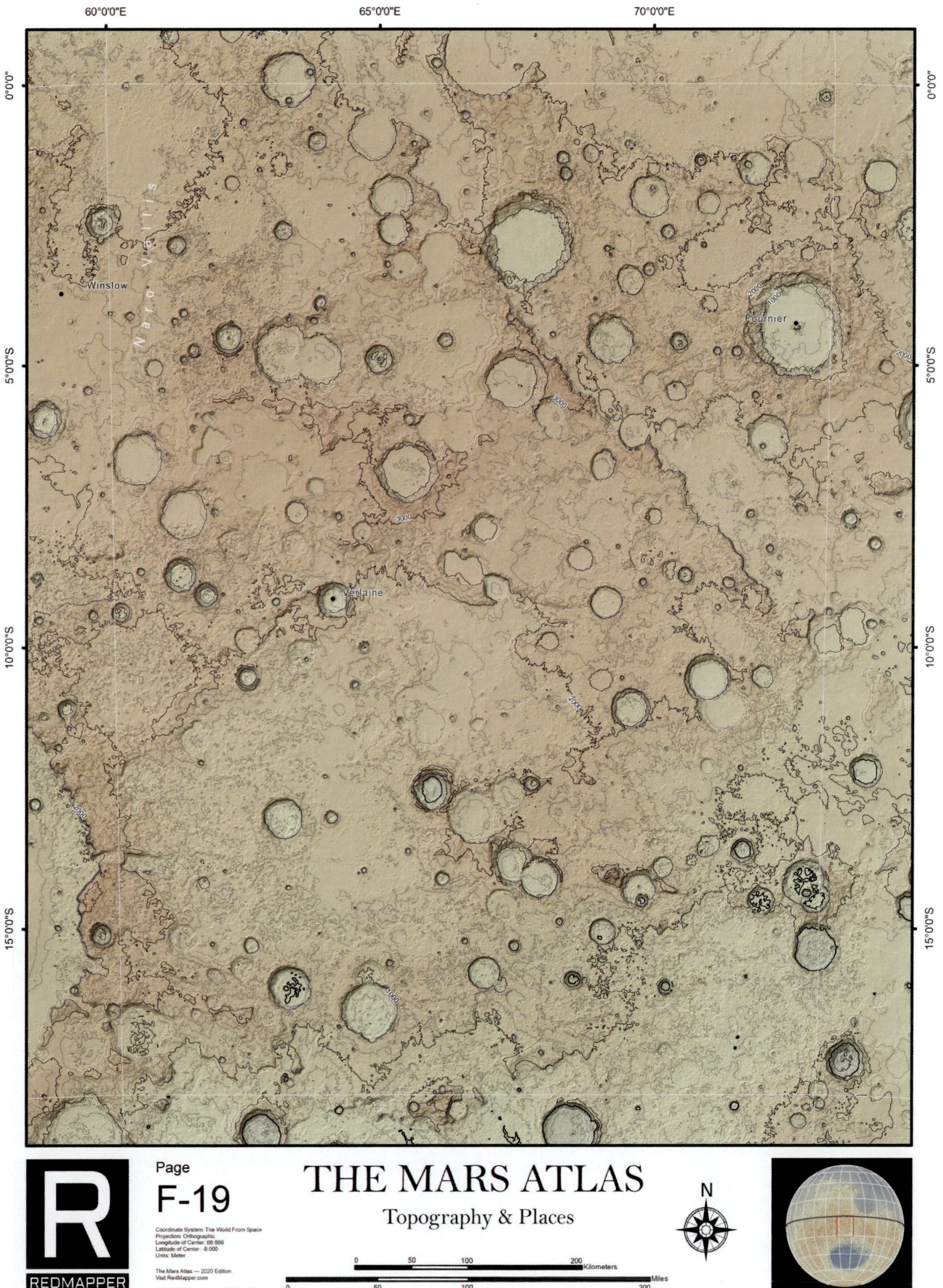
60°0'0"E
65°0'0"E
70°0'0"E
0°0'0"
5°0'0"S
10°0'0"S
15°0'0"S
Naro Vallis
Winslow
Fournier
Verlaine
2000
1000
3000
R
REDMAPPER
Page
F-19
Coordinate System: The World From Space
Projection: Orthographic
Longitude of Center: 66.666
Latitude of Center: -9.000
Units: Meter
The Mars Atlas — 2020 Edition
Visit RedMapper.com
© 2019-2020 RedMapper - All Rights Reserved
THE MARS ATLAS
Topography & Places
N
0 50 100 200 Kilometers
0 50 100 200 Miles

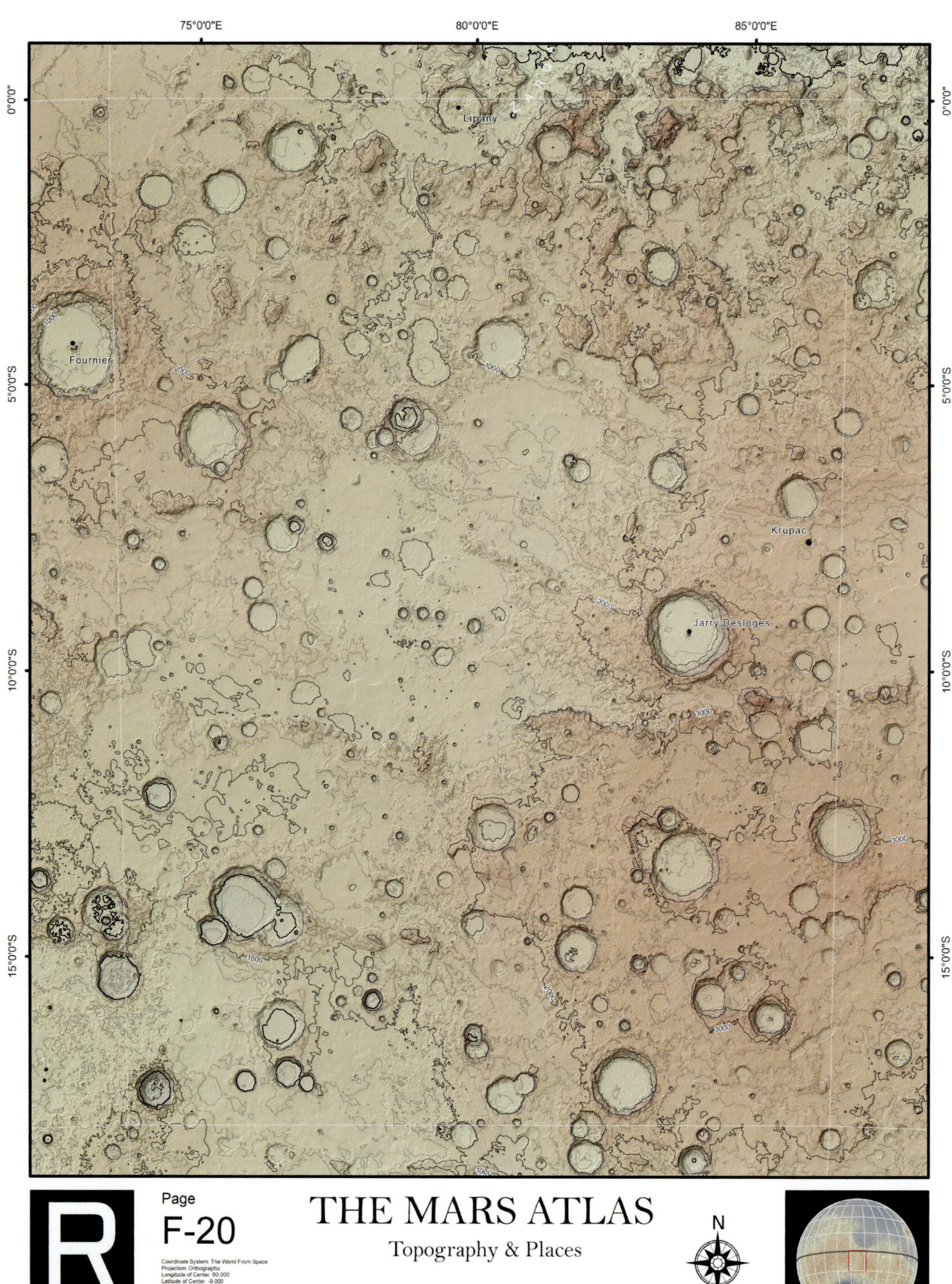

R
REDMAPPER

Page
F-20

Coordinate System: The World From Space
Projection: Orthographic
Longitude of Center: 80.000
Latitude of Center: -9.000
Units: Meter

The Mars Atlas — 2020 Edition.
Visit RedMapper.com

THE MARS ATLAS

Topography & Places

0 50 100 200 Kilometers
0 50 100 200 Miles

N

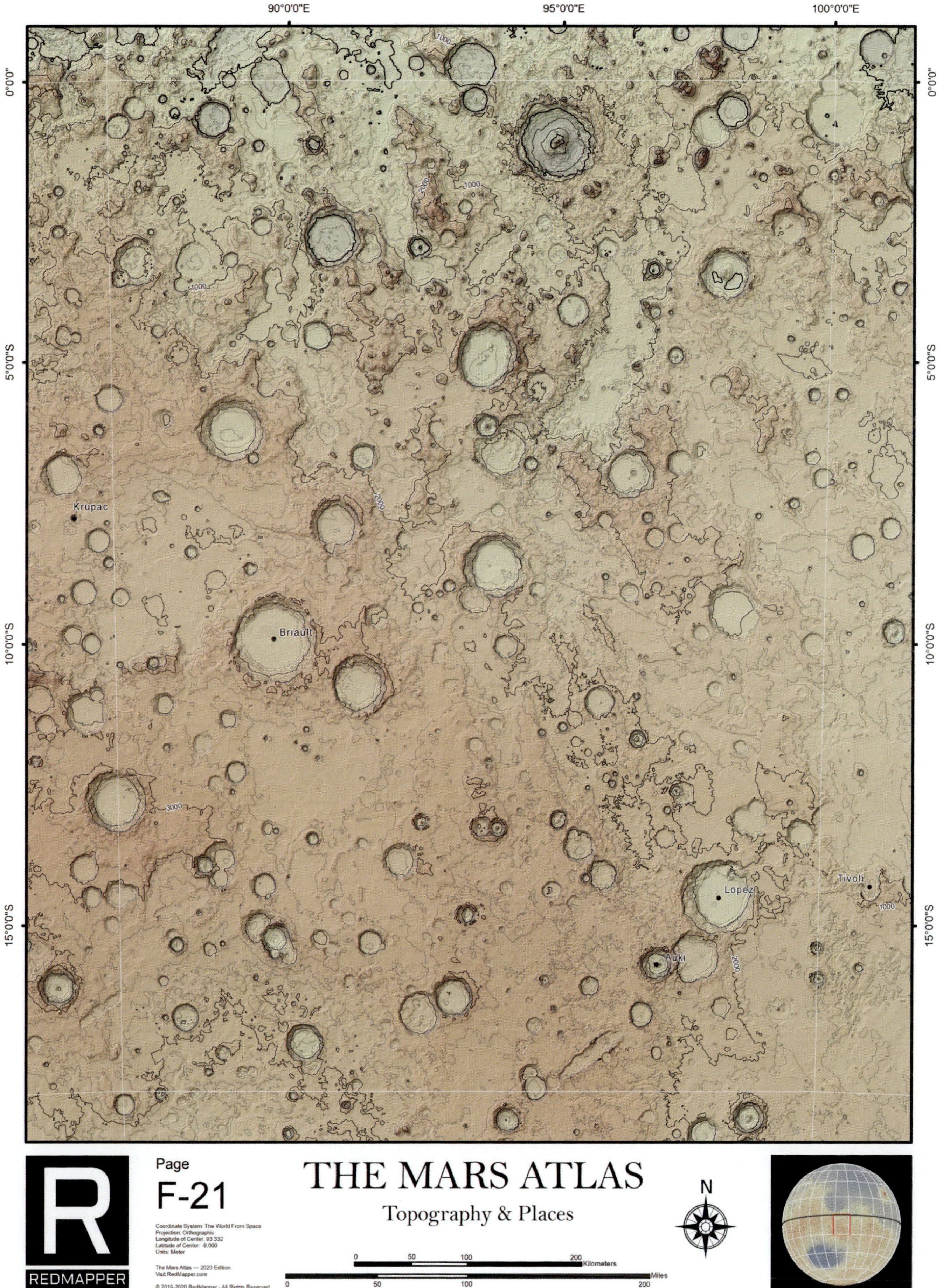
90°0'0"E
95°0'0"E
100°0'0"E
0°0'0"
5°0'0"S
10°0'0"S
15°0'0"S
Krupac
Briault
Lopez
Auki
Tivoli
1000
2000
3000
R
REDMAPPER
Page
F-21
Coordinate System: The World From Space
Projection: Orthographic
Longitude of Center: 93.332
Latitude of Center: -9.000
Units: Meter
The Mars Atlas — 2020 Edition.
Visit RedMapper.com
© 2019-2020 RedMapper - All Rights Reserved
THE MARS ATLAS
Topography & Places
N
0 50 100 200 Kilometers
0 50 100 200 Miles

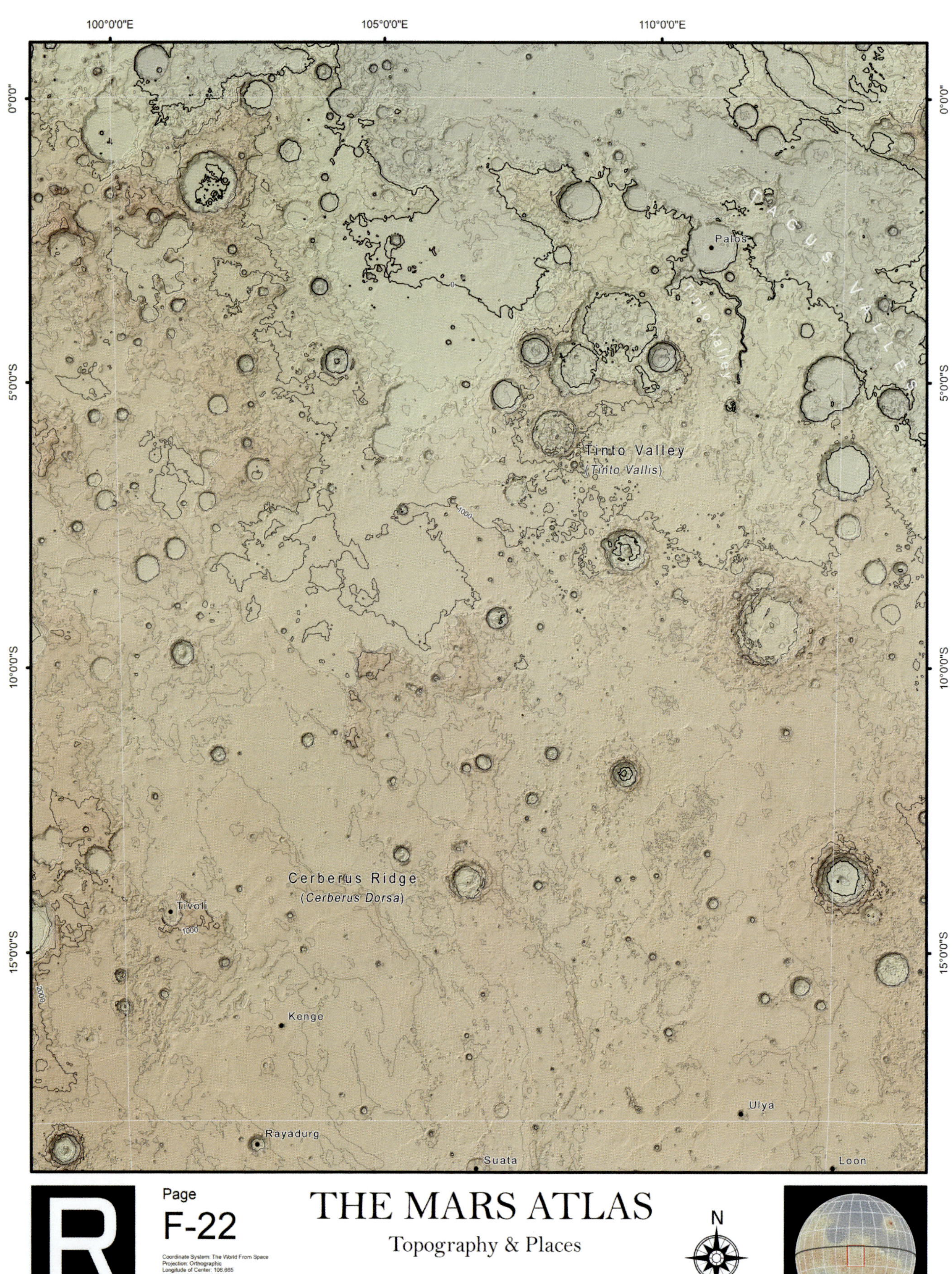

REDMAPPER

Page

F-22

THE MARS ATLAS

Topography & Places

Coordinate System: The World From Space
Projection: Orthographic
Longitude of Center: 106.885
Latitude of Center: -9.000
Units: Meter

The Mars Atlas — 2020 Edition.
Visit RedMapper.com

N

0 50 100 200 Kilometers

0 50 100 200 Miles

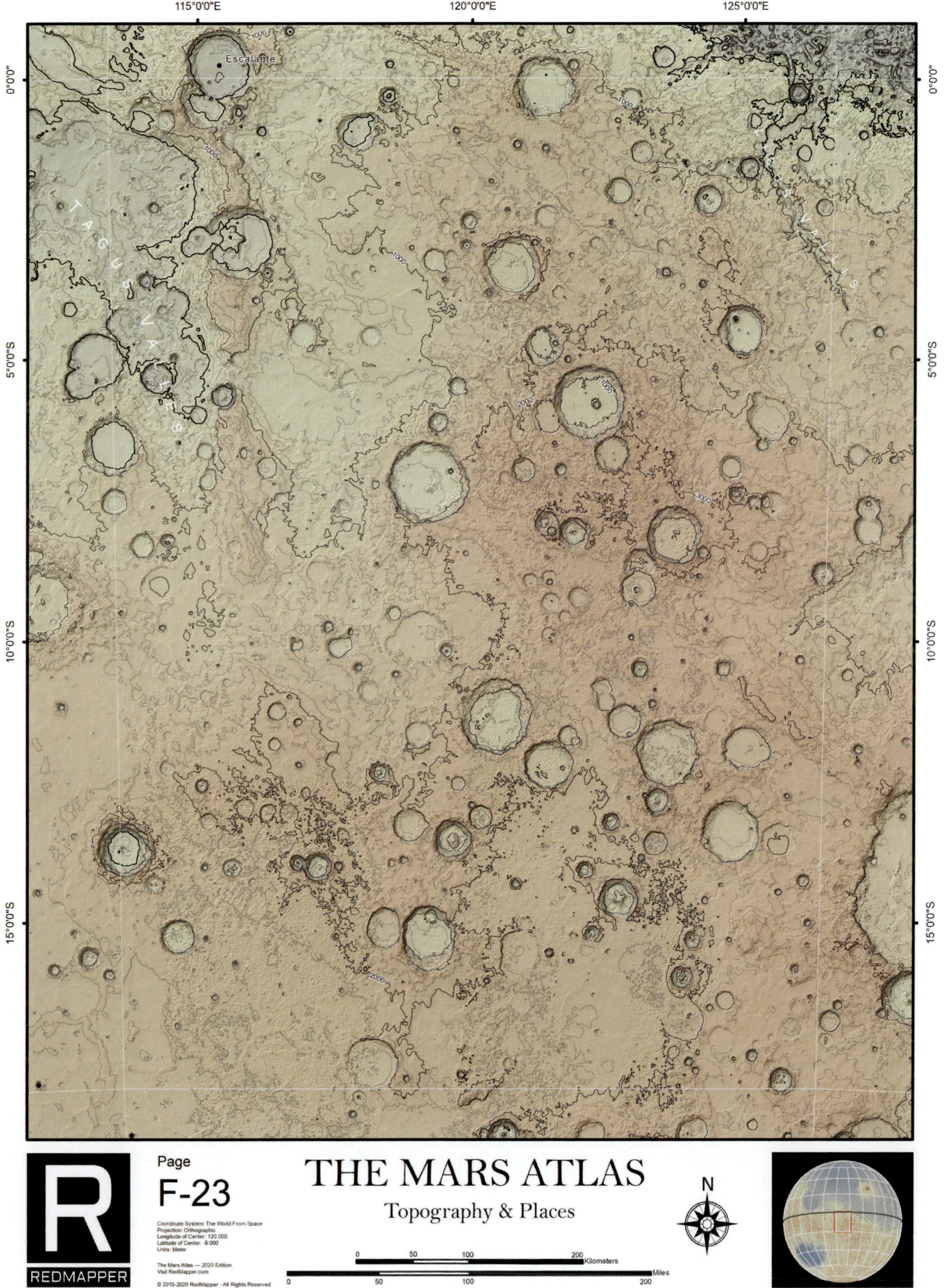
115°0'0"E
120°0'0"E
125°0'0"E
0°0'0"
5°0'0"S
10°0'0"S
15°0'0"S
Escalante
TAGUS VALLES
LICUS VALLIS
1000
2000
3000
Page
F-23
Coordinate System: The World From Space
Projection: Orthographic
Longitude of Center: 120.000
Latitude of Center: -9.000
Units: Meter
The Mars Atlas — 2020 Edition.
Visit RedMapper.com
© 2019-2020 RedMapper - All Rights Reserved
THE MARS ATLAS
Topography & Places
N
0 50 100 200 Kilometers
0 50 100 200 Miles
R
REDMAPPER

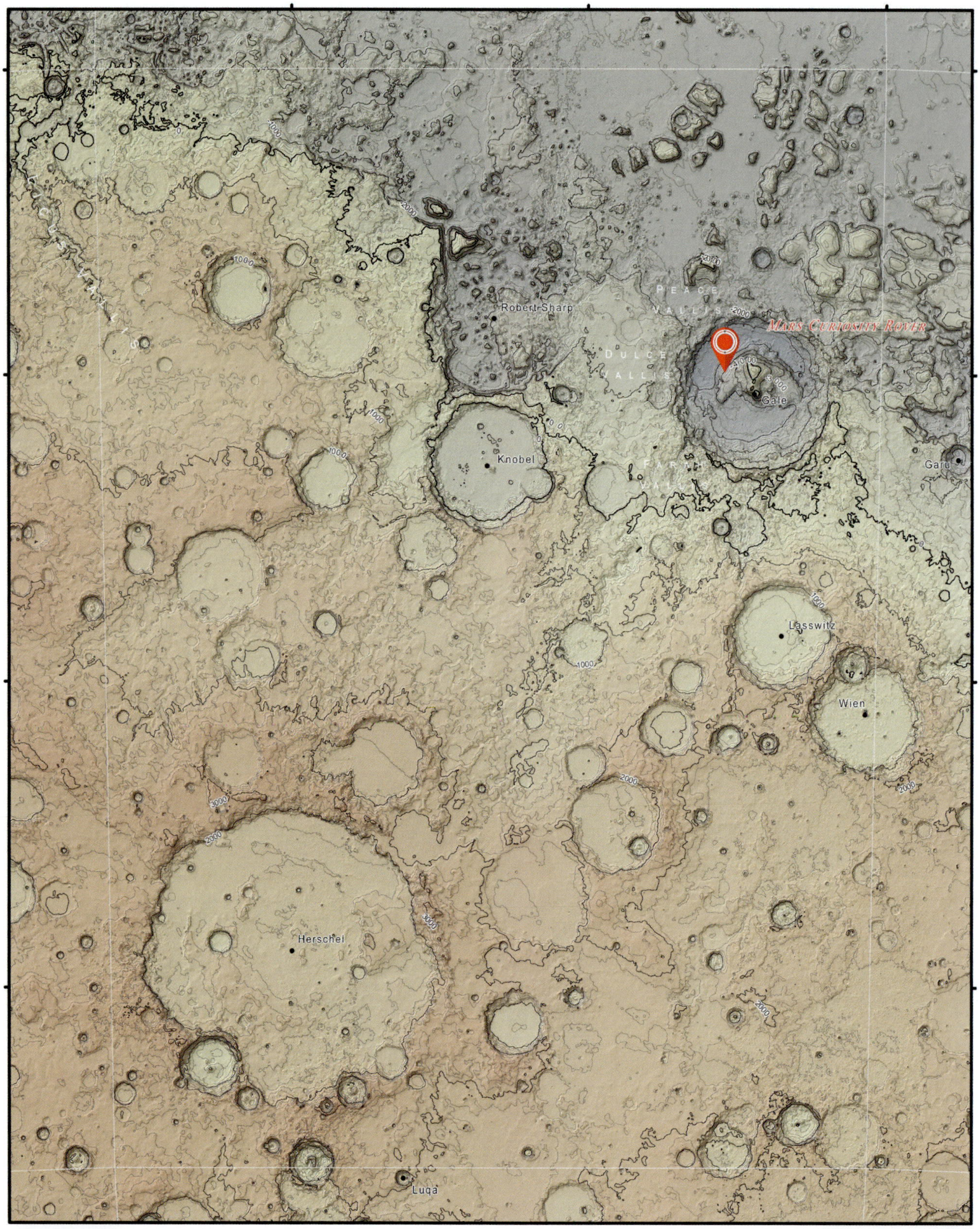

130°0'0"E 135°0'0"E 140°0'0"E

0°0'0" 5°0'0"S 10°0'0"S 15°0'0"S

Robert Sharp

PEACE VALLIS

DULCE VALLIS

FARAH VALLIS

MARS CURIOSITY ROVER

Gale

Knobel

Garu

Lasswitz

Wien

Herschel

Luqa

Page

F-24

Coordinate System: The World From Space
Projection: Orthographic
Longitude of Center: 133.332
Latitude of Center: -9.000
Units: Meter

The Mars Atlas — 2020 Edition.
Visit RedMapper.com

THE MARS ATLAS

Topography & Places

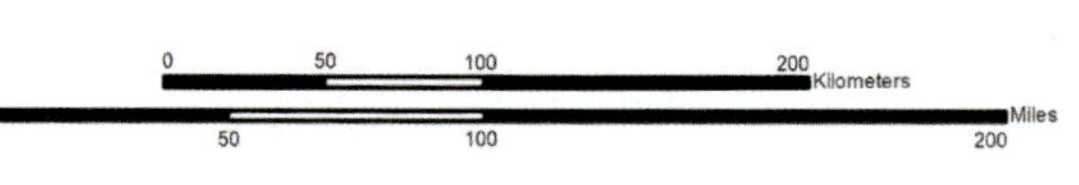

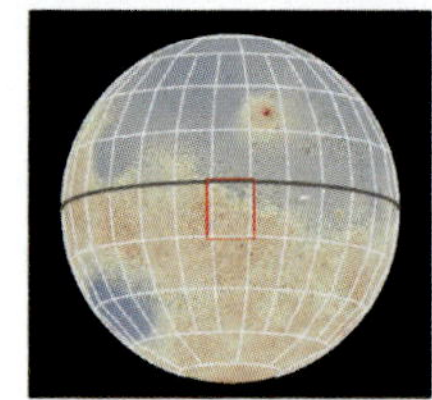

140°0'0"E
145°0'0"E
150°0'0"E
0°0'0"
5°0'0"S
10°0'0"S
15°0'0"S
Quthing
Gunjur
Obock
Neves
Asau
Aeolis Ridge
(Aeolis Dorsa)
Wien
R
REDMAPPER
Page
F-25
Coordinate System: The World From Space
Projection: Orthographic
Longitude of Center: 146.665
Latitude of Center: -9.000
Units: Meter
The Mars Atlas — 2020 Edition.
Visit RedMapper.com
© 2019-2020 RedMapper - All Rights Reserved
THE MARS ATLAS
Topography & Places
N
0 50 100 200 Kilometers
0 50 100 200 Miles

R
REDMAPPER

Page
F-26

Coordinate System: The World From Space
Projection: Orthographic
Longitude of Center: 160.000
Latitude of Center: -9.000
Units: Meter

The Mars Atlas — 2020 Edition.
Visit RedMapper.com

THE MARS ATLAS

Topography & Places

N

0 50 100 200 Kilometers

0 50 100 200 Miles

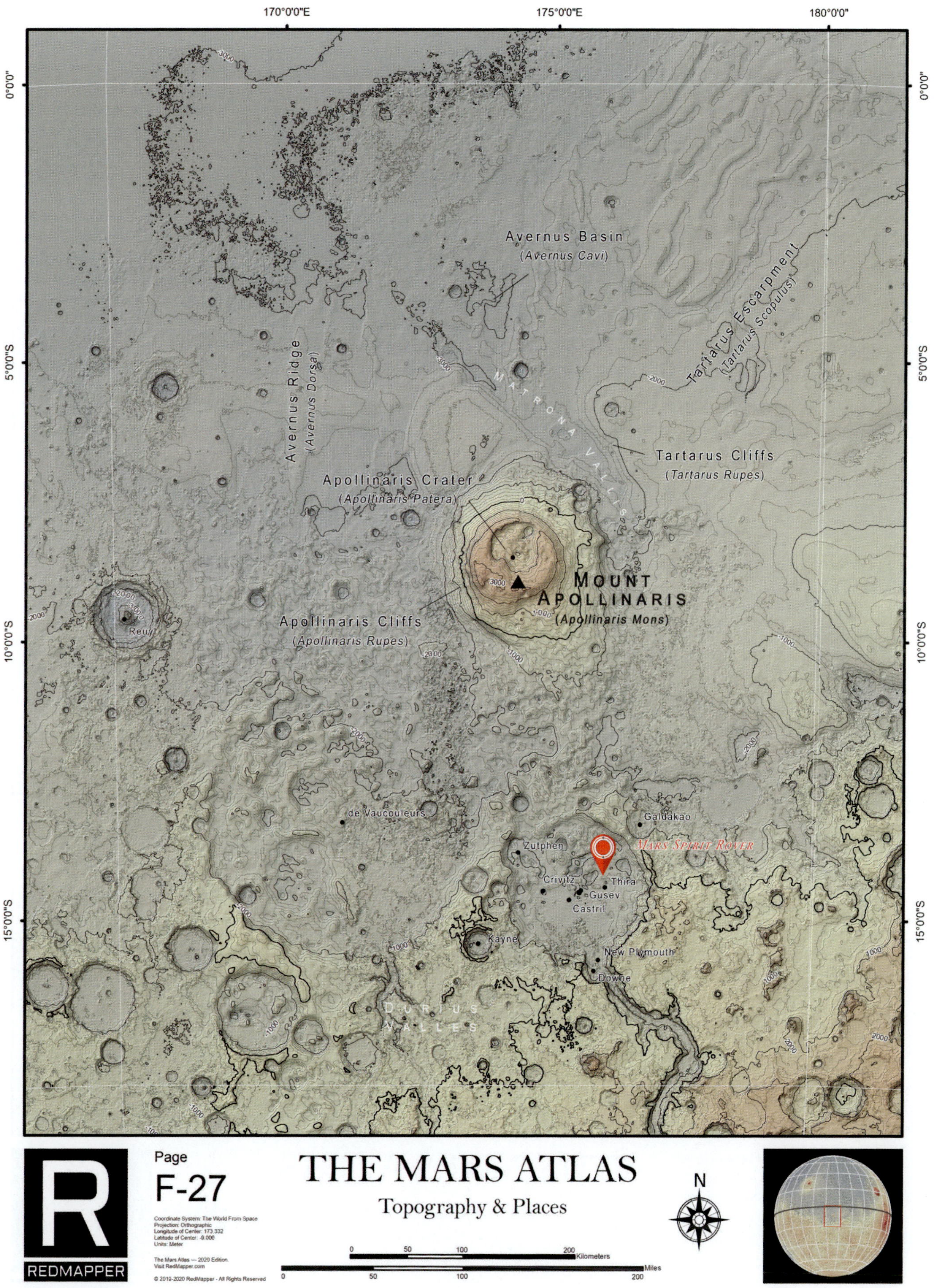
170°0'0"E
175°0'0"E
180°0'0"
0°0'0"
5°0'0"S
10°0'0"S
15°0'0"S
Avernus Basin
(Avernus Cavi)
Tartarus Escarpment
(Tartarus Scopulus)
Avernus Ridge
(Avernus Dorsa)
MATRONA VALLIS
Tartarus Cliffs
(Tartarus Rupes)
Apollinaris Crater
(Apollinaris Patera)
MOUNT APOLLINARIS
(Apollinaris Mons)
Apollinaris Cliffs
(Apollinaris Rupes)
Reuyl
de Vaucouleurs
Galdakao
Zutphen
MARS SPIRIT ROVER
Crivitz
Thira
Gusev
Castril
Kayne
New Plymouth
Downe
DURIUS VALLES
REDMAPPER
Page
F-27
Coordinate System: The World From Space
Projection: Orthographic
Longitude of Center: 173.332
Latitude of Center: -9.000
Units: Meter
The Mars Atlas — 2020 Edition.
Visit RedMapper.com
© 2019-2020 RedMapper - All Rights Reserved
THE MARS ATLAS
Topography & Places
N
Kilometers
Miles

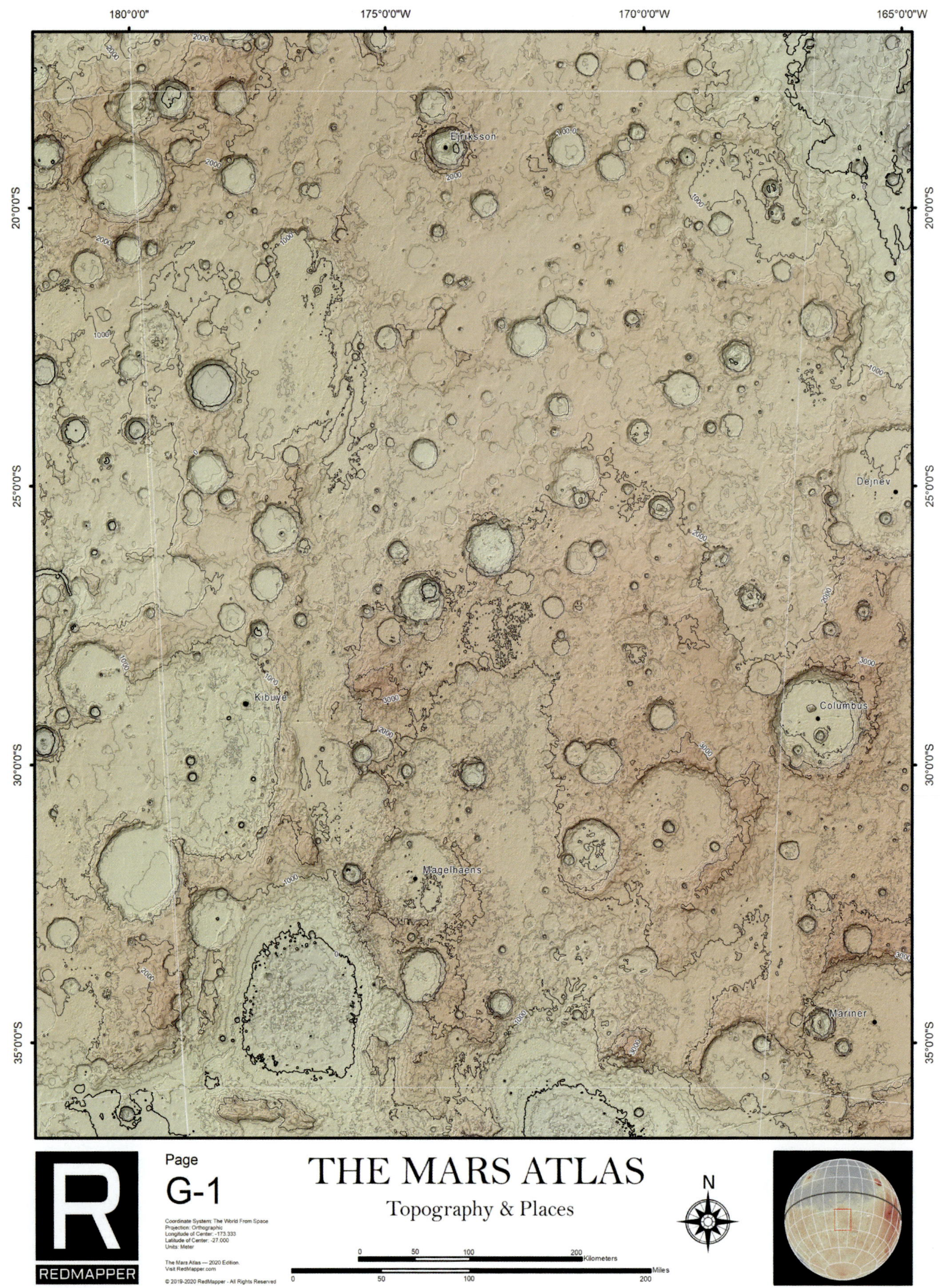

180°0'0"
175°0'0"W
170°0'0"W
165°0'0"W
20°0'0"S
25°0'0"S
30°0'0"S
35°0'0"S
Eiriksson
Dejnev
Kibuye
Columbus
Magelhaens
Mariner
2000
1000
3000
R
REDMAPPER
Page
G-1
Coordinate System: The World From Space
Projection: Orthographic
Longitude of Center: -173.333
Latitude of Center: -27.000
Units: Meter
The Mars Atlas — 2020 Edition.
Visit RedMapper.com
© 2019-2020 RedMapper - All Rights Reserved
THE MARS ATLAS
Topography & Places
N
0
50
100
200
Kilometers
Miles

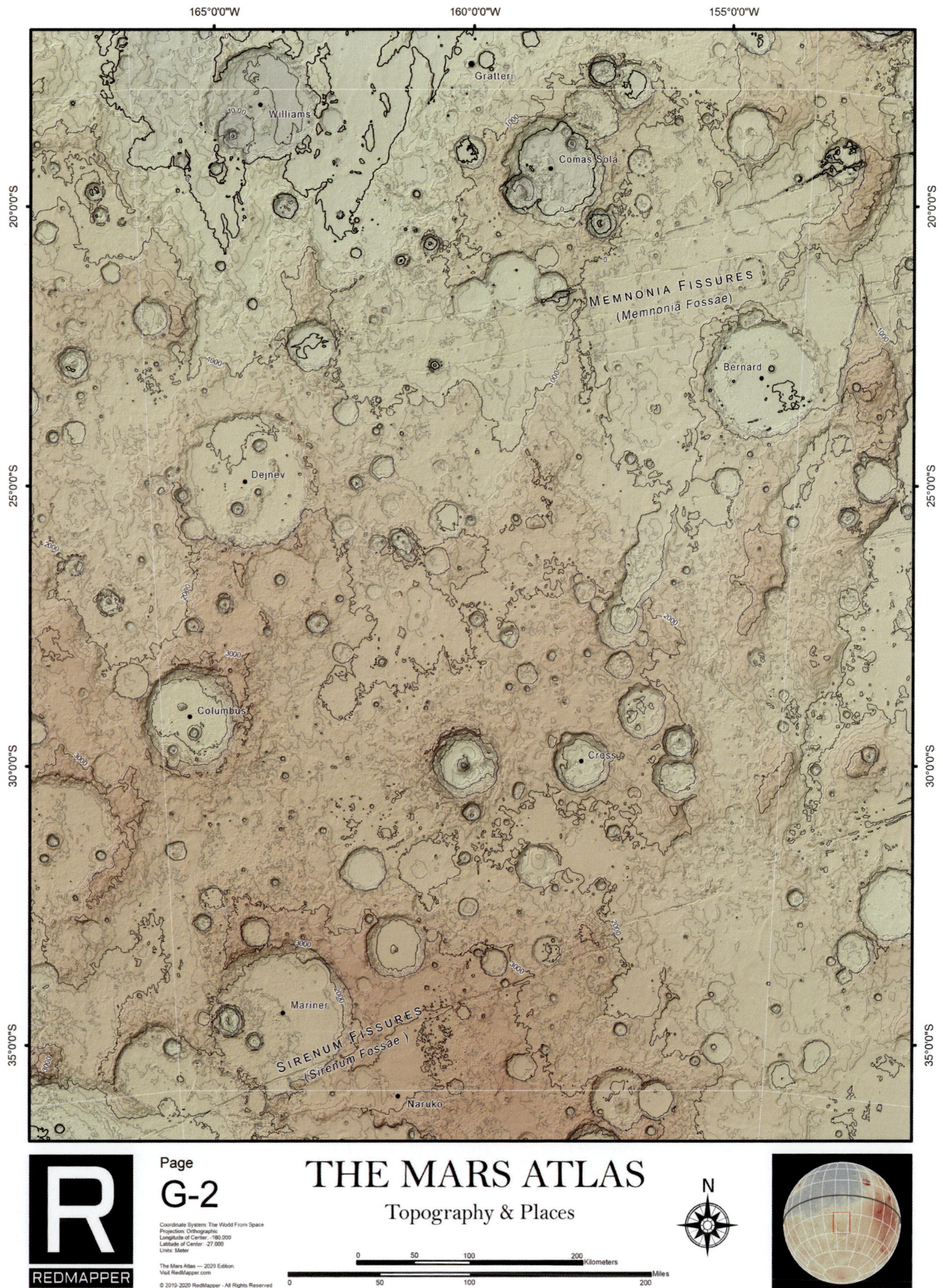

REDMAPPER

Page

G-2

Coordinate System: The World From Space
Projection: Orthographic
Longitude of Center: -160.000
Latitude of Center: -27.000
Units: Meter

The Mars Atlas — 2020 Edition.
Visit RedMapper.com

THE MARS ATLAS

Topography & Places

N

0 50 100 200 Kilometers

0 50 100 200 Miles

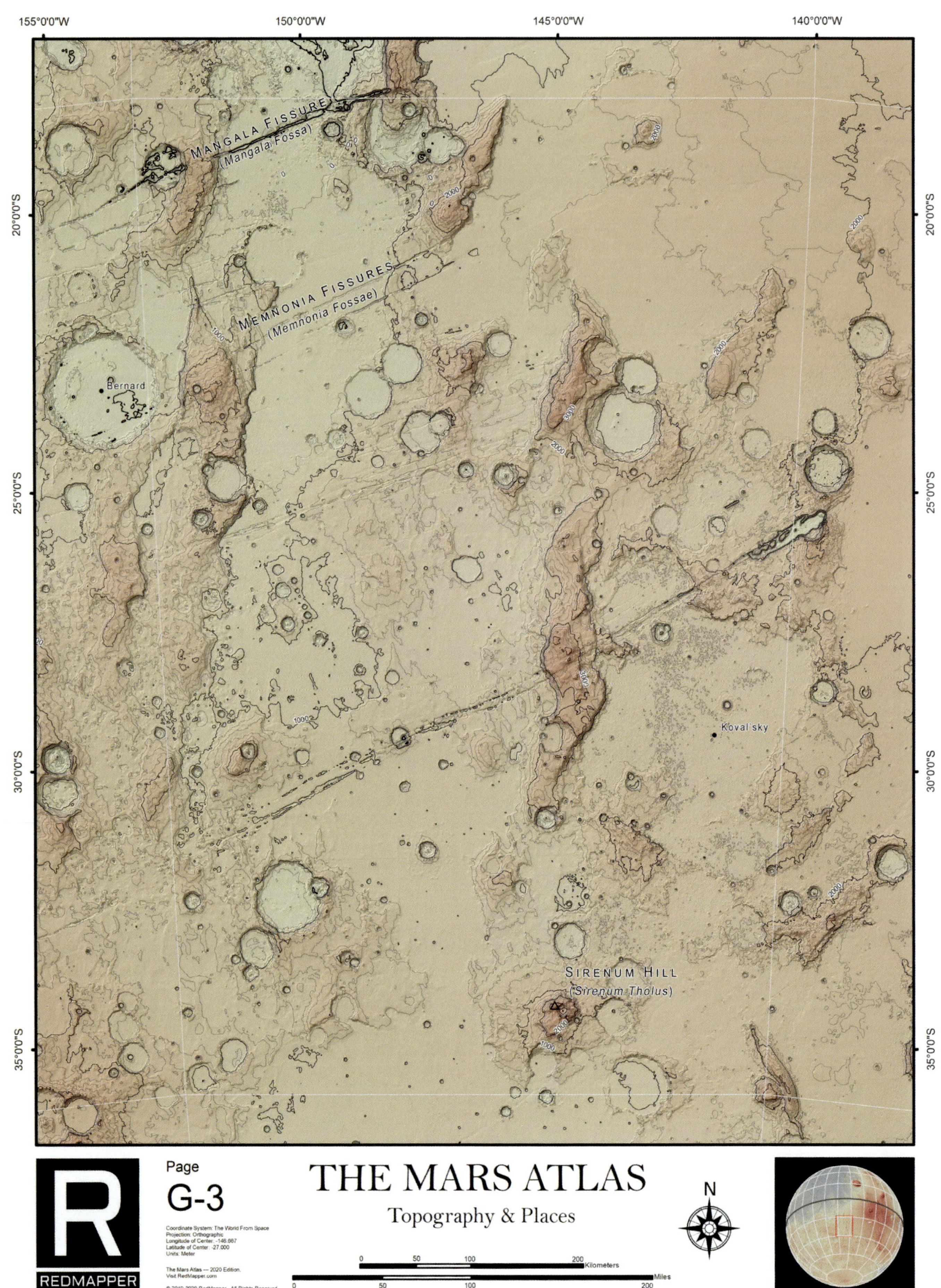
155°0'0"W
150°0'0"W
145°0'0"W
140°0'0"W
20°0'0"S
25°0'0"S
30°0'0"S
35°0'0"S
MANGALA FISSURE
(Mangala Fossa)
MEMNONIA FISSURES
(Memnonia Fossae)
Bernard
Kovalsky
SIRENUM HILL
(Sirenum Tholus)
Page
G-3
THE MARS ATLAS
Topography & Places
N
REDMAPPER
Coordinate System: The World From Space
Projection: Orthographic
Longitude of Center: -146.667
Latitude of Center: -27.000
Units: Meter
The Mars Atlas — 2020 Edition.
Visit RedMapper.com
© 2019-2020 RedMapper - All Rights Reserved
Kilometers
Miles

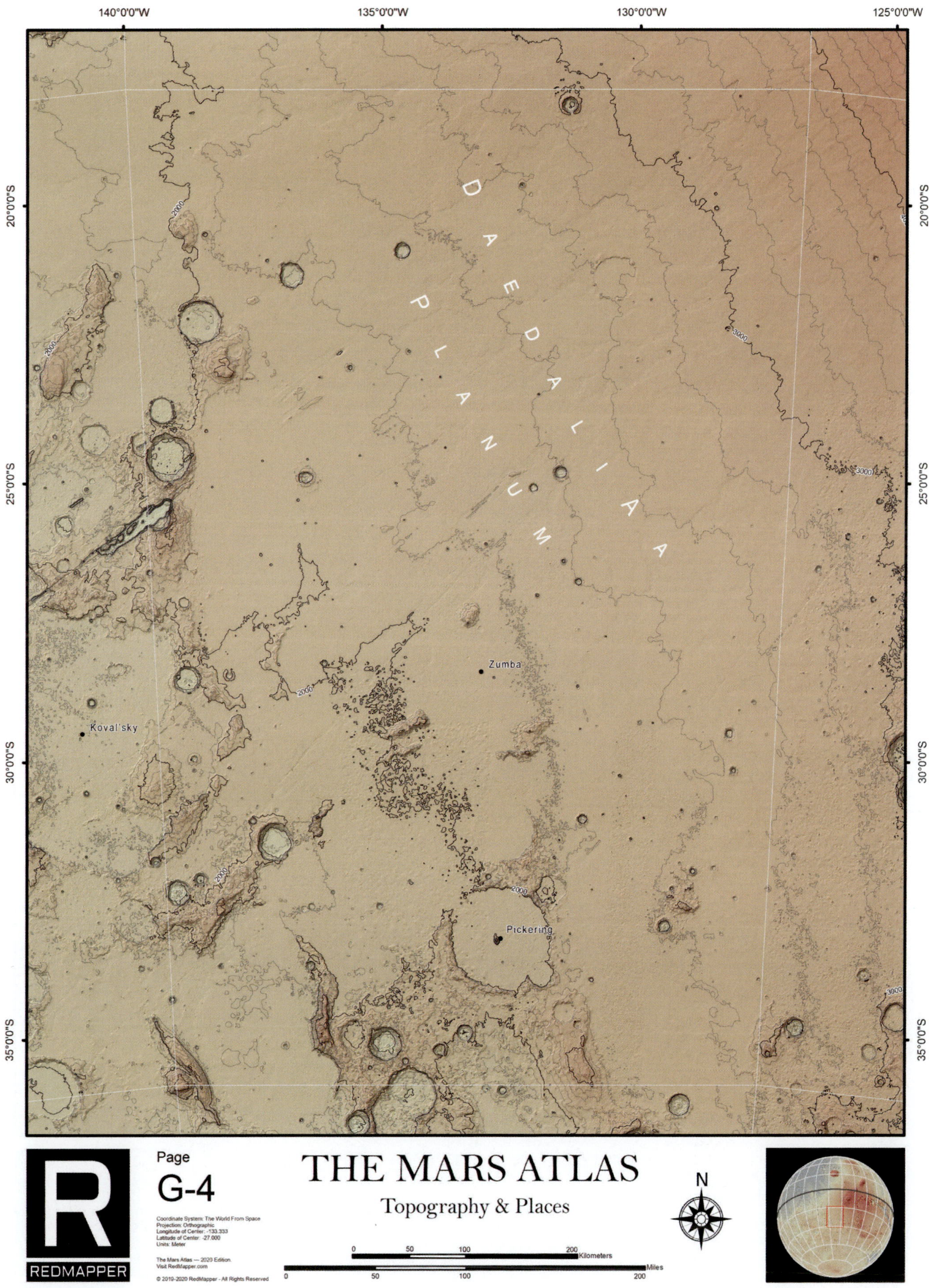

R
REDMAPPER

Page
G-4

Coordinate System: The World From Space
Projection: Orthographic
Longitude of Center: -133.333
Latitude of Center: -27.000
Units: Meter

The Mars Atlas — 2020 Edition.
Visit RedMapper.com

THE MARS ATLAS

Topography & Places

N

0 50 100 200 Kilometers
0 50 100 200 Miles

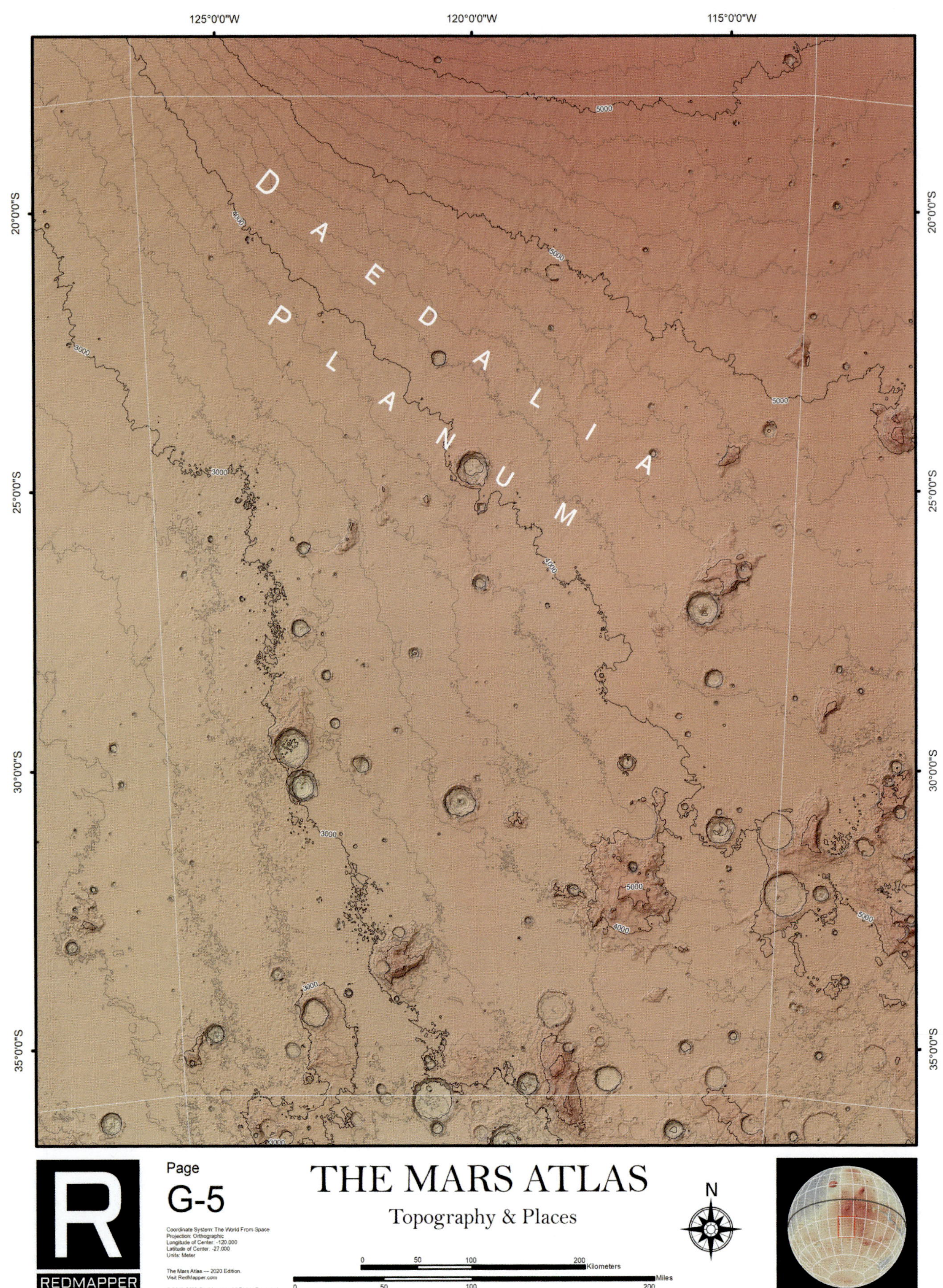
125°0'0"W
120°0'0"W
115°0'0"W
20°0'0"S
25°0'0"S
30°0'0"S
35°0'0"S
DAEDALIA PLANUM
5000
4000
3000
R
REDMAPPER
Page
G-5
Coordinate System: The World From Space
Projection: Orthographic
Longitude of Center: -120.000
Latitude of Center: -27.000
Units: Meter
The Mars Atlas — 2020 Edition.
Visit RedMapper.com
© 2019-2020 RedMapper - All Rights Reserved
THE MARS ATLAS
Topography & Places
N
0
50
100
200
Kilometers
Miles

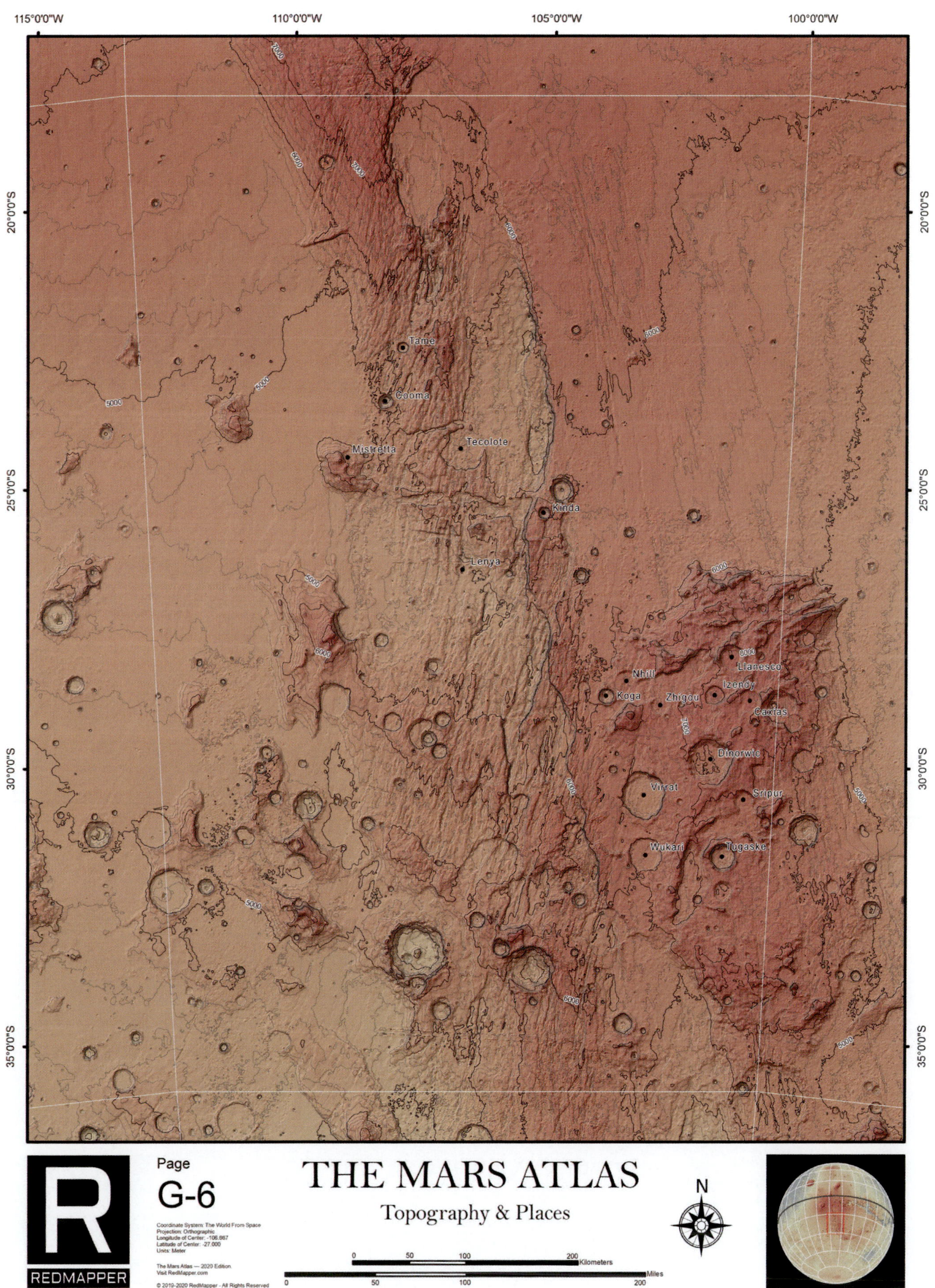

Page

G-6

THE MARS ATLAS

Topography & Places

Coordinate System: The World From Space
Projection: Orthographic
Longitude of Center: -106.667
Latitude of Center: -27.000
Units: Meter

The Mars Atlas — 2020 Edition.
Visit RedMapper.com

0 50 100 200 Kilometers

0 50 100 200 Miles

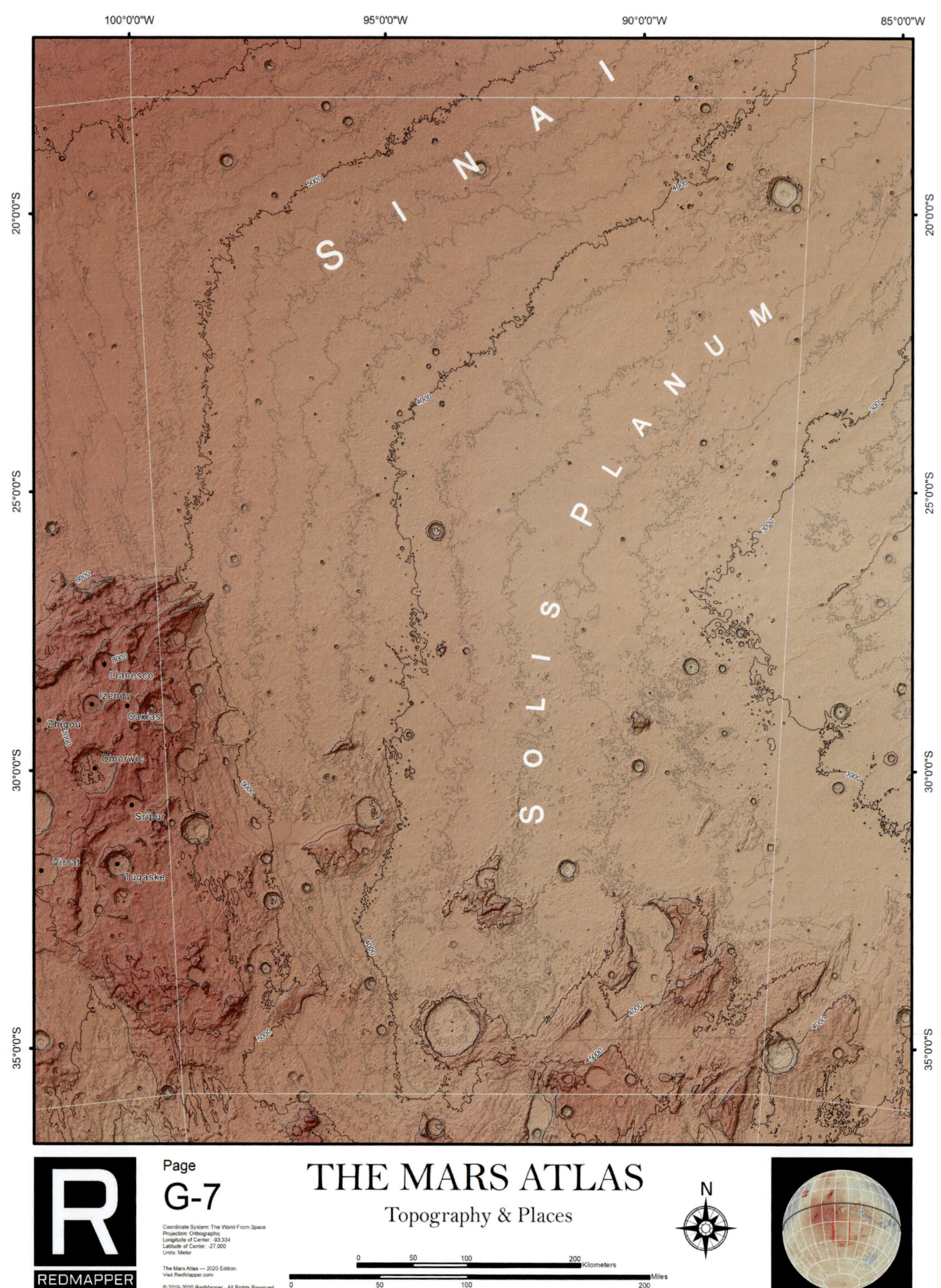
100°0'0"W
95°0'0"W
90°0'0"W
85°0'0"W
20°0'0"S
25°0'0"S
30°0'0"S
35°0'0"S
SINAI
SOLIS PLANUM
Llanesco
Izendy
Caxias
Zhigou
Dinorwic
Sripur
Virrat
Tugaske
5000
4000
3000
6000
7000
REDMAPPER
Page
G-7
Coordinate System: The World From Space
Projection: Orthographic
Longitude of Center: -93.334
Latitude of Center: -27.000
Units: Meter
The Mars Atlas — 2020 Edition.
Visit RedMapper.com
© 2019-2020 RedMapper - All Rights Reserved
THE MARS ATLAS
Topography & Places
0 50 100 200 Kilometers
0 50 100 200 Miles
N

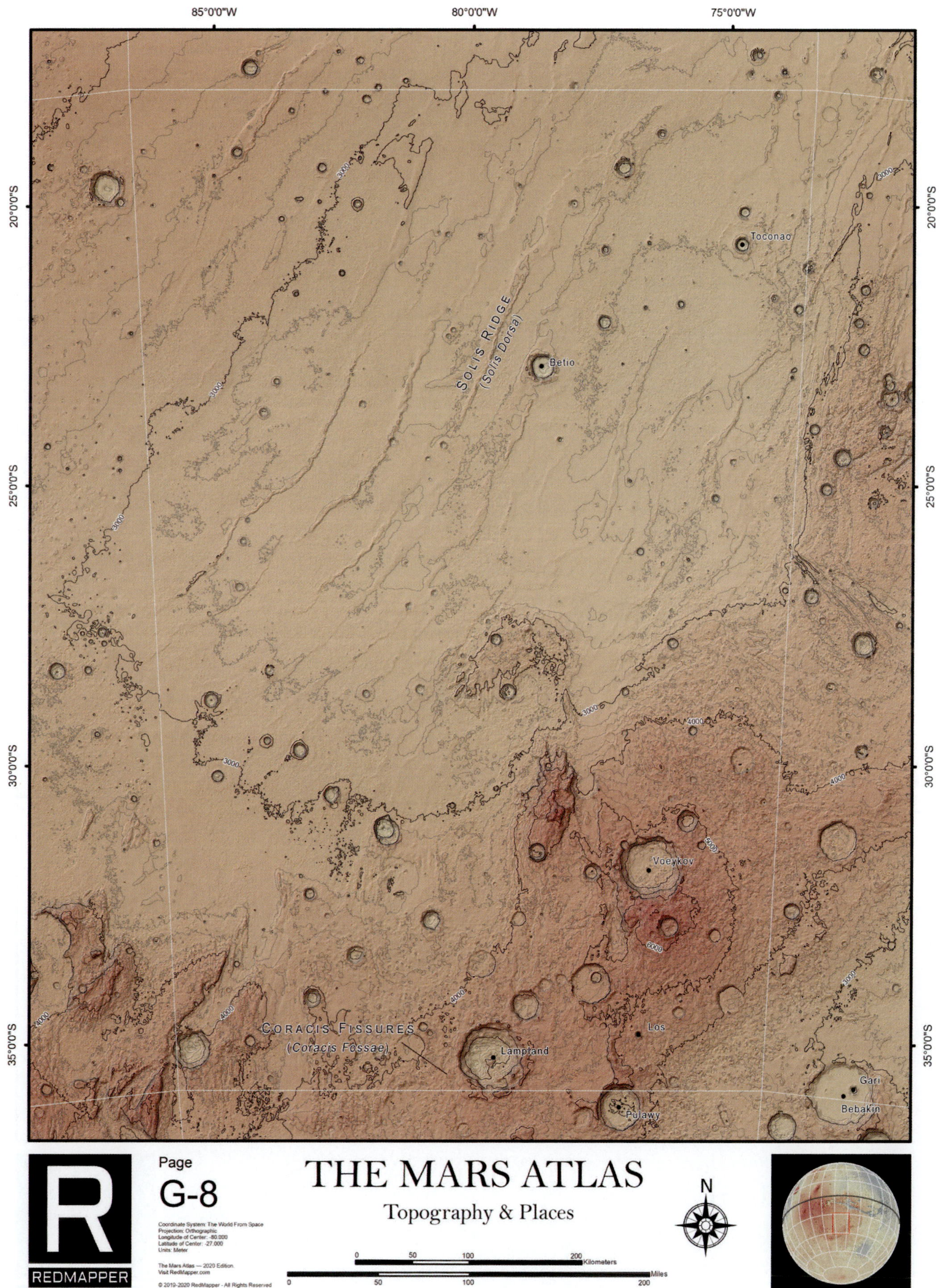
85°0'0"W
80°0'0"W
75°0'0"W
20°0'0"S
25°0'0"S
30°0'0"S
35°0'0"S
SOLIS RIDGE
(Solis Dorsa)
Betio
Toconao
Voeykov
Los
Lamprand
Pulawy
Gari
Bebakin
CORACIS FISSURES
(Coracis Fossae)
3000
4000
5000
6000
R
REDMAPPER
Page
G-8
THE MARS ATLAS
Topography & Places
Coordinate System: The World From Space
Projection: Orthographic
Longitude of Center: -80.000
Latitude of Center: -27.000
Units: Meter
The Mars Atlas — 2020 Edition.
Visit RedMapper.com
© 2019-2020 RedMapper - All Rights Reserved
N
0 50 100 200 Kilometers
0 50 100 200 Miles

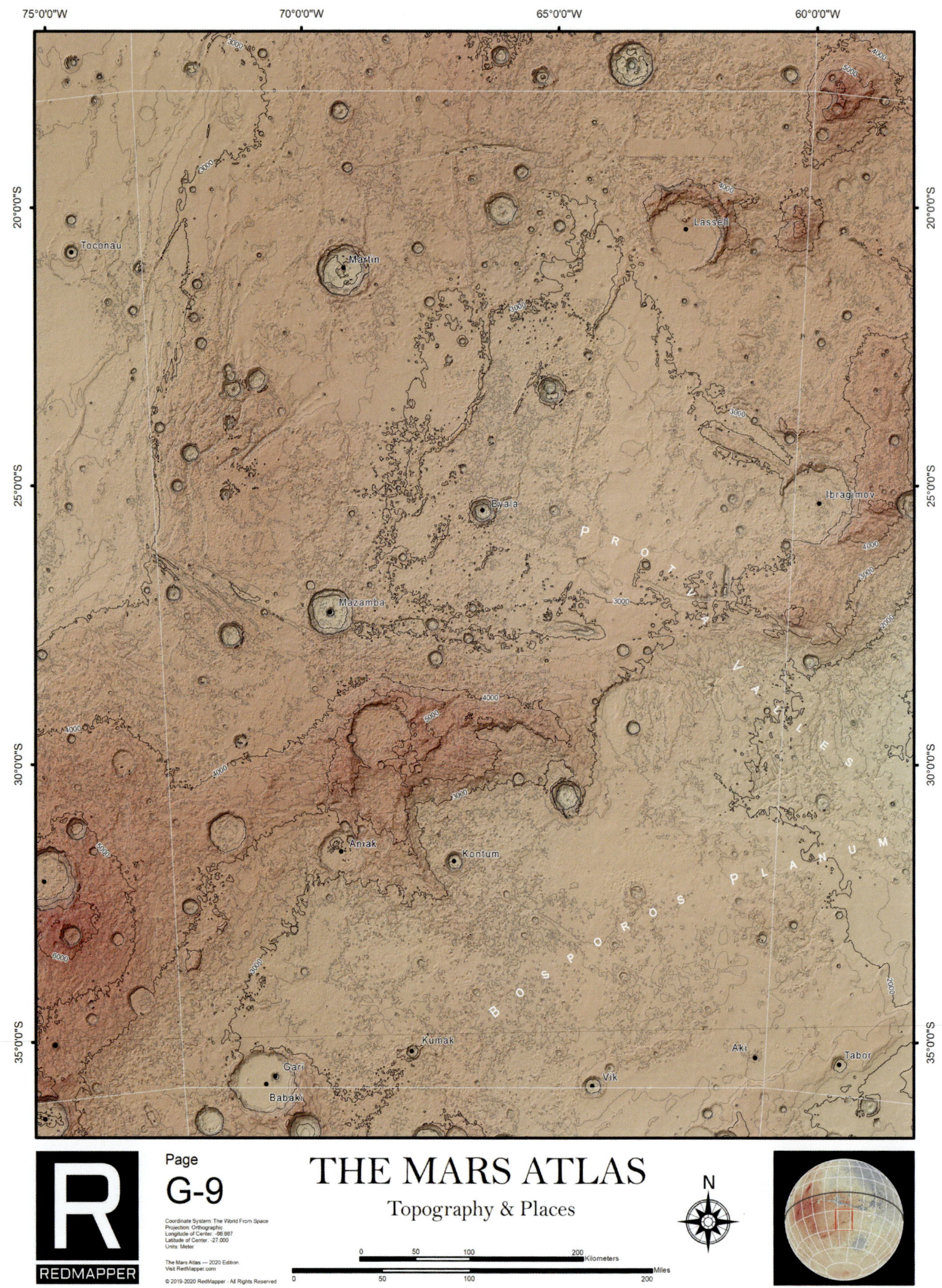

REDMAPPER

Page
G-9

Coordinate System: The World From Space
Projection: Orthographic
Longitude of Center: -88.867
Latitude of Center: -27.000
Units: Meter

The Mars Atlas — 2020 Edition
Visit RedMapper.com

THE MARS ATLAS

Topography & Places

N

0 50 100 200 Kilometers

0 50 100 200 Miles

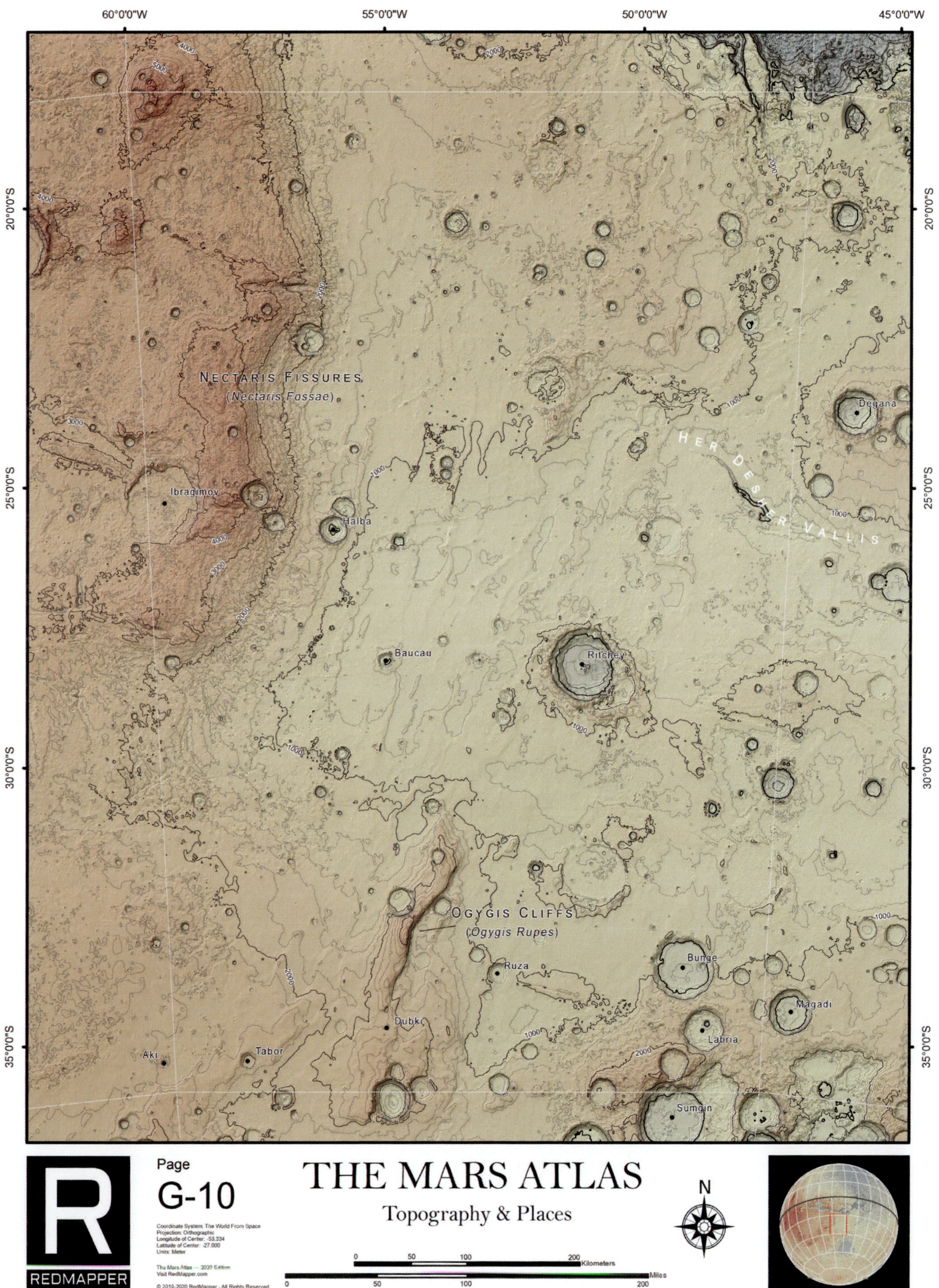

R
REDMAPPER

Page
G-10

Coordinate System: The World From Space
Projection: Orthographic
Longitude of Center: -53.334
Latitude of Center: -27.000
Units: Meter

The Mars Atlas — 2020 Edition
Visit RedMapper.com

THE MARS ATLAS

Topography & Places

N

0 50 100 200 Kilometers

0 50 100 200 Miles

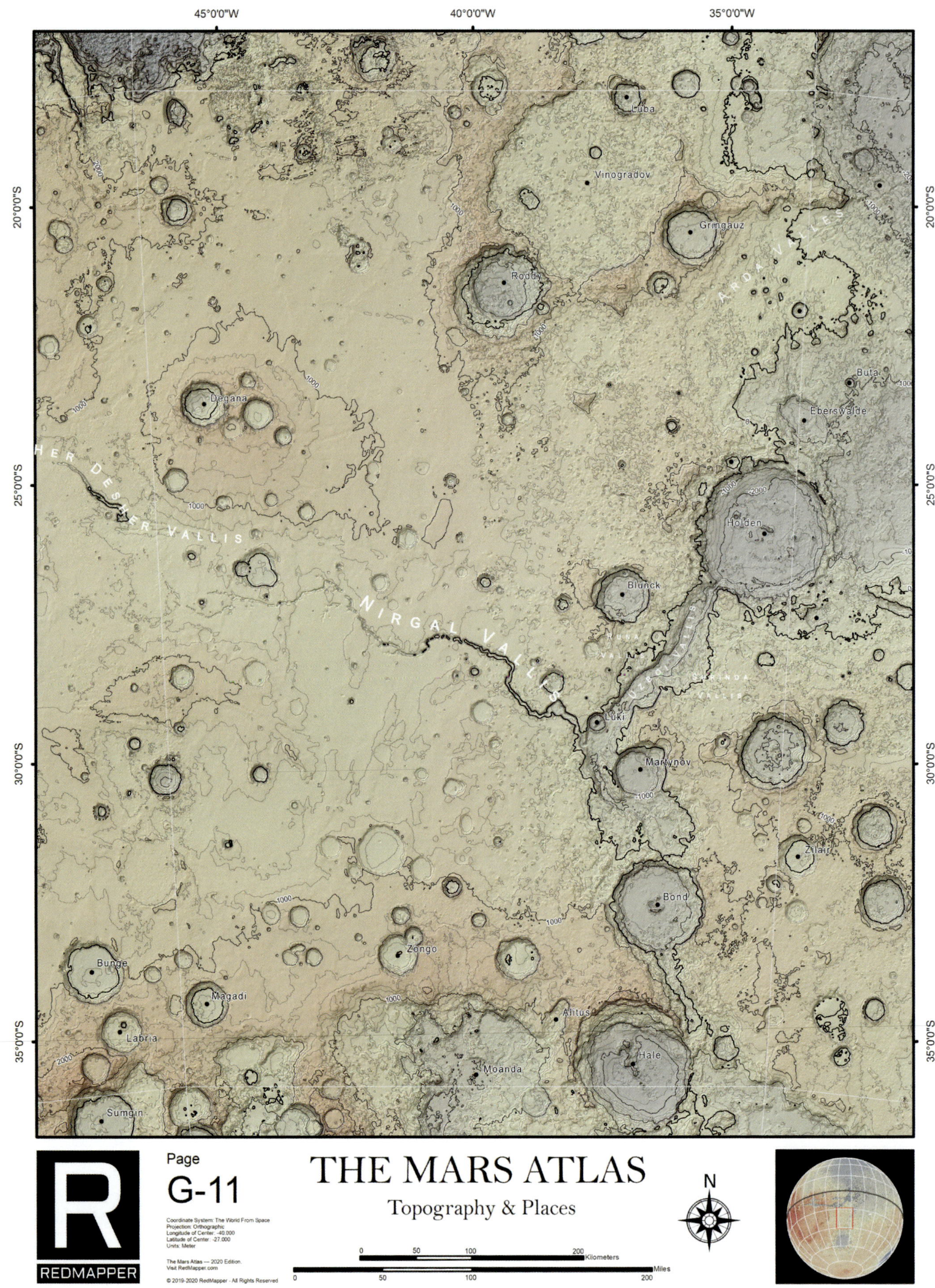

45°0'0"W
40°0'0"W
35°0'0"W
20°0'0"S
25°0'0"S
30°0'0"S
35°0'0"S
Luba
Vinogradov
Gringauz
Roddy
ARDA VALLES
Buta
Eberswalde
Degana
Holden
VALLIS
NIRGAL VALLIS
Blunck
Luki
Martynov
Zilair
Bond
Zongo
Bunge
Magadi
Labria
Alitus
Moanda
Hale
Sumgin
1000
2000
REDMAPPER
Page
G-11
Coordinate System: The World From Space
Projection: Orthographic
Longitude of Center: -40.000
Latitude of Center: -27.000
Units: Meter
The Mars Atlas — 2020 Edition.
Visit RedMapper.com
© 2019-2020 RedMapper - All Rights Reserved
THE MARS ATLAS
Topography & Places
N
0 50 100 200 Kilometers
0 50 100 200 Miles

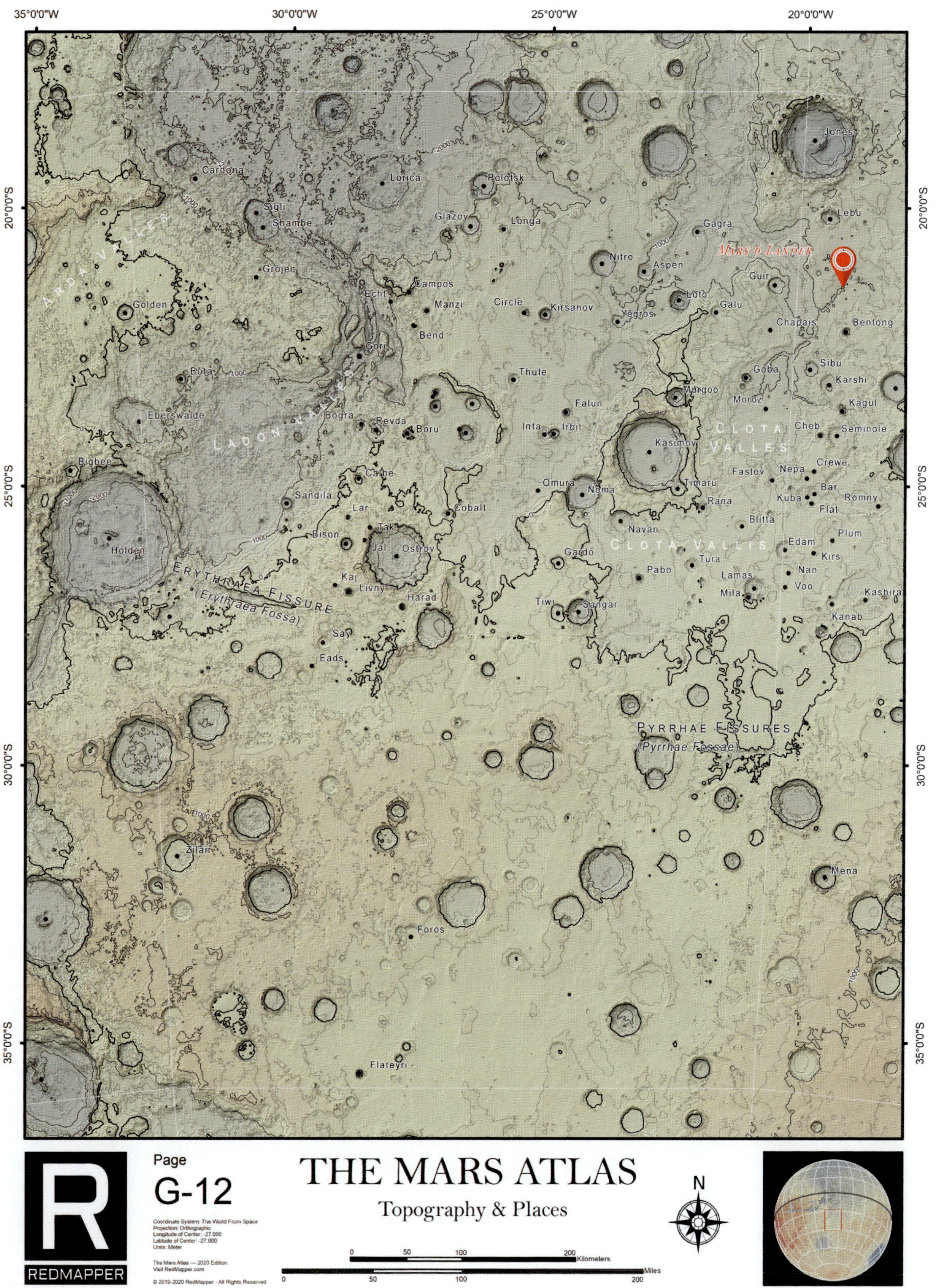
35°0'0"W
30°0'0"W
25°0'0"W
20°0'0"W
20°0'0"S
25°0'0"S
30°0'0"S
35°0'0"S
Jones
Lebu
Mars 6 Lander
Gagra
Aspen
Nitro
Guir
Lolo
Galu
Yegros
Chapais
Bentong
Cardona
Lorica
Polotsk
Glazov
Longa
Sigli
Shambe
Ardo Valles
Grojec
Campos
Echt
Manzi
Circle
Kirsanov
Golden
Bend
Gori
Bulla
Thule
Goba
Sibu
Karshi
Margoo
Moroz
Kagul
Falun
Eberswalde
Ladon Valles
Bogra
Revda
Boru
Inta
Irbit
Clota Valles
Cheb
Seminole
Kasimov
Bigbee
Crewe
Fastov
Nepa
Calbe
Omura
Noma
Timaru
Bar
Sandila
Rana
Kuba
Romny
Lar
Cobalt
Flat
Blitta
Tak
Navan
Bison
Plum
Holden
Jal
Ostrov
Clota Vallis
Edam
Kirs
Gardo
Tura
Erythraea Fissure
(Erythraea Fossa)
Pabo
Nan
Kaj
Lamas
Livny
Voo
Mila
Kashira
Harad
Tiwi
Sangar
Kanab
Say
Eads
Pyrrhae Fissures
(Pyrrhae Fossae)
Zilair
Mena
Foros
Flateyri
1000
2000
R
REDMAPPER
Page
G-12
Coordinate System: The World From Space
Projection: Orthographic
Longitude of Center: -27.000
Latitude of Center: -27.000
Units: Meter
The Mars Atlas — 2020 Edition.
Visit RedMapper.com
© 2019-2020 RedMapper - All Rights Reserved
THE MARS ATLAS
Topography & Places
N
0
50
100
200
Kilometers
0
50
100
200
Miles

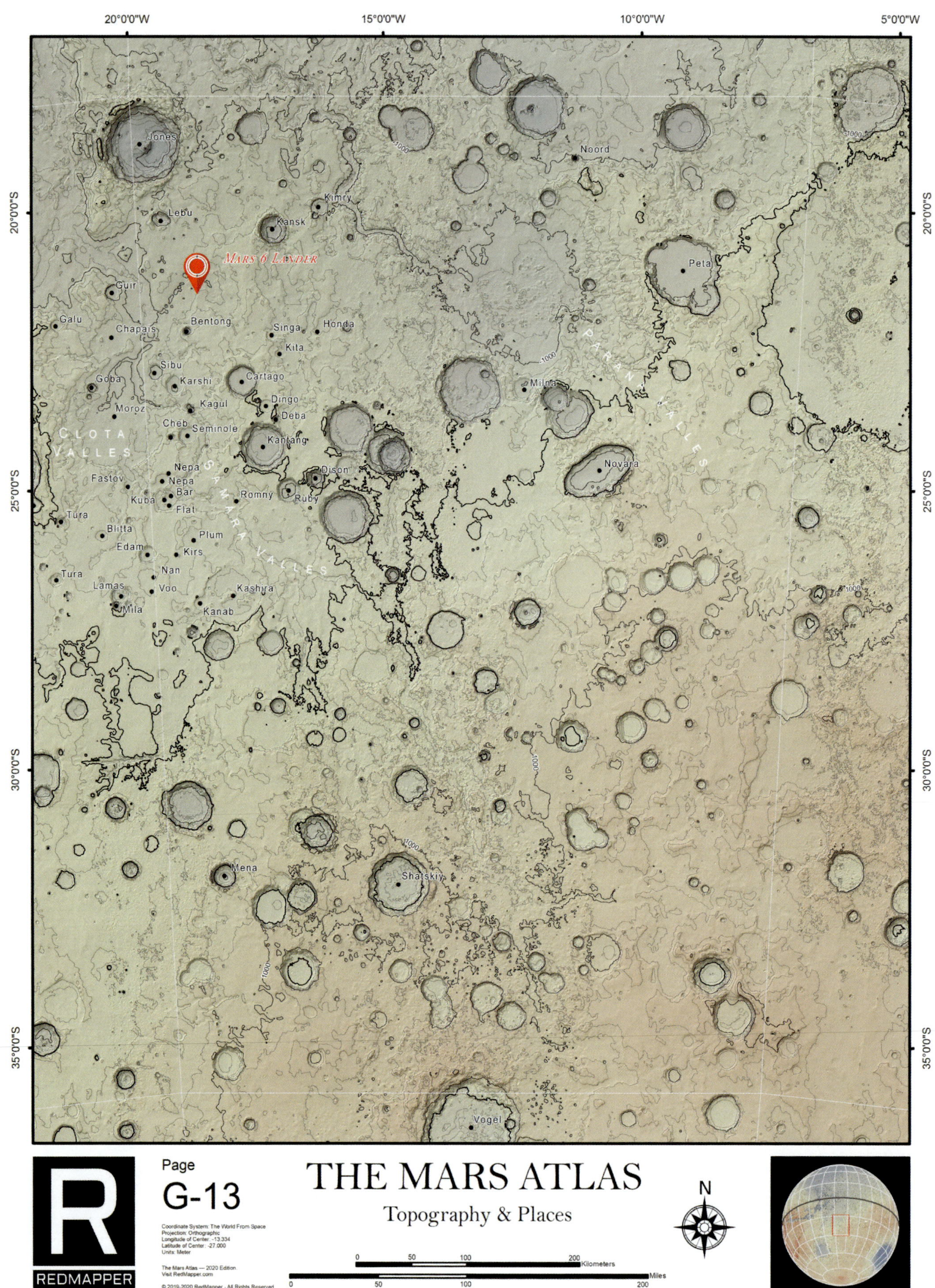

REDMAPPER

Page
G-13

Coordinate System: The World From Space
Projection: Orthographic
Longitude of Center: -13.334
Latitude of Center: -27.000
Units: Meter

The Mars Atlas — 2020 Edition
Visit RedMapper.com

THE MARS ATLAS

Topography & Places

N

0 50 100 200 Kilometers

0 50 100 200 Miles

5°0'0"W
0°0'0"
5°0'0"E
20°0'0"S
25°0'0"S
30°0'0"S
35°0'0"S
1000
2000
Dollfus
Newcomb
MARKHAVALLIS
Greeley
R
REDMAPPER
Page
G-14
Coordinate System: The World From Space
Projection: Orthographic
Longitude of Center: 0.000
Latitude of Center: -9.000
Units: Meter
The Mars Atlas — 2020 Edition.
Visit RedMapper.com
© 2019-2020 RedMapper - All Rights Reserved
THE MARS ATLAS
Topography & Places
N
0
50
100
200
Kilometers
Miles

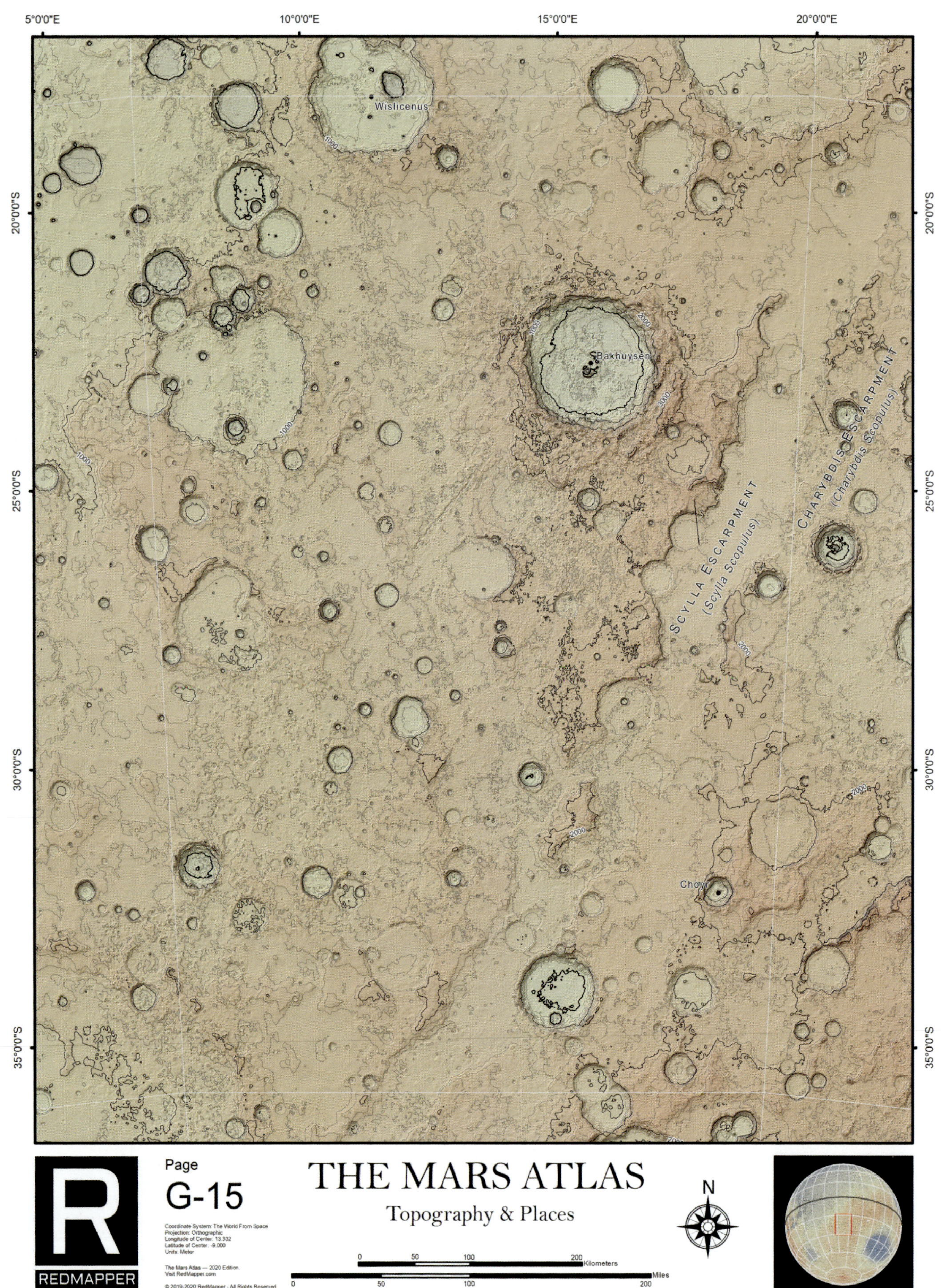
5°0'0"E
10°0'0"E
15°0'0"E
20°0'0"E
20°0'0"S
25°0'0"S
30°0'0"S
35°0'0"S
Wislicenus
Bakhuysen
SCYLLA ESCARPMENT
(Scylla Scopulus)
CHARYBDIS ESCARPMENT
(Charybdis Scopulus)
Choyr
1000
2000
3000
R
REDMAPPER
Page
G-15
Coordinate System: The World From Space
Projection: Orthographic
Longitude of Center: 13.332
Latitude of Center: -9.000
Units: Meter
The Mars Atlas — 2020 Edition
Visit RedMapper.com
© 2019-2020 RedMapper - All Rights Reserved
THE MARS ATLAS
Topography & Places
N
0 50 100 200 Kilometers
0 50 100 200 Miles

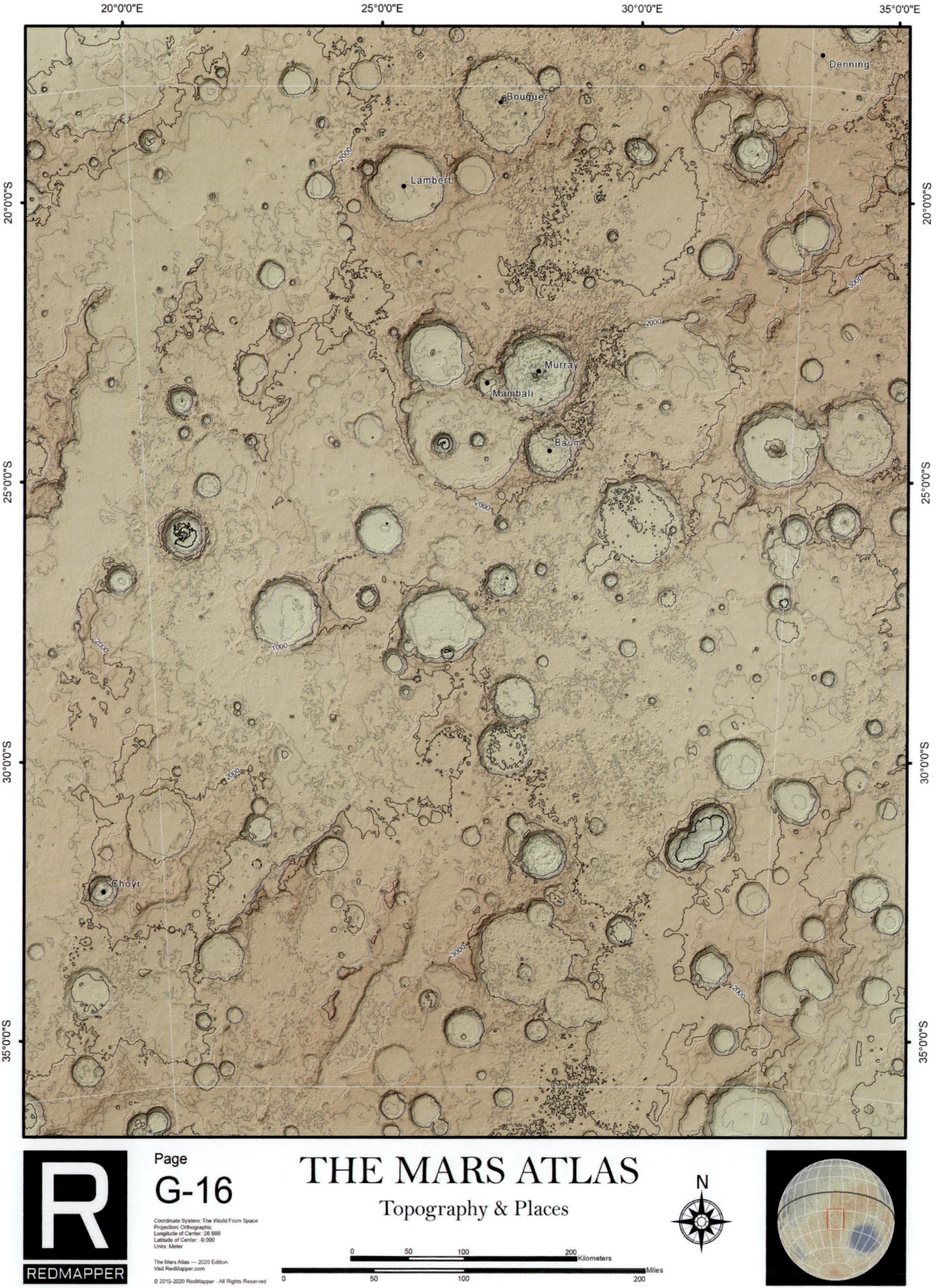

Page

G-16

Coordinate System: The World From Space
Projection: Orthographic
Longitude of Center: 26.666
Latitude of Center: -9.000
Units: Meter

The Mars Atlas — 2020 Edition.
Visit RedMapper.com

THE MARS ATLAS

Topography & Places

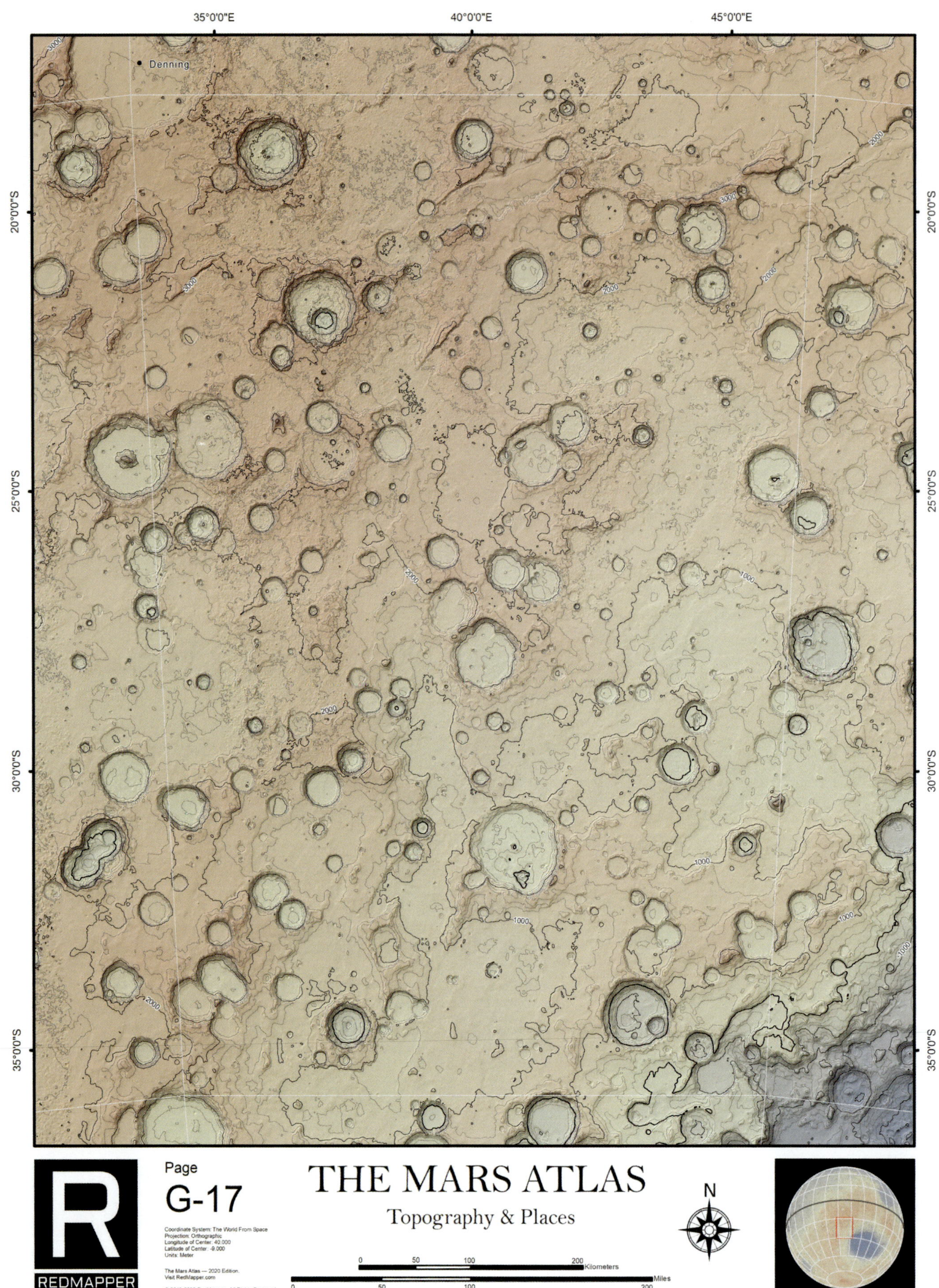
35°0'0"E
40°0'0"E
45°0'0"E
20°0'0"S
25°0'0"S
30°0'0"S
35°0'0"S
Denning
3000
2000
1000
R
REDMAPPER
Page
G-17
Coordinate System: The World From Space
Projection: Orthographic
Longitude of Center: 40.000
Latitude of Center: -9.000
Units: Meter
The Mars Atlas — 2020 Edition.
Visit RedMapper.com
© 2019-2020 RedMapper - All Rights Reserved
THE MARS ATLAS
Topography & Places
N
0 50 100 200 Kilometers
0 50 100 200 Miles

REDMAPPER

Page
G-18

THE MARS ATLAS

Topography & Places

Coordinate System: The World From Space
Projection: Orthographic
Longitude of Center: 53.332
Latitude of Center: -27.000
Units: Meter

The Mars Atlas — 2020 Edition.
Visit RedMapper.com

0 50 100 200 Kilometers
0 50 100 200 Miles

N

60°0'0"E
65°0'0"E
70°0'0"E
75°0'0"E
20°0'0"S
25°0'0"S
30°0'0"S
35°0'0"S
Harris
Sahek
Terby
-1000
-2000
-3000
-4000
-5000
-6000
-7000
CORONAE PLA
S PALUS
Badwater
CORONAE ESCARPMENT
(Coronae Scopulus)
HELLAS CANYON
(Hellas Chasma)
Kufstein
Talus
REDMAPPER
Page
G-19
Coordinate System: The World From Space
Projection: Orthographic
Longitude of Center: 66.666
Latitude of Center: -27.000
Units: Meter
The Mars Atlas — 2020 Edition
Visit RedMapper.com
© 2019-2020 RedMapper - All Rights Reserved
THE MARS ATLAS
Topography & Places
N
0 50 100 200 Kilometers
0 50 100 200 Miles

75°0'0"E
80°0'0"E
85°0'0"E
20°0'0"S
25°0'0"S
30°0'0"S
35°0'0"S
Saheki
Millochau
Okotoks
Jumla
Vichada Valles
Napo Vallis
Huallaga Vallis
Runanga
Gokwe
Auce
Jorn
Terby
Hadriacus Basin
(Hadriacus Cavi)
McCauley
Jori
Suzhi
Isil
Kasabi
Mount Anseris
(Anseris Mons)
Batson
Nako
Salkhad
Tignish
Mount Peraea
(Peraea Mons)
Navua Valles
Nae Planum
Majuro
Coronae Mountains
(Coronae Montes)
Njesko
Taejin
Talas
Poti
R
REDMAPPER
Page
G-20
Coordinate System: The World From Space
Projection: Orthographic
Longitude of Center: 80.000
Latitude of Center: -27.000
Units: Meter
The Mars Atlas — 2020 Edition
Visit RedMapper.com
© 2019-2020 RedMapper - All Rights Reserved
THE MARS ATLAS
Topography & Places
N
Kilometers
Miles

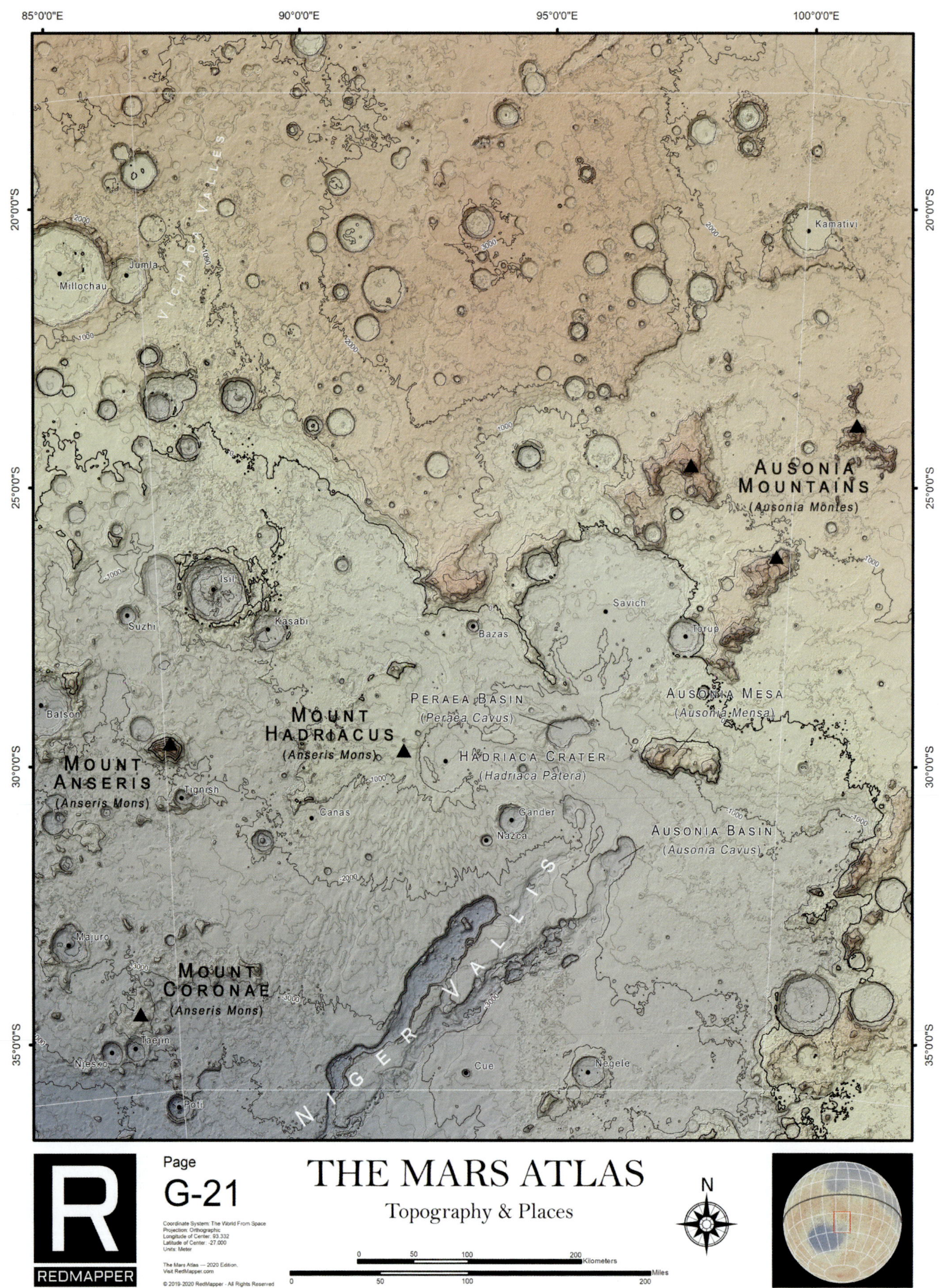

R
REDMAPPER

Page
G-21

Coordinate System: The World From Space
Projection: Orthographic
Longitude of Center: 93.332
Latitude of Center: -27.000
Units: Meter

The Mars Atlas — 2020 Edition.
Visit RedMapper.com

THE MARS ATLAS

Topography & Places

N

0 50 100 200 Kilometers

0 50 100 200 Miles

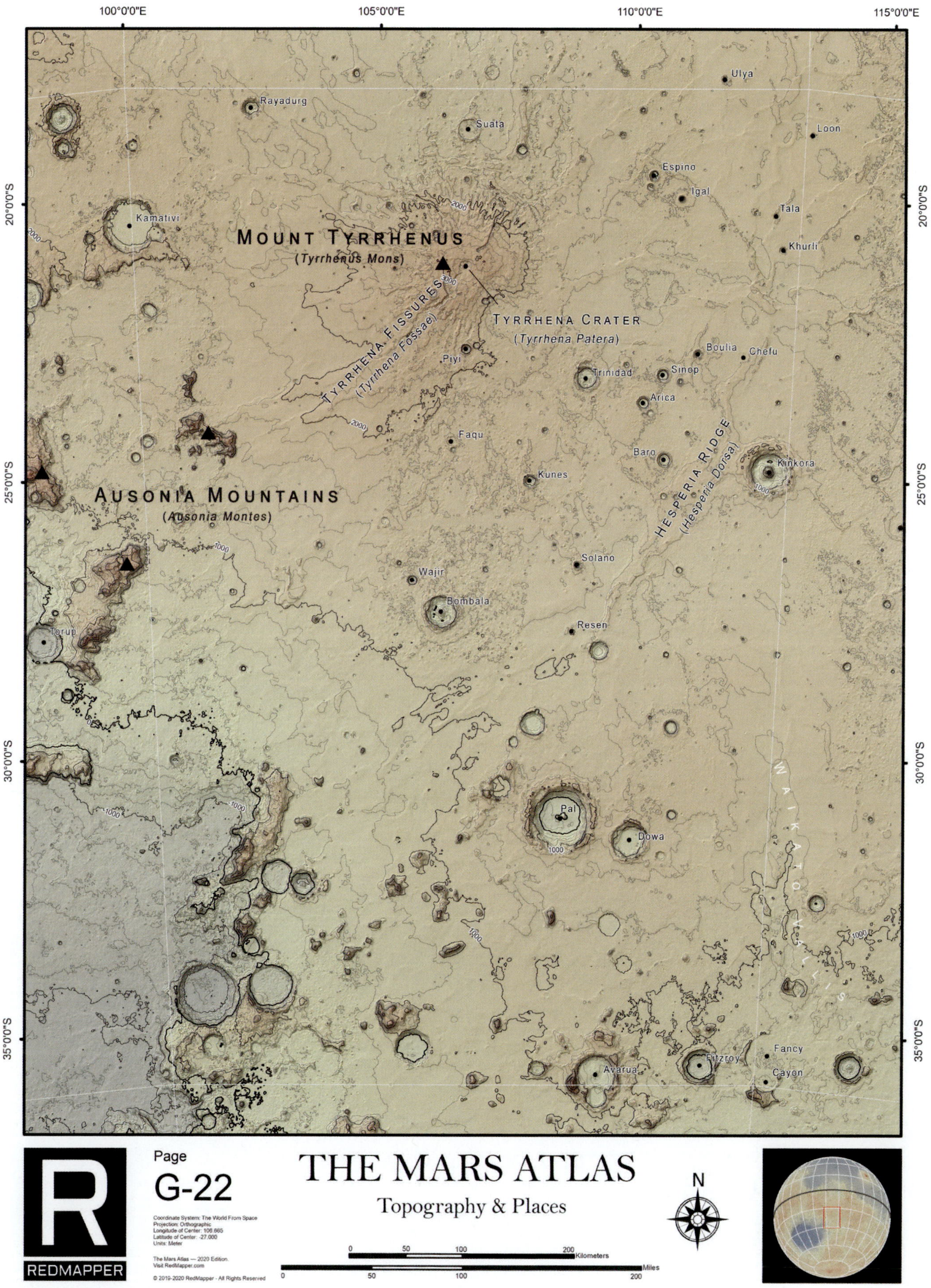
100°0'0"E
105°0'0"E
110°0'0"E
115°0'0"E
20°0'0"S
25°0'0"S
30°0'0"S
35°0'0"S
Ulya
Rayadurg
Suata
Loon
Espino
Igal
Tala
Kamativi
Khurli
MOUNT TYRRHENUS
(Tyrrhenus Mons)
TYRRHENA FISSURES
(Tyrrhena Fossae)
TYRRHENA CRATER
(Tyrrhena Patera)
Piyi
Boulia
Chefu
Trinidad
Sinop
Arica
Faqu
Baro
Kinkora
Kunes
HESPERIA RIDGE
(Hesperia Dorsa)
AUSONIA MOUNTAINS
(Ausonia Montes)
Solano
Wajir
Bombala
Resen
Torup
Pal
Dowa
WAIKATO VALLIS
Fancy
Fitzroy
Avarua
Cayon
2000
3000
1000
R
REDMAPPER
Page
G-22
Coordinate System: The World From Space
Projection: Orthographic
Longitude of Center: 106.665
Latitude of Center: -27.000
Units: Meter
The Mars Atlas — 2020 Edition.
Visit RedMapper.com
© 2019-2020 RedMapper - All Rights Reserved
THE MARS ATLAS
Topography & Places
N
0
50
100
200
Kilometers
Miles

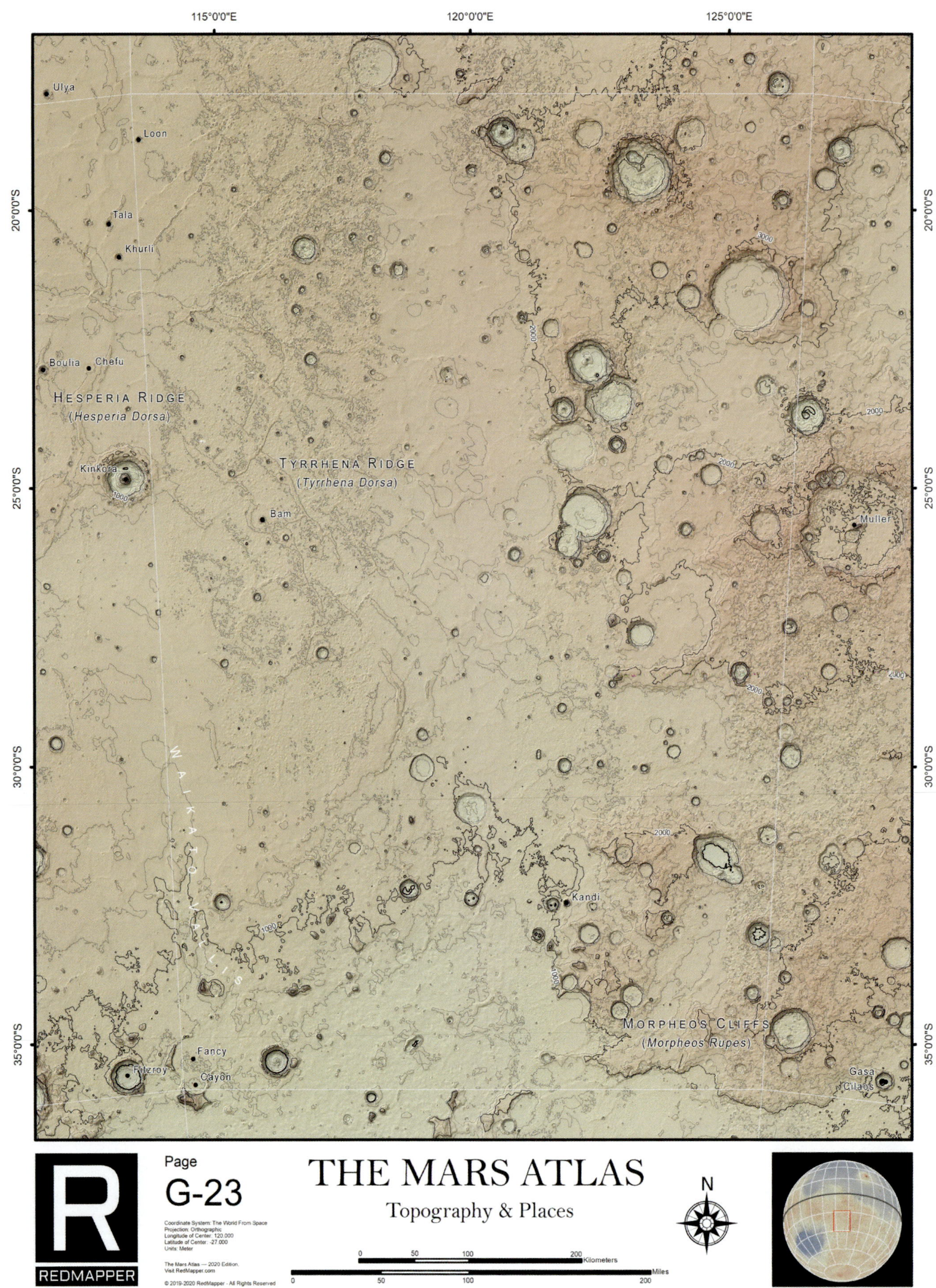

115°0'0"E
120°0'0"E
125°0'0"E
20°0'0"S
25°0'0"S
30°0'0"S
35°0'0"S
Ulya
Loon
Tala
Khurli
Boulia
Chefu
HESPERIA RIDGE
(Hesperia Dorsa)
Kinkora
TYRRHENA RIDGE
(Tyrrhena Dorsa)
Bam
Muller
WAIKATO VALLIS
Kandi
MORPHEOS CLIFFS
(Morpheos Rupes)
Fancy
Fitzroy
Cayon
Gasa
Cilaos
3000
2000
1000
REDMAPPER
Page
G-23
Coordinate System: The World From Space
Projection: Orthographic
Longitude of Center: 120.000
Latitude of Center: -27.000
Units: Meter
The Mars Atlas — 2020 Edition.
Visit RedMapper.com
© 2019-2020 RedMapper - All Rights Reserved
THE MARS ATLAS
Topography & Places
N
0 50 100 200 Kilometers
0 50 100 200 Miles

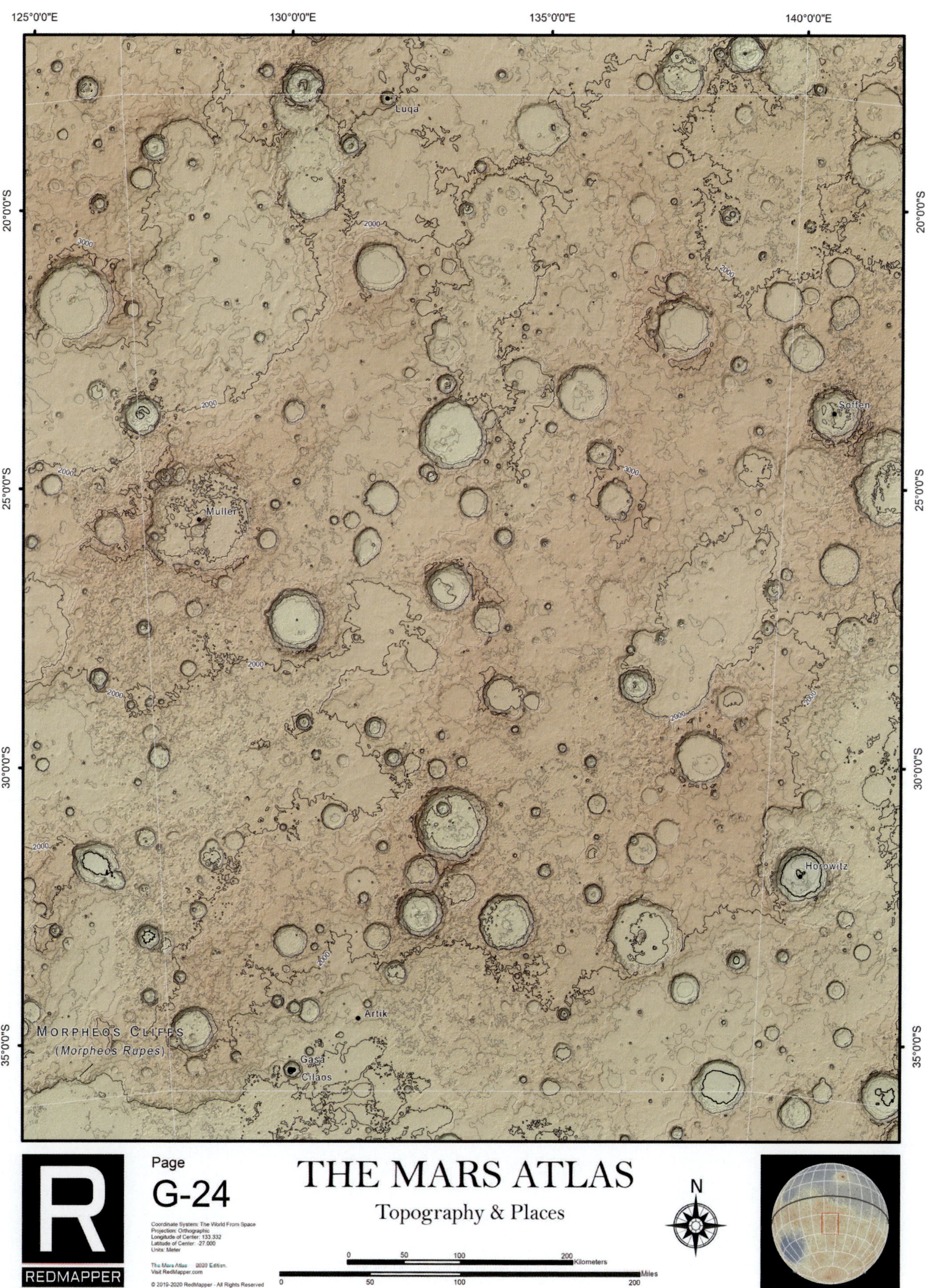

R
REDMAPPER

Page
G-24

Coordinate System: The World From Space
Projection: Orthographic
Longitude of Center: 133.332
Latitude of Center: -27.000
Units: Meter

The Mars Atlas 2020 Edition.
Visit RedMapper.com

THE MARS ATLAS

Topography & Places

N

0 50 100 200 Kilometers

0 50 100 200 Miles

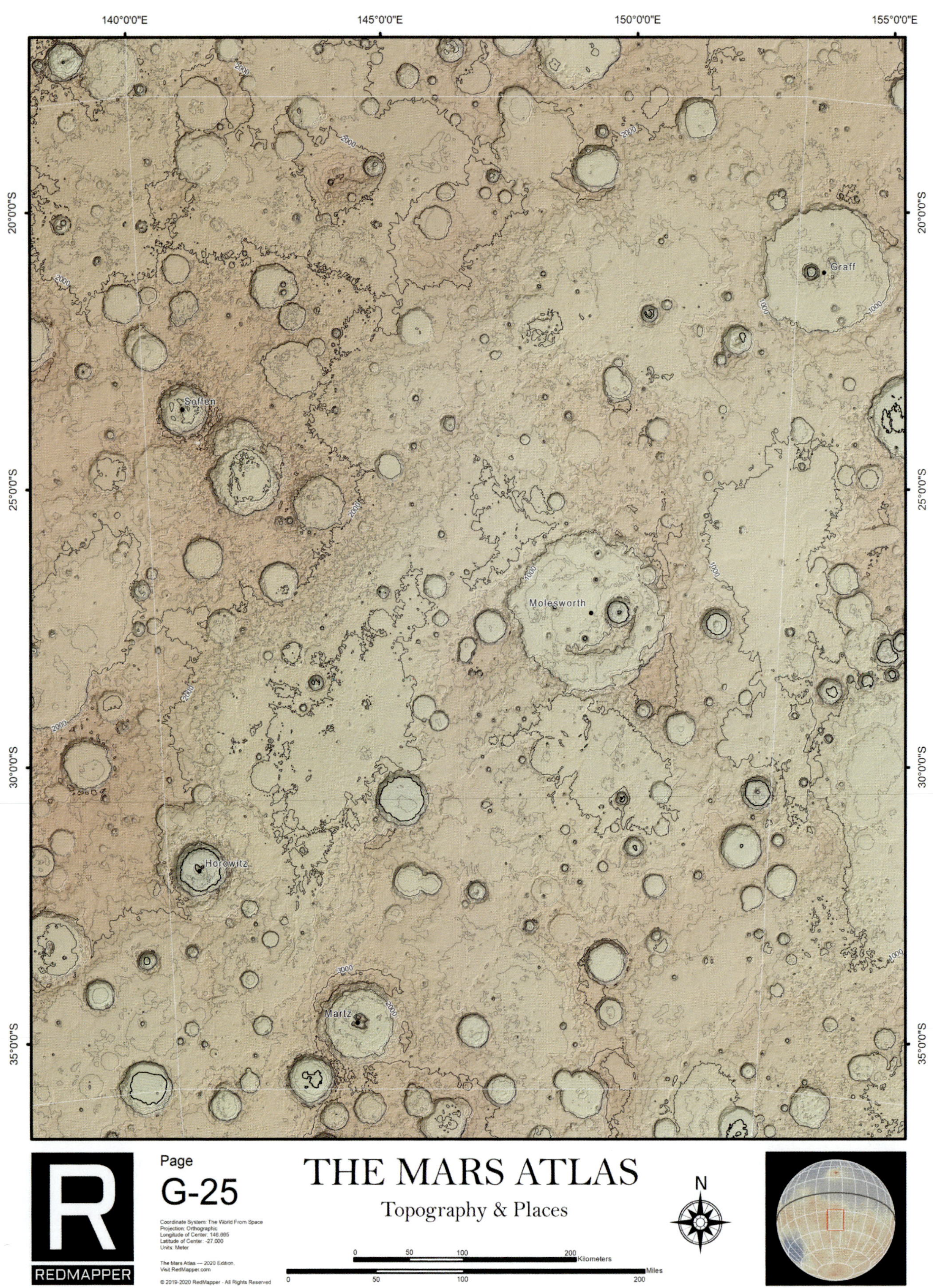

REDMAPPER

Page

G-25

Coordinate System: The World From Space
Projection: Orthographic
Longitude of Center: 146.665
Latitude of Center: -27.000
Units: Meter

The Mars Atlas — 2020 Edition.
Visit RedMapper.com

THE MARS ATLAS

Topography & Places

N

0 50 100 200 Kilometers

0 50 100 200 Miles

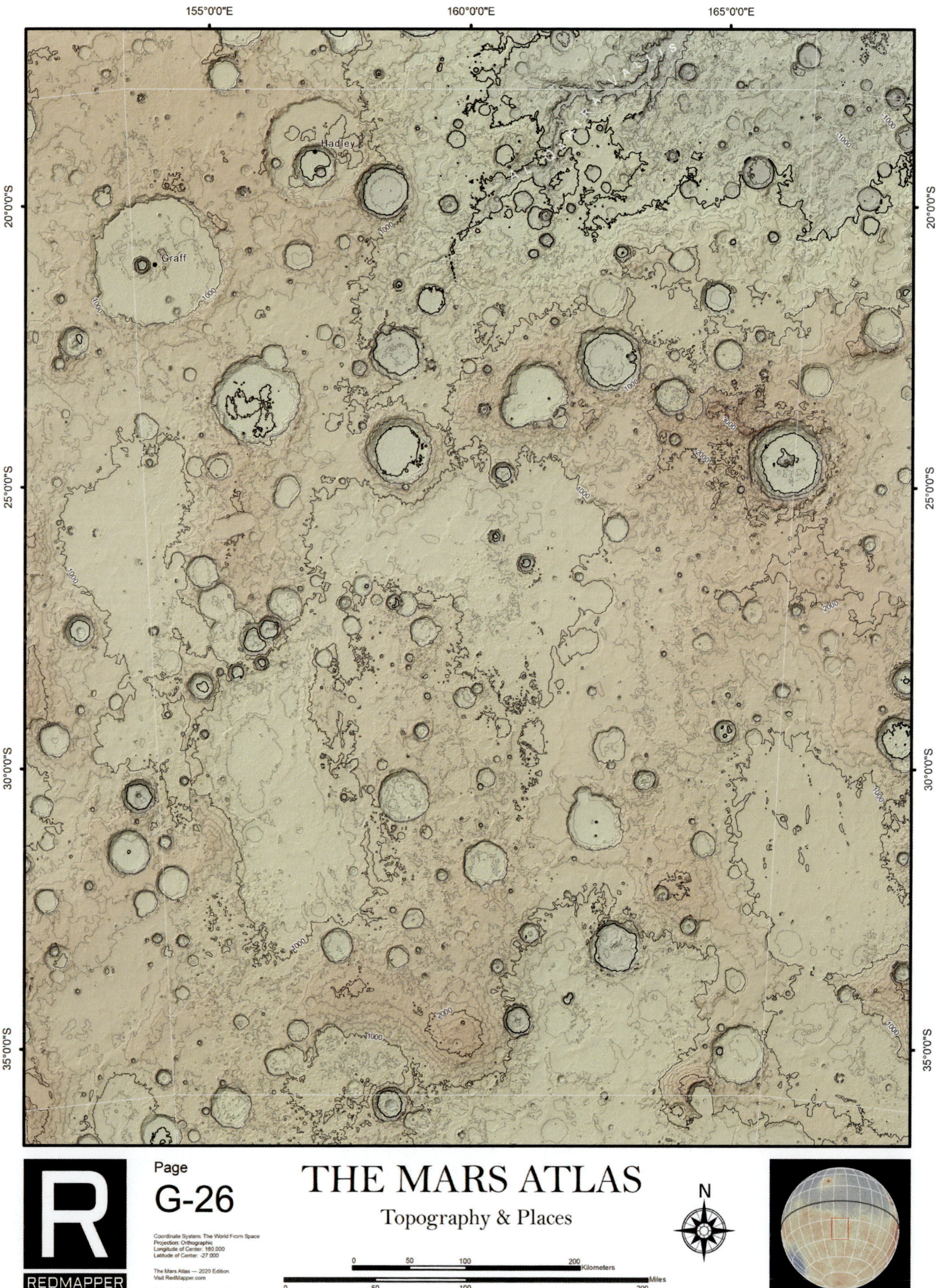

R
REDMAPPER

Page
G-26

Coordinate System: The World From Space
Projection: Orthographic
Longitude of Center: 160.000
Latitude of Center: -27.000

The Mars Atlas — 2020 Edition.
Visit RedMapper.com

THE MARS ATLAS

Topography & Places

0 50 100 200 Kilometers

0 50 100 200 Miles

N

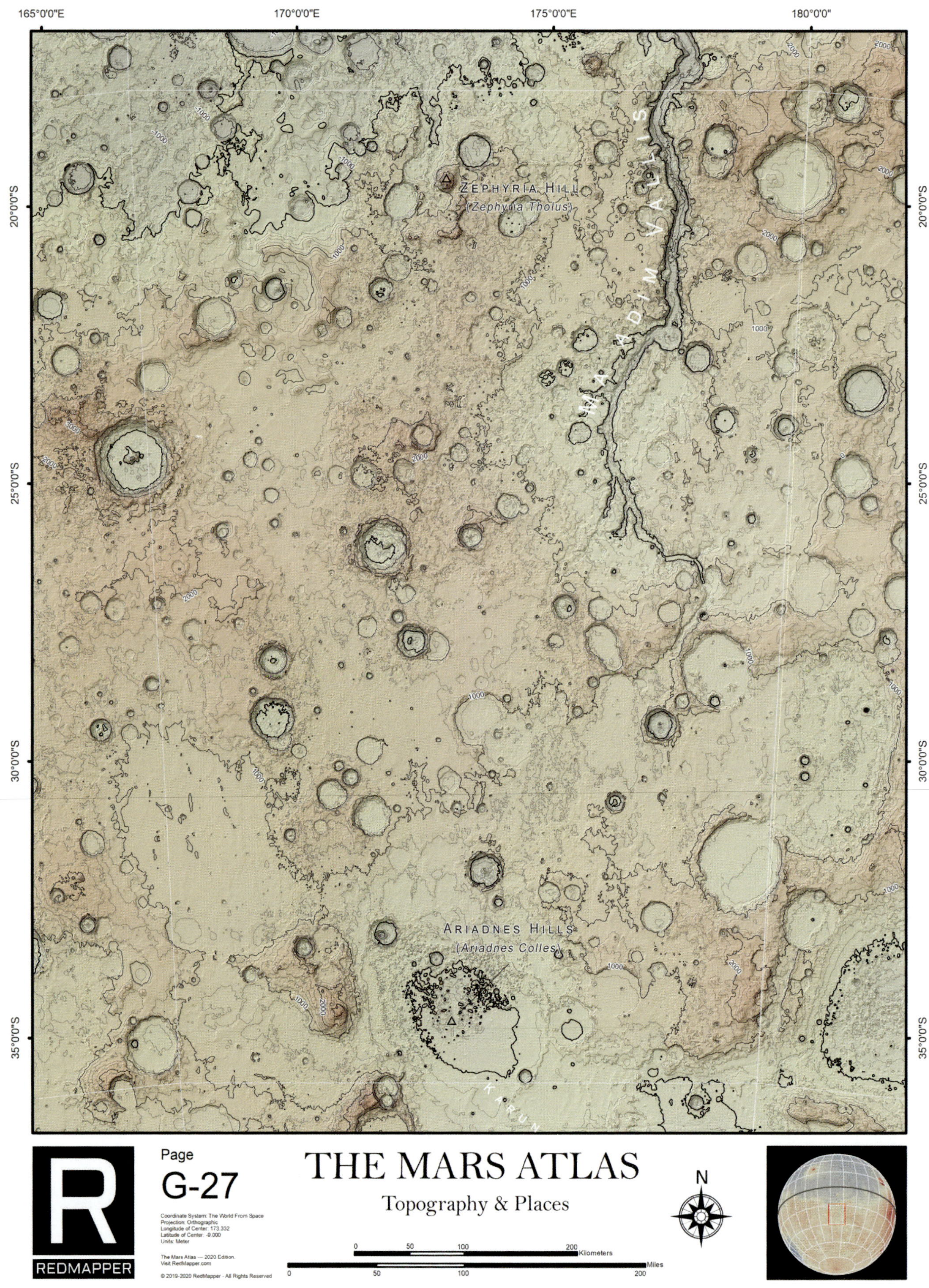
165°0'0"E
170°0'0"E
175°0'0"E
180°0'0"
20°0'0"S
25°0'0"S
30°0'0"S
35°0'0"S
ZEPHYRIA HILL
(Zephyria Tholus)
MA'ADIM VALLIS
ARIADNES HILLS
(Ariadnes Colles)
KARUN
1000
2000
Page
G-27
THE MARS ATLAS
Topography & Places
N
0 50 100 200 Kilometers
0 50 100 200 Miles
REDMAPPER
Coordinate System: The World From Space
Projection: Orthographic
Longitude of Center: 173.332
Latitude of Center: -9.000
Units: Meter
The Mars Atlas — 2020 Edition.
Visit RedMapper.com

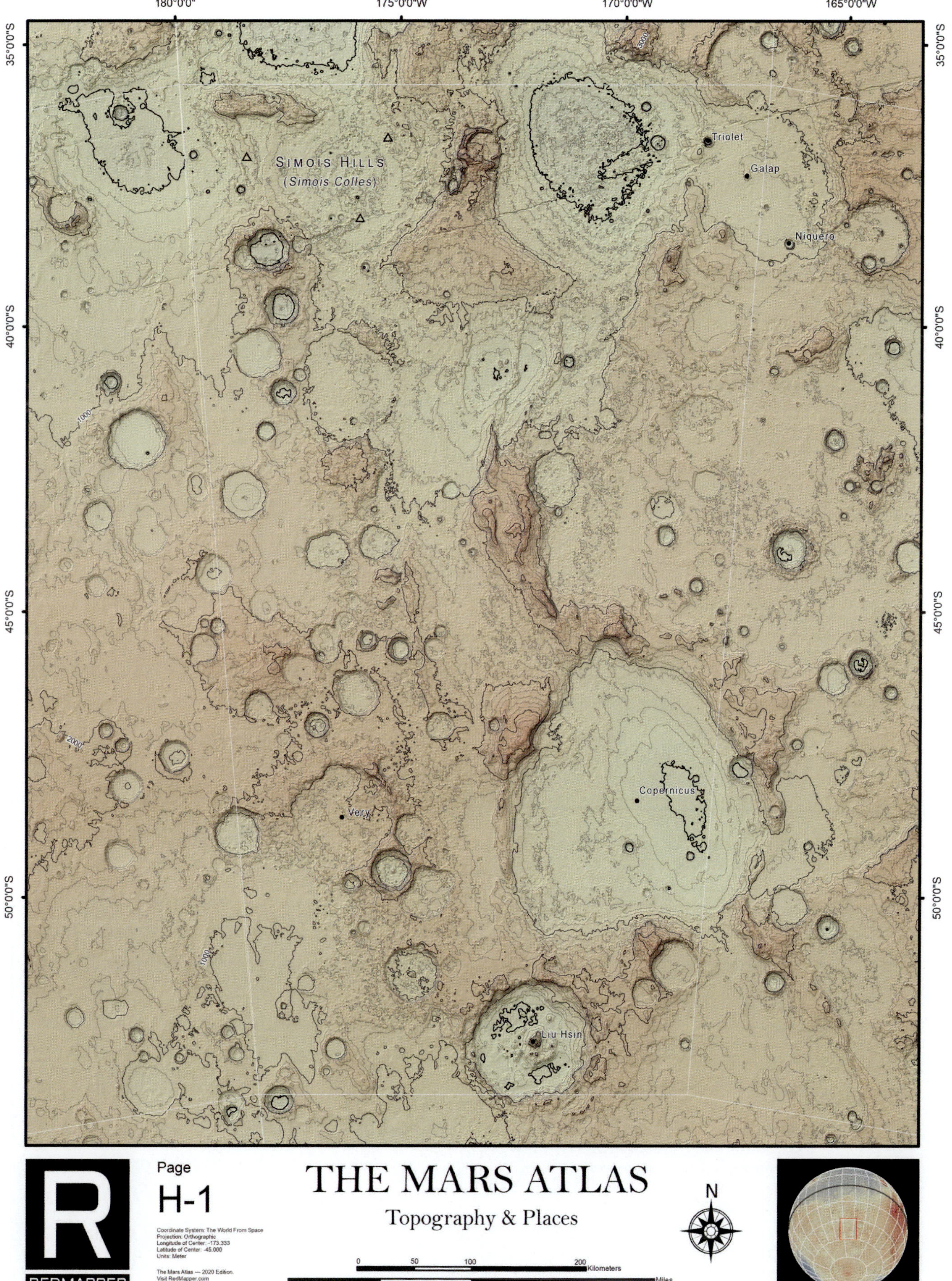

R
REDMAPPER

Page
H-1

Coordinate System: The World From Space
Projection: Orthographic
Longitude of Center: -173.333
Latitude of Center: -45.000
Units: Meter

The Mars Atlas — 2020 Edition.
Visit RedMapper.com

THE MARS ATLAS

Topography & Places

N

0 50 100 200 Kilometers

0 50 100 200 Miles

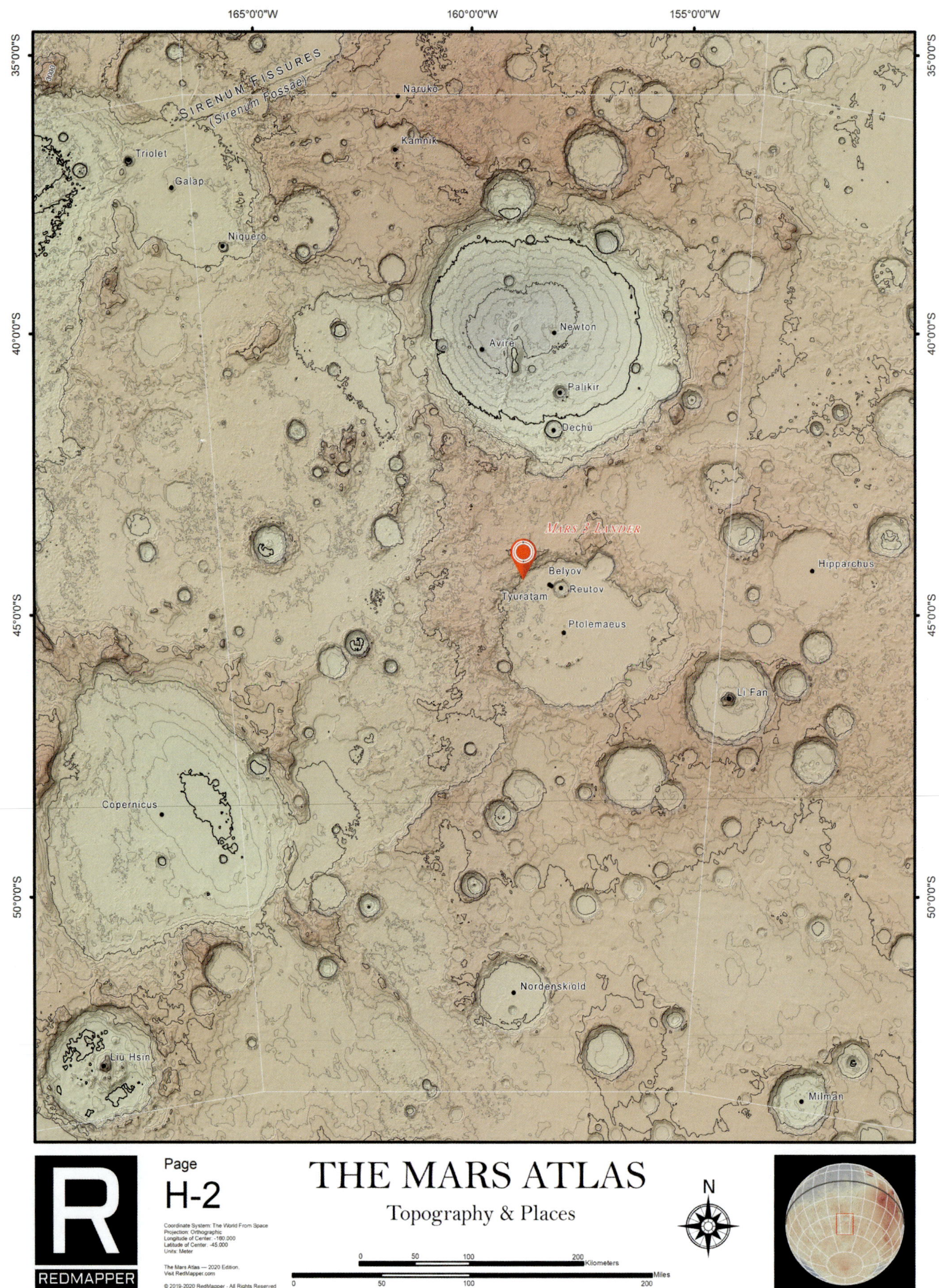

165°0'0"W
160°0'0"W
155°0'0"W
35°0'0"S
40°0'0"S
45°0'0"S
50°0'0"S
SIRENUM FISSURES
(Sirenum Fossae)
Naruko
Kamnik
Triolet
Galap
Niquero
Newton
Avire
Palikir
Dechu
MARS 3 LANDER
Belyov
Reutov
Tyuratam
Ptolemaeus
Hipparchus
Li Fan
Copernicus
Nordenskiold
Liu Hsin
Milman
R
REDMAPPER
Page
H-2
Coordinate System: The World From Space
Projection: Orthographic
Longitude of Center: -160.000
Latitude of Center: -45.000
Units: Meter
The Mars Atlas — 2020 Edition.
Visit RedMapper.com
© 2019-2020 RedMapper - All Rights Reserved
THE MARS ATLAS
Topography & Places
0 50 100 200 Kilometers
0 50 100 200 Miles
N

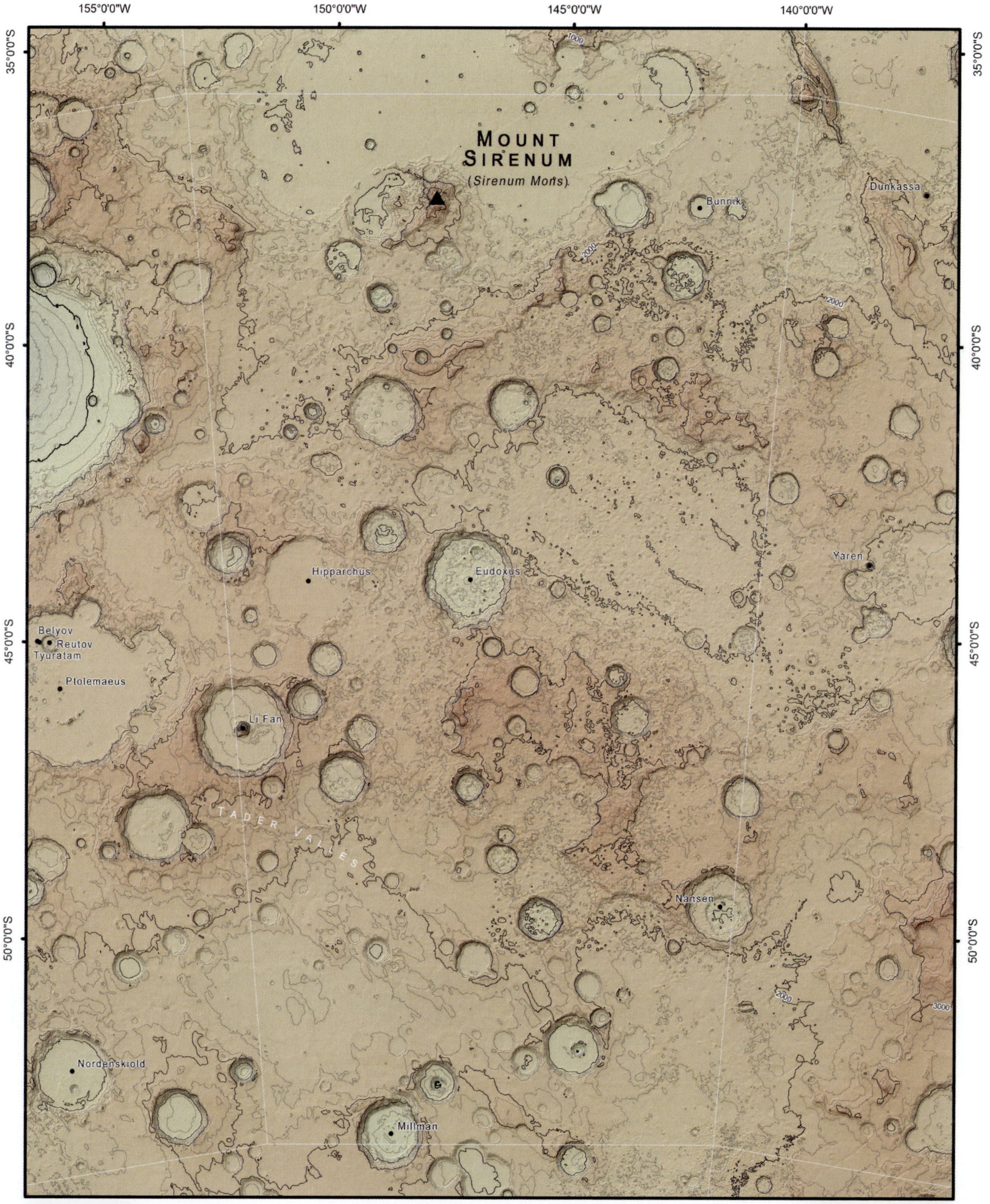

155°0'0"W
150°0'0"W
145°0'0"W
140°0'0"W

35°0'0"S
40°0'0"S
45°0'0"S
50°0'0"S

MOUNT SIRENUM
(Sirenum Mons)
Bunnik
Dunkassa
Hipparchus
Eudoxus
Yaren
Belyov
Reutov
Tyuratam
Ptolemaeus
Li Fan
TADER VALLES
Nansen
Nordenskiold
Millman
1000
2000
2000
2000
3000

Page
H-3

Coordinate System: The World From Space
Projection: Orthographic
Longitude of Center: -146.667
Latitude of Center: -45.000
Units: Meter

The Mars Atlas — 2020 Edition.
Visit RedMapper.com

THE MARS ATLAS

Topography & Places

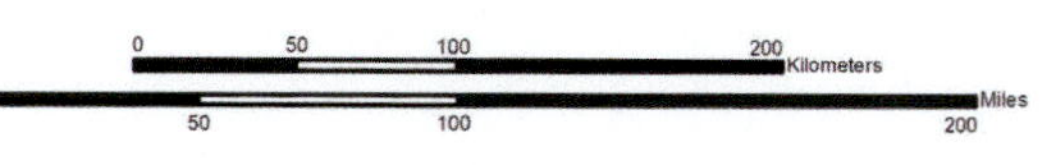

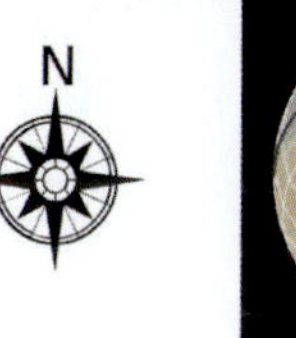

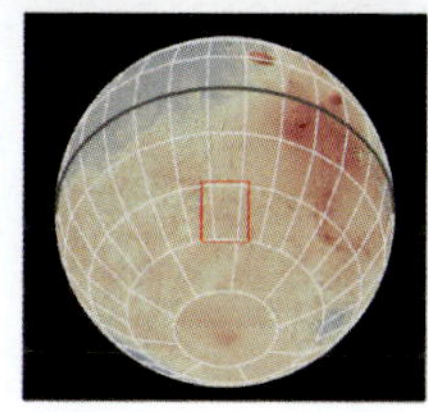

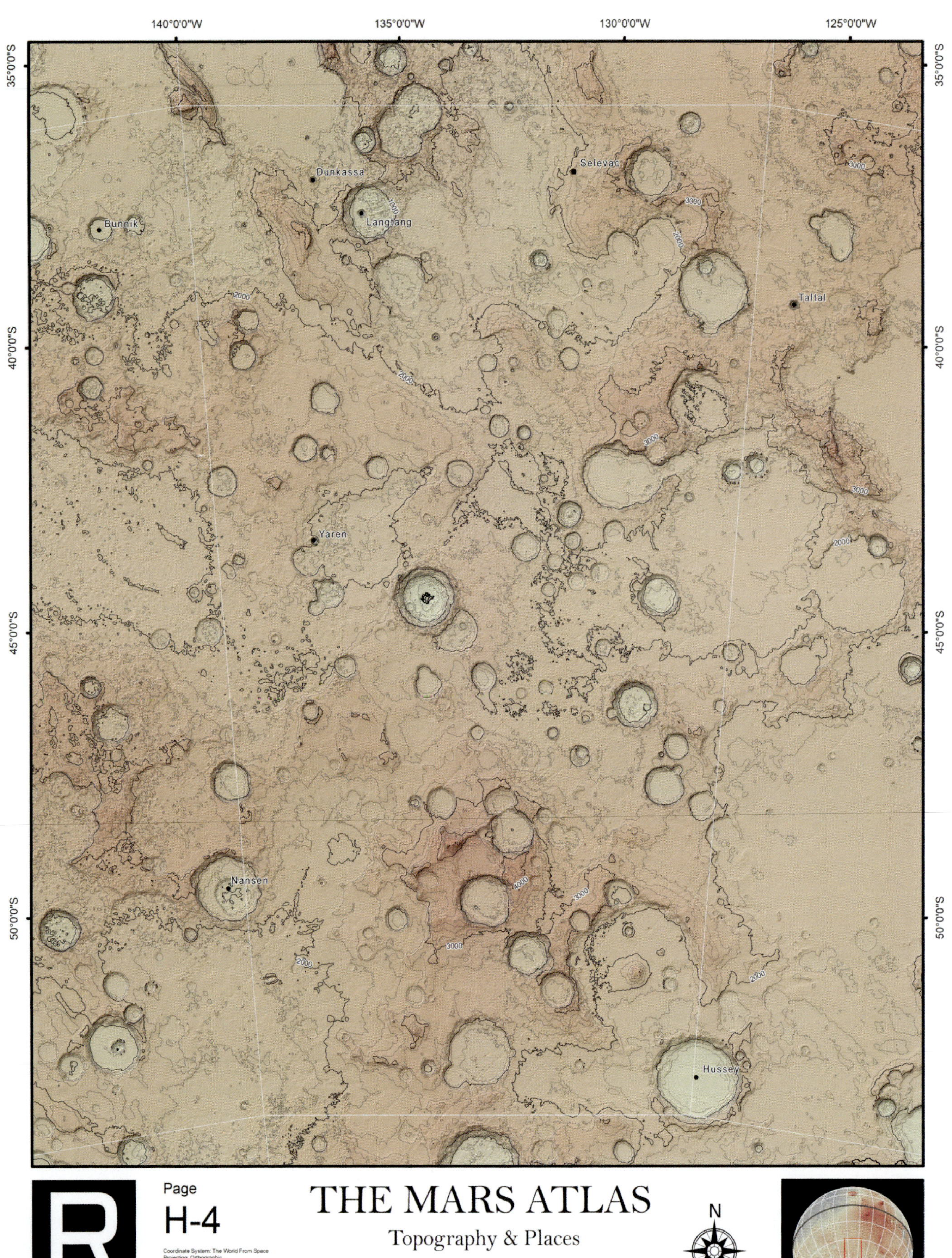

R
REDMAPPER

Page
H-4

Coordinate System: The World From Space
Projection: Orthographic
Longitude of Center: -133.333
Latitude of Center: -45.000
Units: Meter

The Mars Atlas — 2020 Edition.
Visit RedMapper.com

THE MARS ATLAS
Topography & Places

N

0 50 100 200 Kilometers
0 50 100 200 Miles

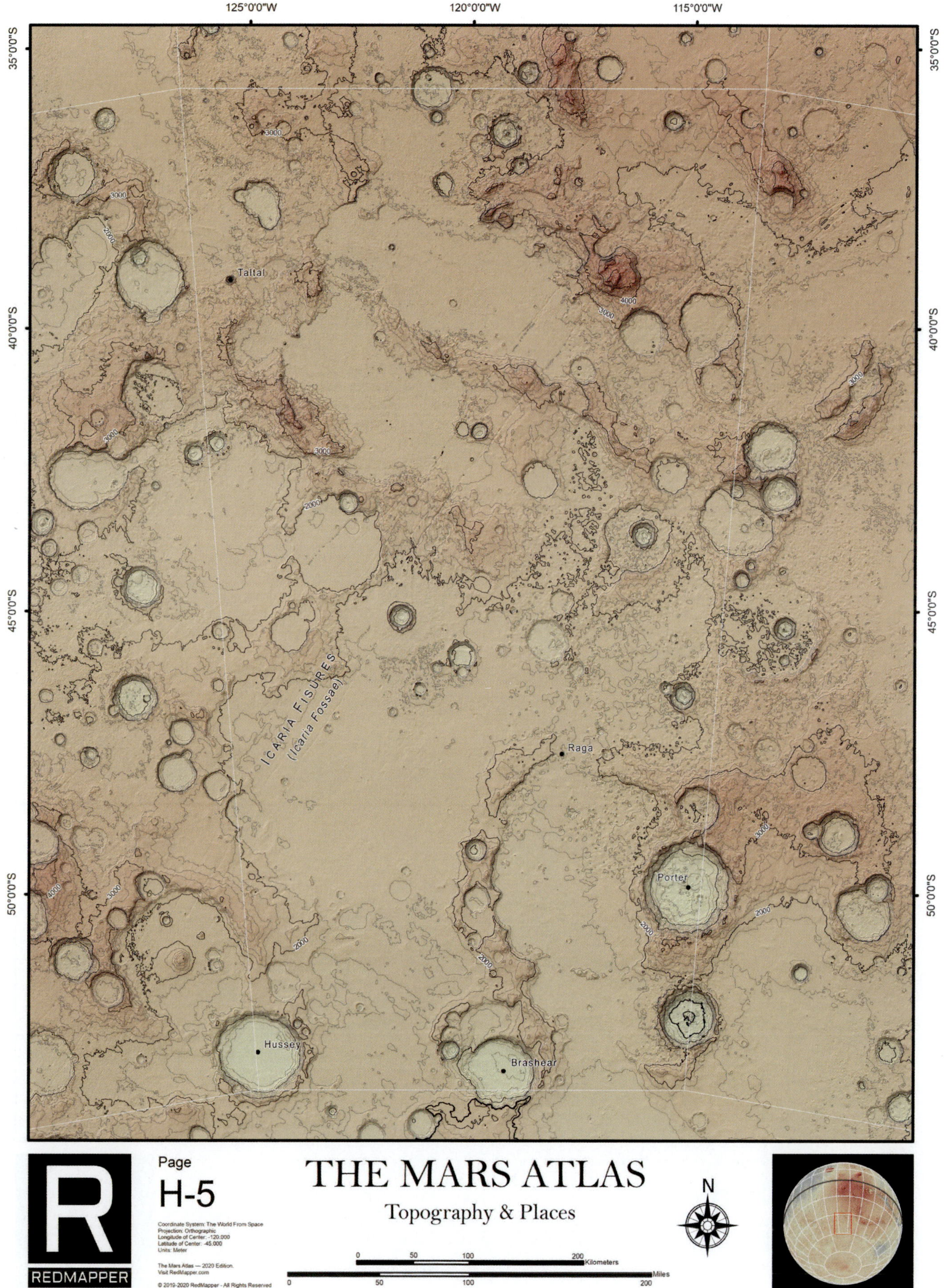
125°0'0"W
120°0'0"W
115°0'0"W
35°0'0"S
40°0'0"S
45°0'0"S
50°0'0"S
Taltal
Raga
ICARIA FISURES
(Icaria Fossae)
Porter
Hussey
Brashear
3000
2000
4000
R
REDMAPPER
Page
H-5
Coordinate System: The World From Space
Projection: Orthographic
Longitude of Center: -120.000
Latitude of Center: -45.000
Units: Meter
The Mars Atlas — 2020 Edition.
Visit RedMapper.com
© 2019-2020 RedMapper - All Rights Reserved
THE MARS ATLAS
Topography & Places
N
0 50 100 200 Kilometers
0 50 100 200 Miles

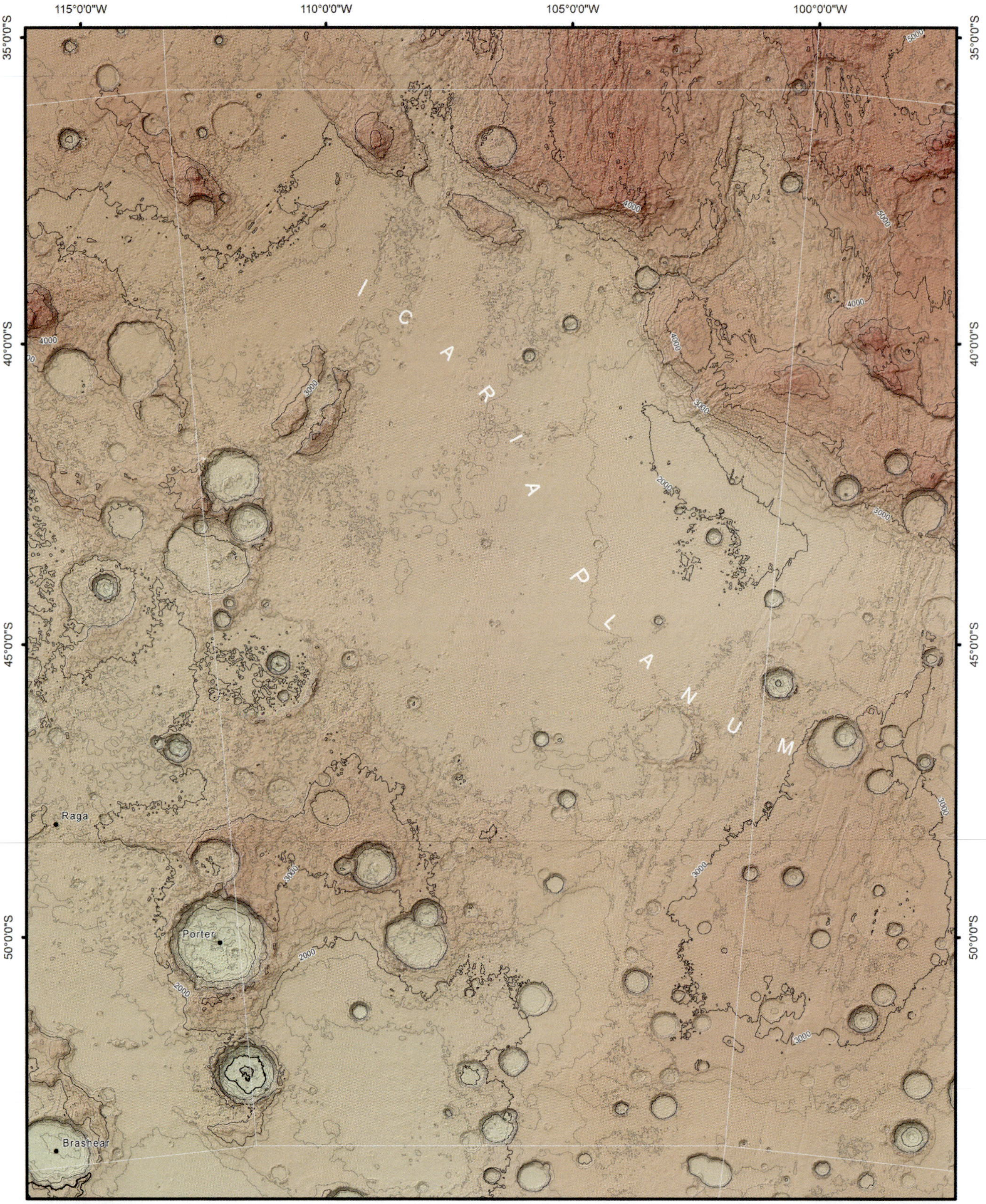

115°0'0"W 110°0'0"W 105°0'0"W 100°0'0"W

35°0'0"S

40°0'0"S

45°0'0"S

50°0'0"S

ICARIA PLANUM

Raga

Porter

Brashear

4000

3000

2000

5000

Page

H-6

Coordinate System: The World From Space
Projection: Orthographic
Longitude of Center: -106.667
Latitude of Center: -45.000
Units: Meter

The Mars Atlas — 2020 Edition.
Visit RedMapper.com

THE MARS ATLAS

Topography & Places

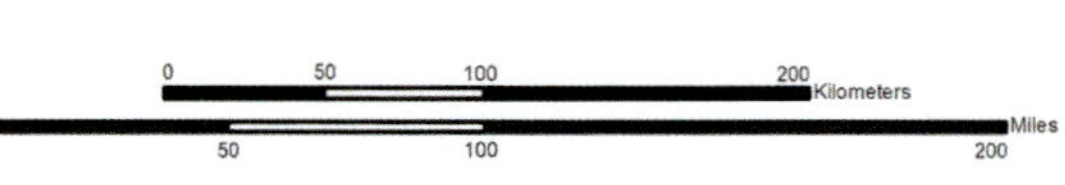

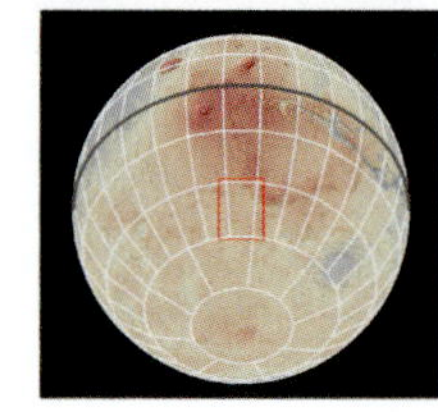

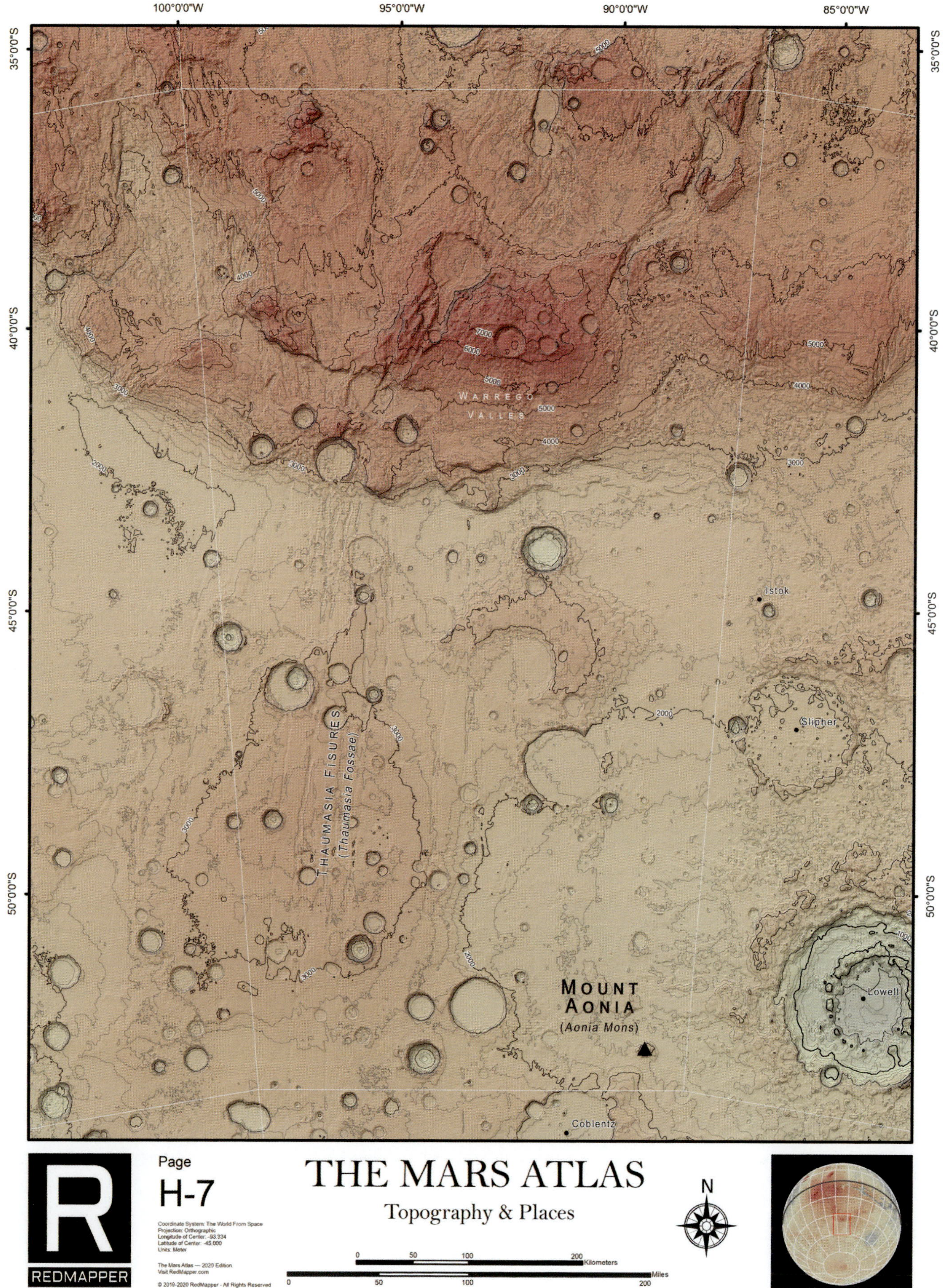

Page
H-7

Coordinate System: The World From Space
Projection: Orthographic
Longitude of Center: -93.334
Latitude of Center: -45.000
Units: Meter

The Mars Atlas — 2020 Edition
Visit RedMapper.com

REDMAPPER

THE MARS ATLAS

Topography & Places

0 50 100 200 Kilometers
0 50 100 200 Miles

N

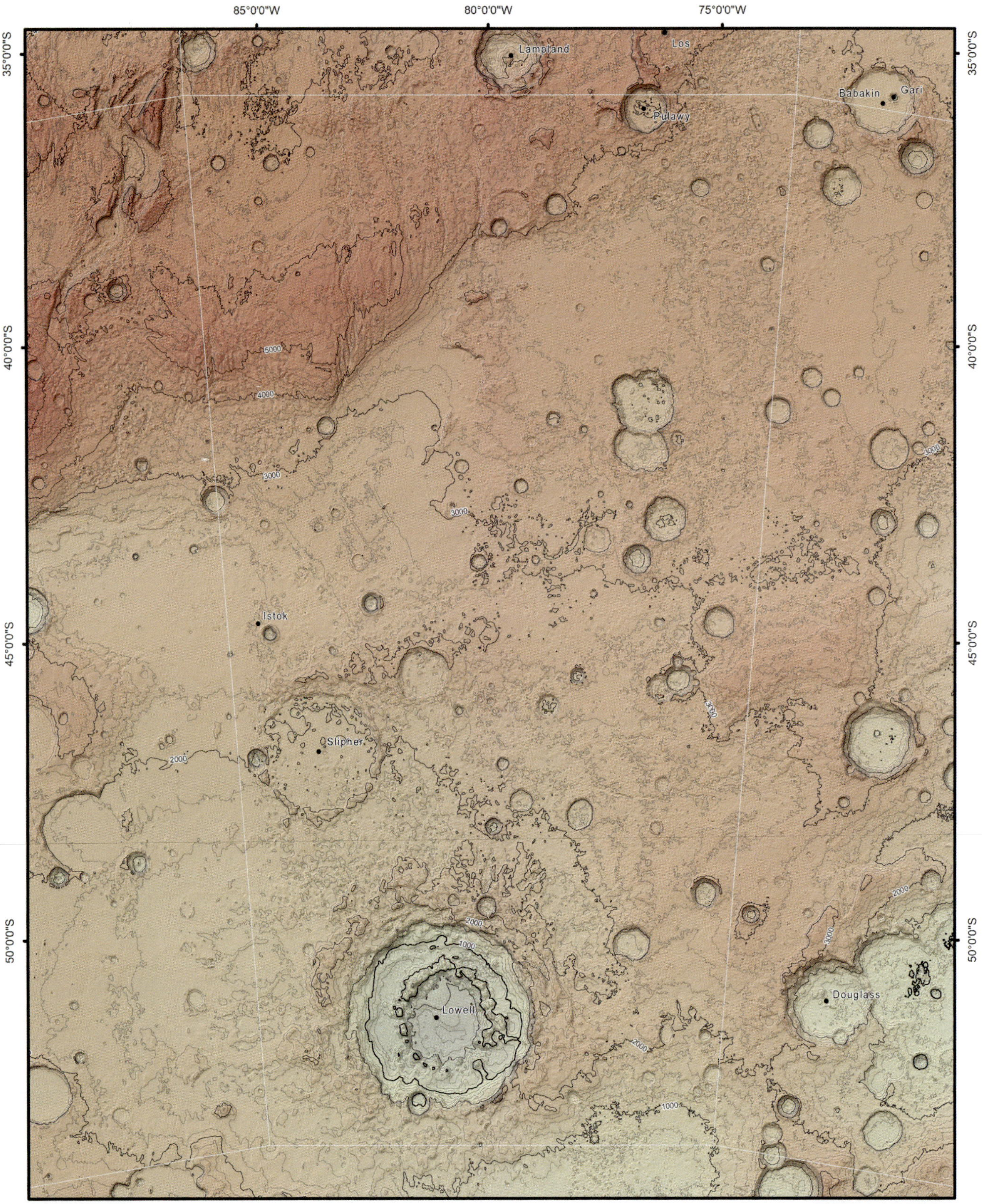

85°0'0"W
80°0'0"W
75°0'0"W

35°0'0"S
40°0'0"S
45°0'0"S
50°0'0"S

Lampland
Los
Babakin
Gari
Pulawy
Istok
Slipher
Lowell
Douglass

5000
4000
3000
2000
1000

Page
H-8

Coordinate System: The World From Space
Projection: Orthographic
Longitude of Center: -80.000
Latitude of Center: -45.000
Units: Meter

The Mars Atlas — 2020 Edition.
Visit RedMapper.com

THE MARS ATLAS

Topography & Places

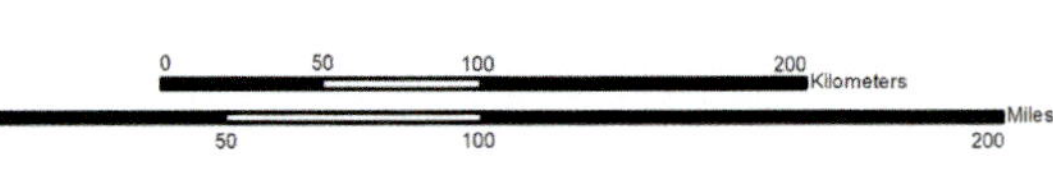

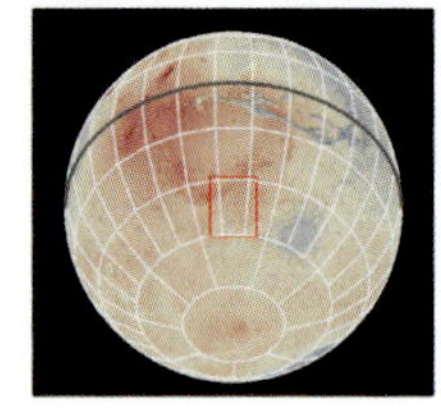

75°0'0"W
70°0'0"W
65°0'0"W
60°0'0"W
35°0'0"S
40°0'0"S
45°0'0"S
50°0'0"S
Los
Kumak
Gari
Babakin
Vik
Aki
Tabor
Pulawy
Halley
OGYGIS DUNEFIELD
(Ogygis Undae)
Douglass
3000
2000
1000
Page
H-9
Coordinate System: The World From Space
Projection: Orthographic
Longitude of Center: -66.667
Latitude of Center: -45.000
Units: Meter
The Mars Atlas — 2020 Edition.
Visit RedMapper.com
© 2019-2020 RedMapper - All Rights Reserved
R
REDMAPPER
THE MARS ATLAS
Topography & Places
N
0
50
100
200
Kilometers
Miles

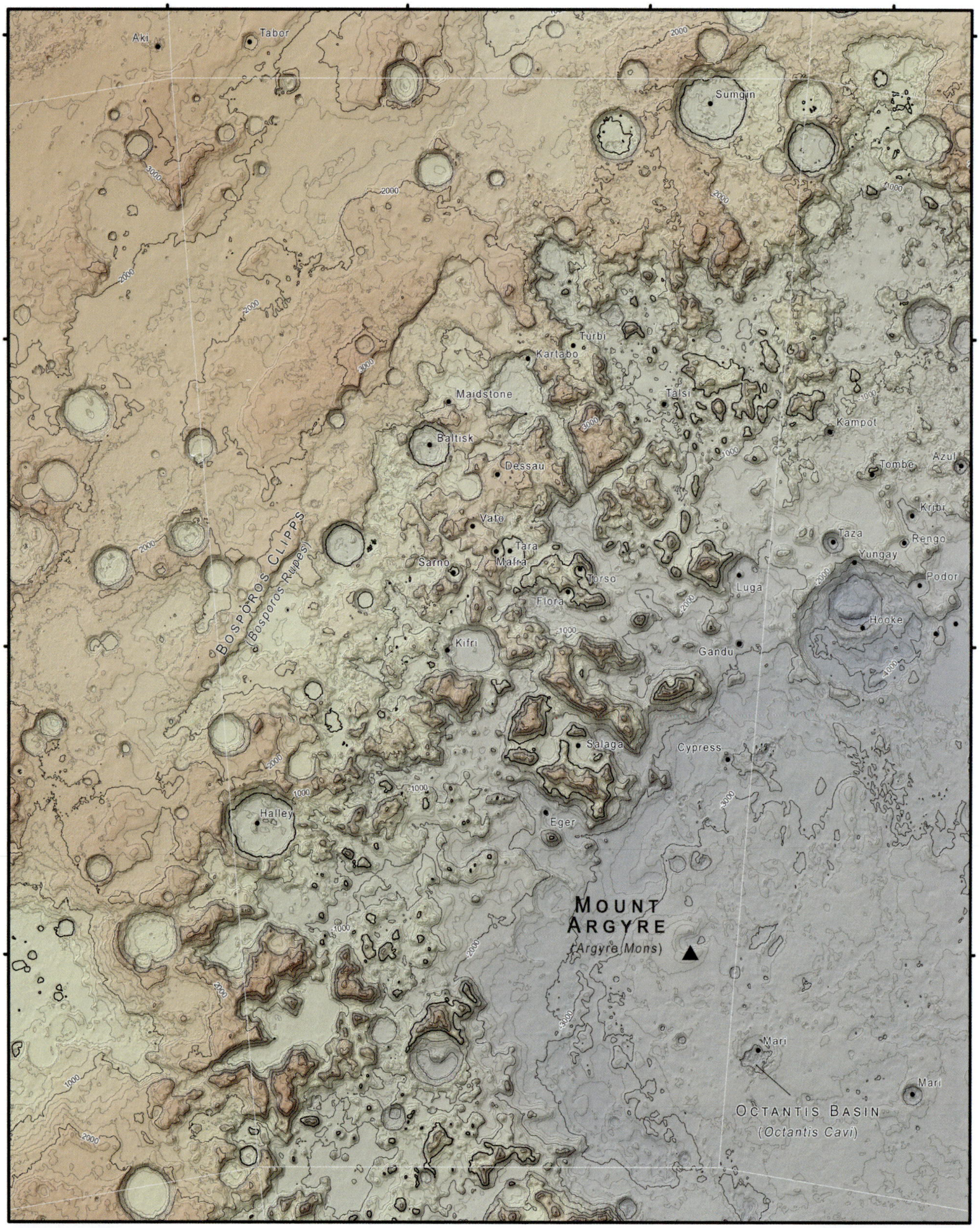

60°0'0"W 55°0'0"W 50°0'0"W 45°0'0"W

35°0'0"S
40°0'0"S
45°0'0"S
50°0'0"S

Aki
Tabor
Sumgin
Turbi
Kartabo
Maidstone
Talsi
Kampot
Baltisk
Dessau
Tombe
Azul
Kribi
Vato
Taza
Rengo
Tara
Malta
Sarno
Yungay
Torso
Podor
Luga
Flora
Hooke
Kifri
Gandu
BOSPOROS CLIPPS
(Bosporos Rupes)
Cypress
Salaga
Halley
Eger
MOUNT ARGYRE
(Argyre Mons)
Mari
Mari
OCTANTIS BASIN
(Octantis Cavi)

Page
H-10

Coordinate System: The World From Space
Projection: Orthographic
Longitude of Center: -53.334
Latitude of Center: -45.000
Units: Meter

The Mars Atlas — 2020 Edition.
Visit RedMapper.com

THE MARS ATLAS

Topography & Places

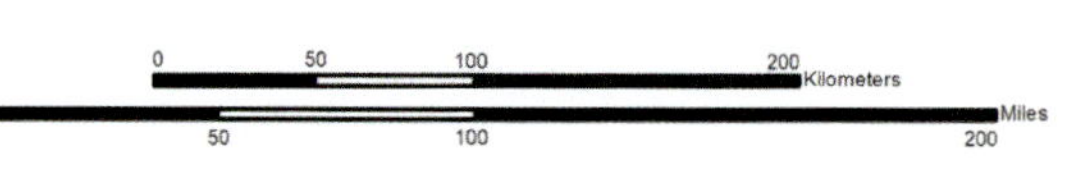

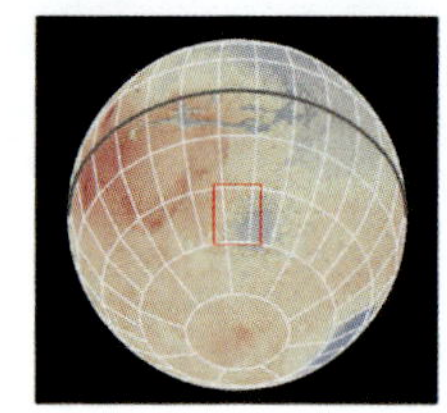

R
REDMAPPER

Page
H-11

Coordinate System: The World From Space
Projection: Orthographic
Longitude of Center: -40.000
Latitude of Center: -45.000
Units: Meter

The Mars Atlas — 2020 Edition.
Visit RedMapper.com

THE MARS ATLAS

Topography & Places

0 50 100 200 Kilometers

0 50 100 200 Miles

N

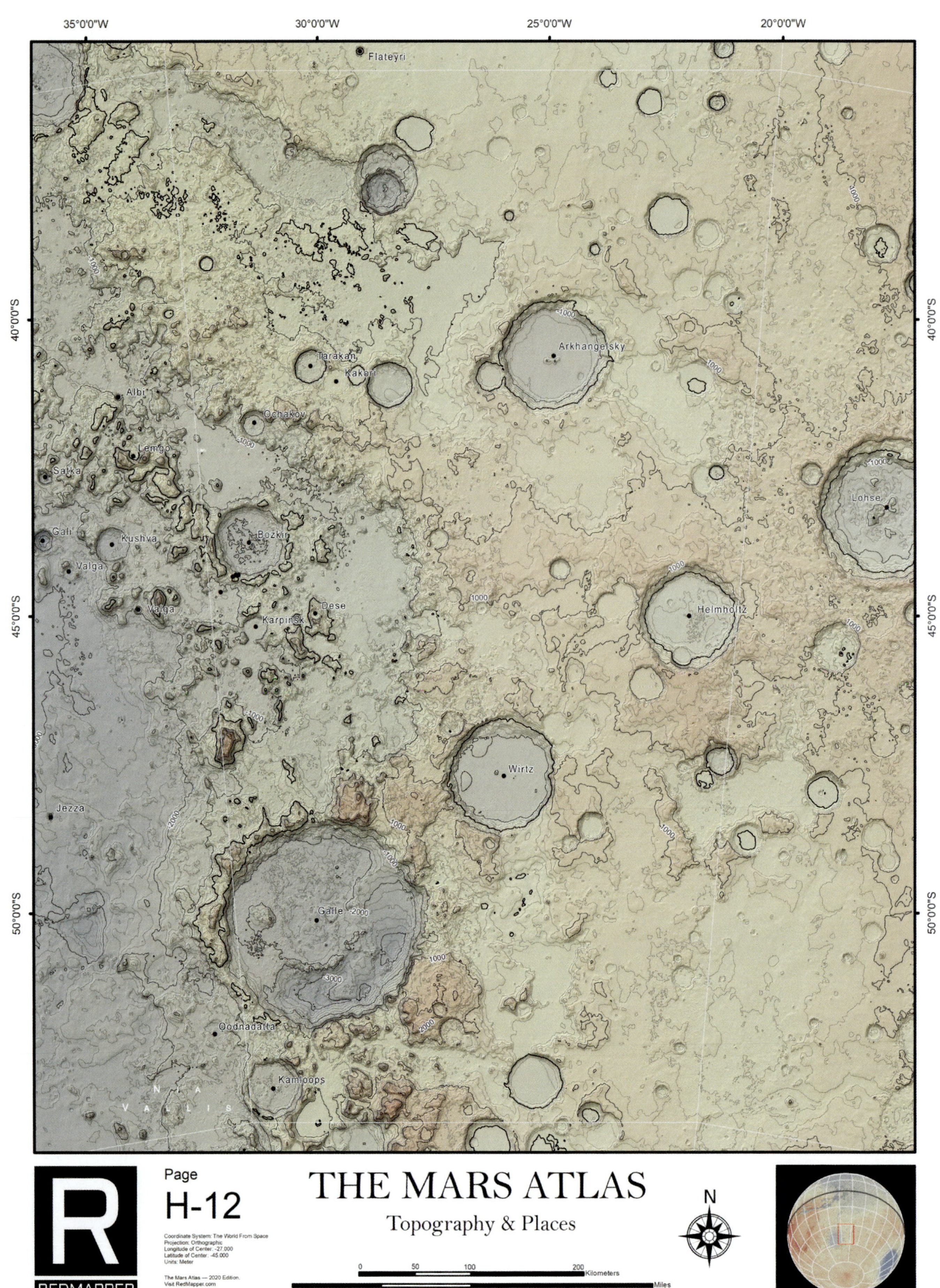

REDMAPPER

Page
H-12

Coordinate System: The World From Space
Projection: Orthographic
Longitude of Center: -27.000
Latitude of Center: -45.000
Units: Meter

The Mars Atlas — 2020 Edition.
Visit RedMapper.com

THE MARS ATLAS

Topography & Places

0 50 100 200 Kilometers

0 50 100 200 Miles

N

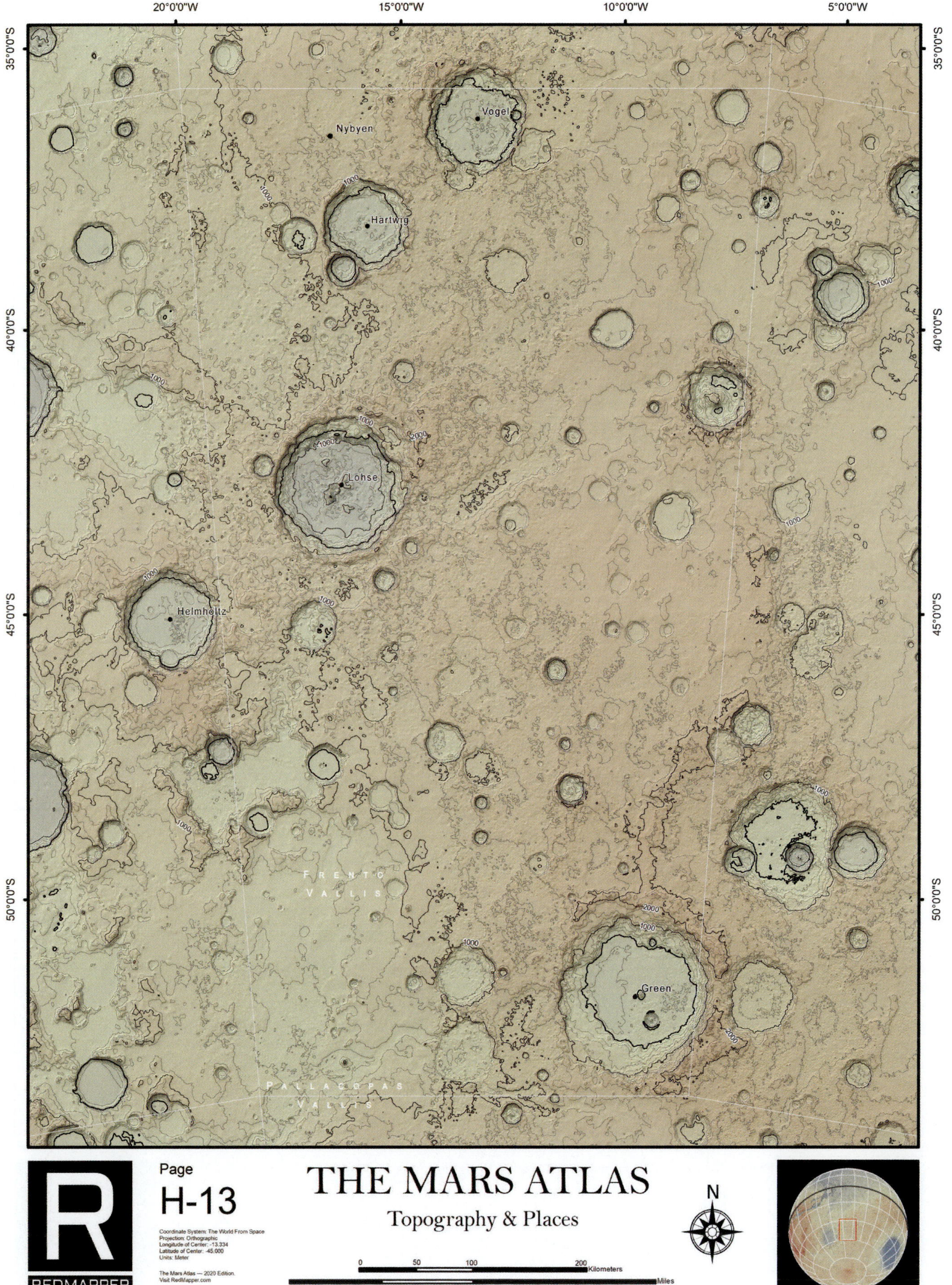

REDMAPPER

Page

H-13

Coordinate System: The World From Space
Projection: Orthographic
Longitude of Center: -13.334
Latitude of Center: -45.000
Units: Meter

The Mars Atlas — 2020 Edition.
Visit RedMapper.com

THE MARS ATLAS

Topography & Places

N

0 50 100 200 Kilometers

0 50 100 200 Miles

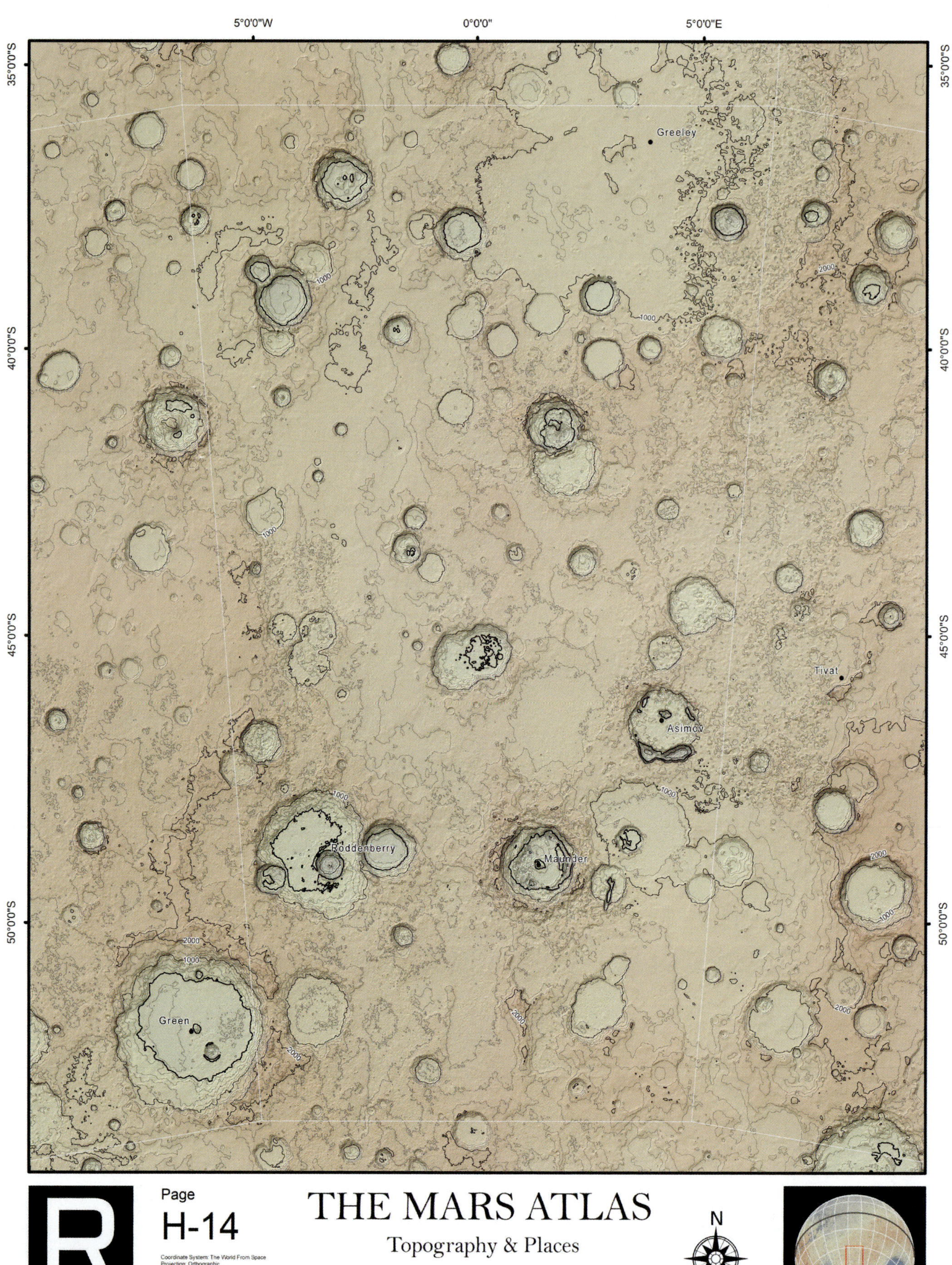

Page

H-14

Coordinate System: The World From Space
Projection: Orthographic
Longitude of Center: 0.000
Latitude of Center: -45.000
Units: Meter

The Mars Atlas — 2020 Edition.
Visit RedMapper.com

THE MARS ATLAS

Topography & Places

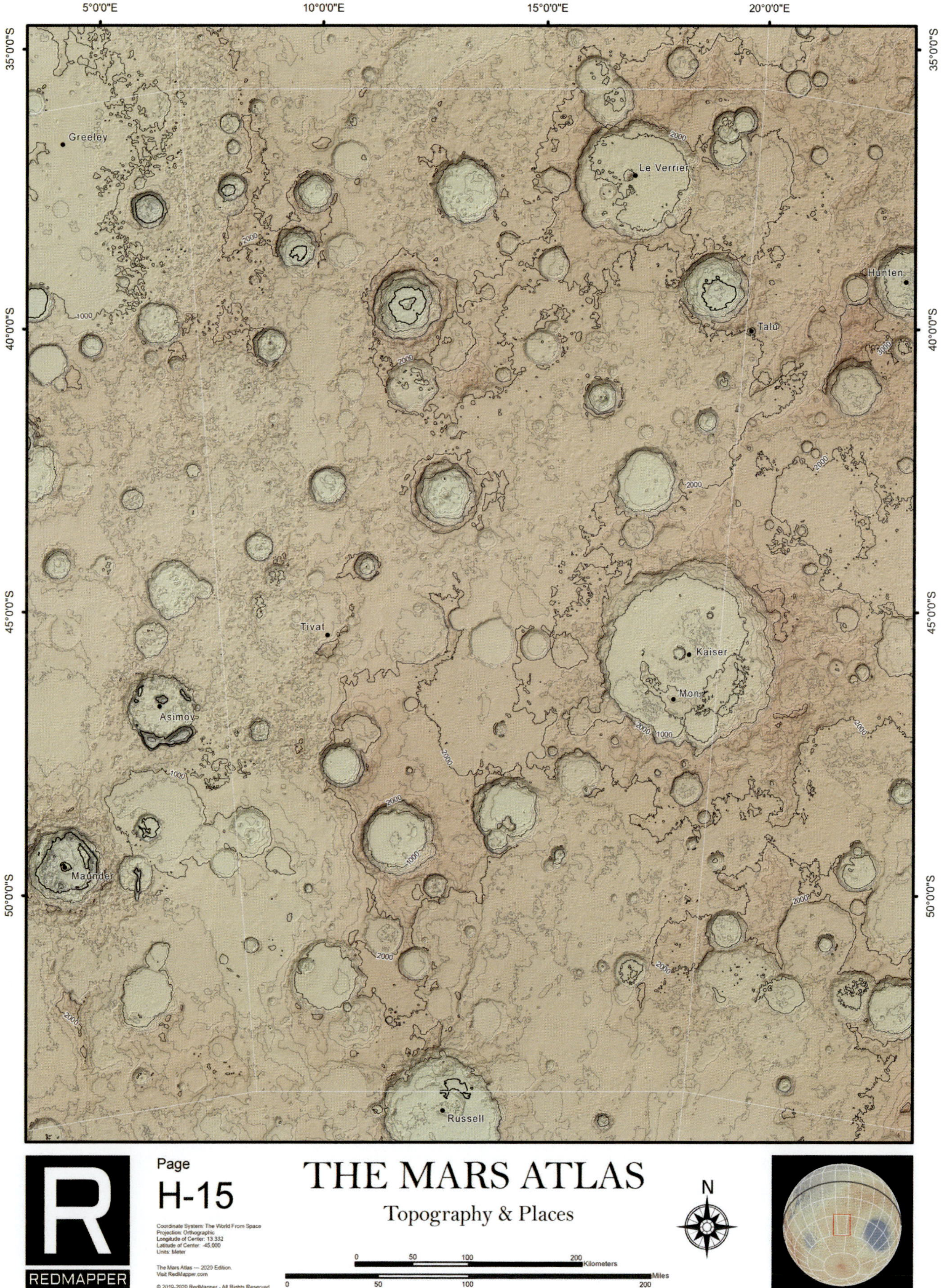
5°0'0"E
10°0'0"E
15°0'0"E
20°0'0"E
35°0'0"S
40°0'0"S
45°0'0"S
50°0'0"S
Greeley
Le Verrier
Hunten
Talu
Tivat
Kaiser
Moni
Asimov
Maunder
Russell
R
REDMAPPER
Page
H-15
Coordinate System: The World From Space
Projection: Orthographic
Longitude of Center: 13.332
Latitude of Center: -45.000
Units: Meter
The Mars Atlas — 2020 Edition.
Visit RedMapper.com
© 2019-2020 RedMapper - All Rights Reserved
THE MARS ATLAS
Topography & Places
N
0 50 100 200 Kilometers
0 50 100 200 Miles

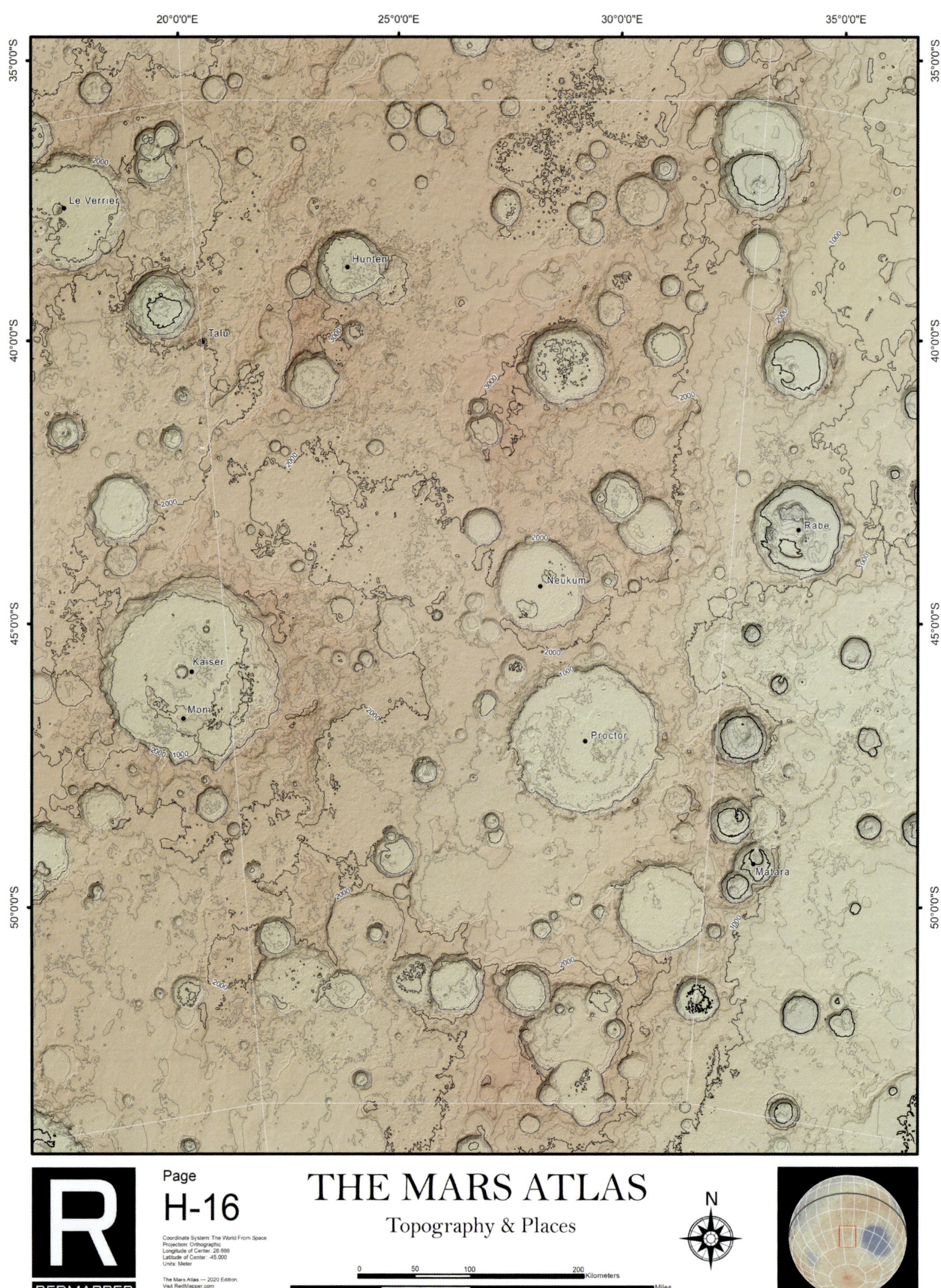

R
REDMAPPER

Page
H-16

Coordinate System: The World From Space
Projection: Orthographic
Longitude of Center: 26.666
Latitude of Center: -45.000
Units: Meter

The Mars Atlas — 2020 Edition.
Visit RedMapper.com

THE MARS ATLAS
Topography & Places

Page

H-17

THE MARS ATLAS

Topography & Places

Coordinate System: The World From Space
Projection: Orthographic
Longitude of Center: 40.000
Latitude of Center: -45.000
Units: Meter

The Mars Atlas — 2020 Edition.
Visit RedMapper.com

REDMAPPER

N

0 50 100 200 Kilometers

0 50 100 200 Miles

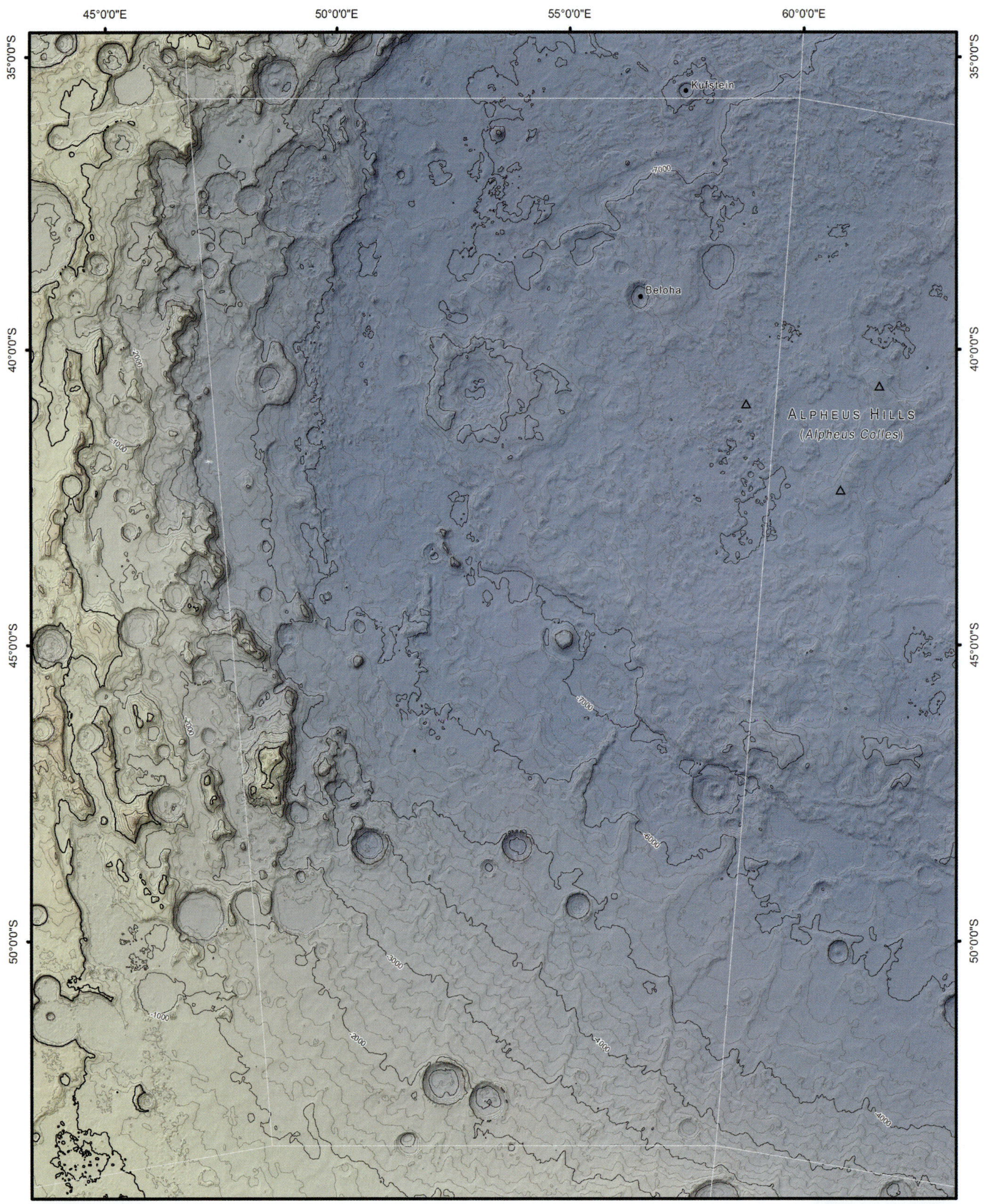

45°0'0"E 50°0'0"E 55°0'0"E 60°0'0"E

35°0'0"S 40°0'0"S 45°0'0"S 50°0'0"S

Kufstein

Beloha

ALPHEUS HILLS
(*Alpheus Colles*)

-7000 -6000 -4000 -3000 -2000 -1000

Page
H-18

Coordinate System: The World From Space
Projection: Orthographic
Longitude of Center: 53.332
Latitude of Center: -45.000
Units: Meter

The Mars Atlas — 2020 Edition.
Visit RedMapper.com

THE MARS ATLAS

Topography & Places

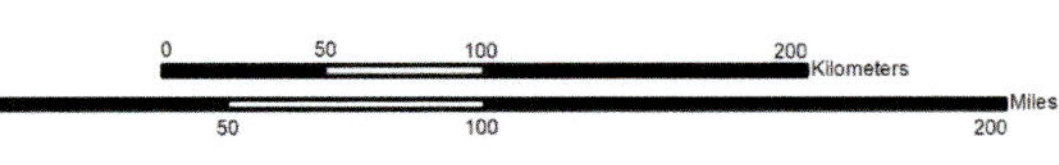

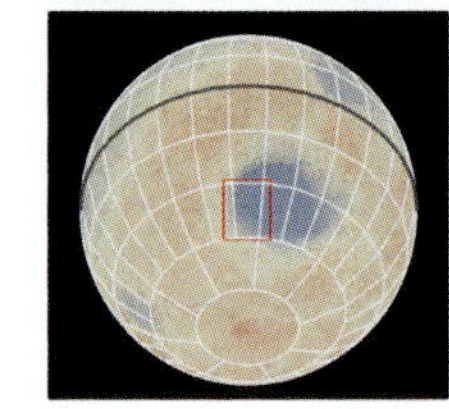

R
REDMAPPER

Page
H-19

Coordinate System: The World From Space
Projection: Orthographic
Longitude of Center: 66.666
Latitude of Center: -45.000
Units: Meter

The Mars Atlas — 2020 Edition.
Visit RedMapper.com

THE MARS ATLAS

Topography & Places

N

0 50 100 200 Kilometers

0 50 100 200 Miles

75°0'0"E
80°0'0"E
85°0'0"E
35°0'0"S
40°0'0"S
45°0'0"S
50°0'0"S
Talas
Njesko
Taejin
Poti
Apia
Mliba
Tungla
Santaca
Daan
LAS PLANITIA
Bogia
ZEA RIDGE
(Zea Dorsa)
Gledhill
-5000
-6000
-4000
-3000
-2000
-1000
Page
H-20
Coordinate System: The World From Space
Projection: Orthographic
Longitude of Center: 80.000
Latitude of Center: -45.000
Units: Meter
The Mars Atlas — 2020 Edition.
Visit RedMapper.com
© 2019-2020 RedMapper - All Rights Reserved
THE MARS ATLAS
Topography & Places
N
0 50 100 200 Kilometers
0 50 100 200 Miles
R
REDMAPPER

85°0'0"E
90°0'0"E
95°0'0"E
100°0'0"E
35°0'0"S
40°0'0"S
45°0'0"S
50°0'0"S
Taejin
Njesko
Poti
Apia
NIGER
Cue
Negele
Penticton
Ayr
Mliba
SUNGARI VALLIS
HARMAKHIS VALLIS
Tungla
Santaca
Quines
Daan
Thom
CENTAURI MOUNTAINS
(Centauri Montes)
Sebec
Gunison
Krishtofovich
Tikhov
Gledhill
R
REDMAPPER
Page
H-21
Coordinate System: The World From Space
Projection: Orthographic
Longitude of Center: 93.332
Latitude of Center: -45.000
Units: Meter
The Mars Atlas — 2020 Edition.
Visit RedMapper.com
© 2019-2020 RedMapper - All Rights Reserved
THE MARS ATLAS
Topography & Places
N
0 50 100 200 Kilometers
0 50 100 200 Miles

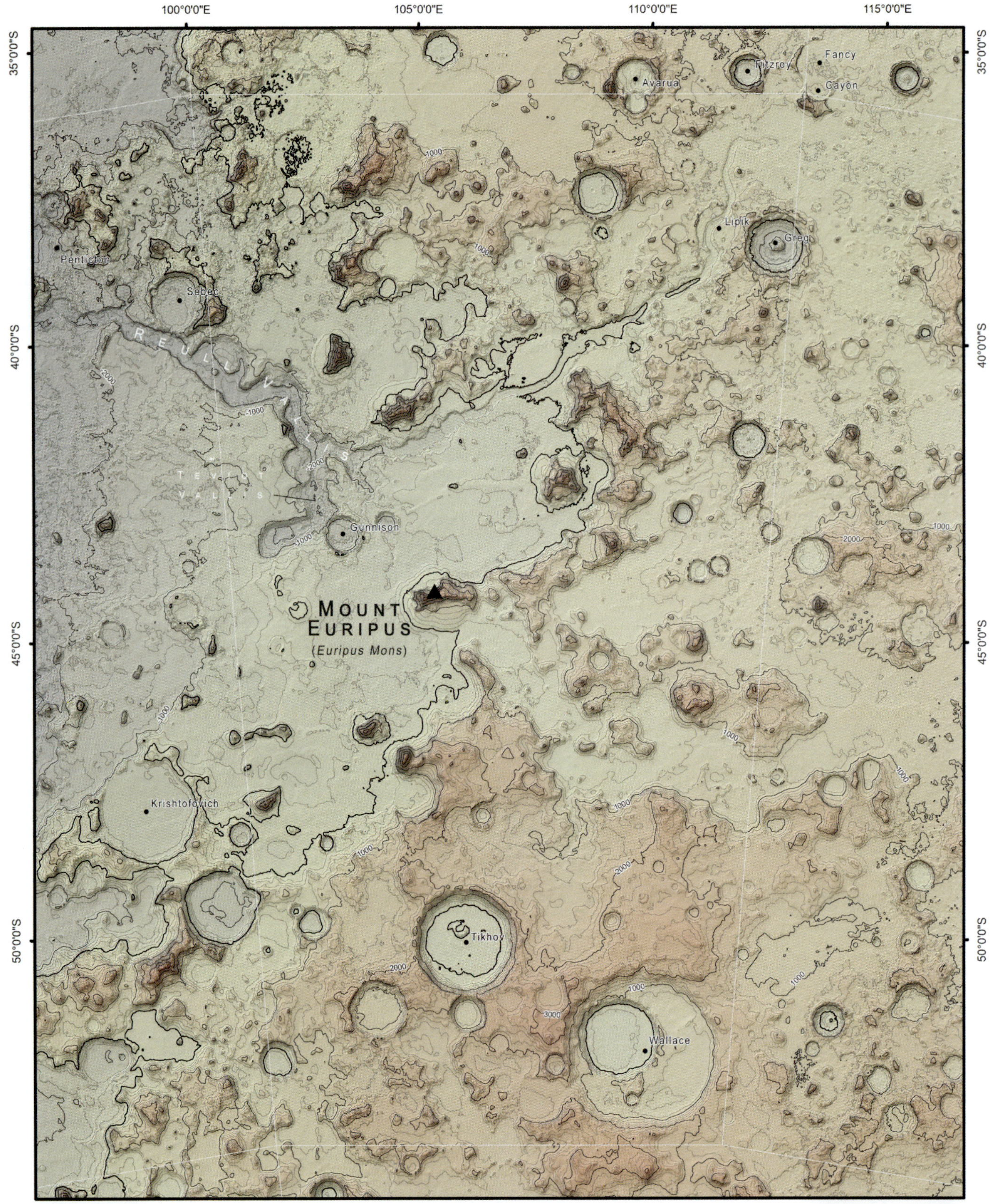

100°0'0"E 105°0'0"E 110°0'0"E 115°0'0"E

35°0'0"S 40°0'0"S 45°0'0"S 50°0'0"S

Avarua
Fitzroy
Fancy
Cayon
Lipik
Gregi
Penticton
Sebec
REULL VALLIS
TEVIOT VALLIS
Gunnison

MOUNT EURIPUS
(Euripus Mons)

Krishtofovich
Tikhov
Wallace

Page
H-22

Coordinate System: The World From Space
Projection: Orthographic
Longitude of Center: 106.885
Latitude of Center: -45.000
Units: Meter

The Mars Atlas — 2020 Edition.
Visit RedMapper.com

THE MARS ATLAS
Topography & Places

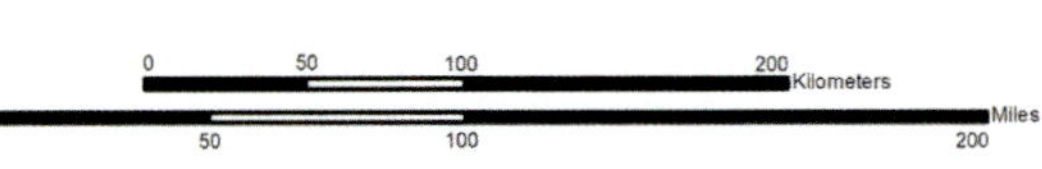

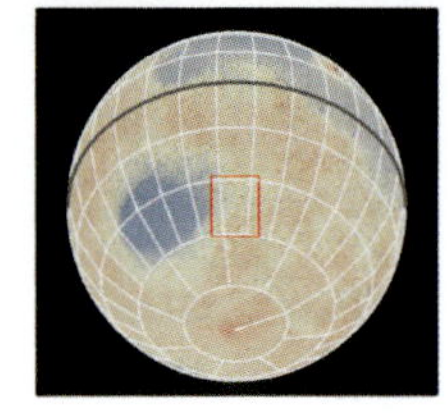

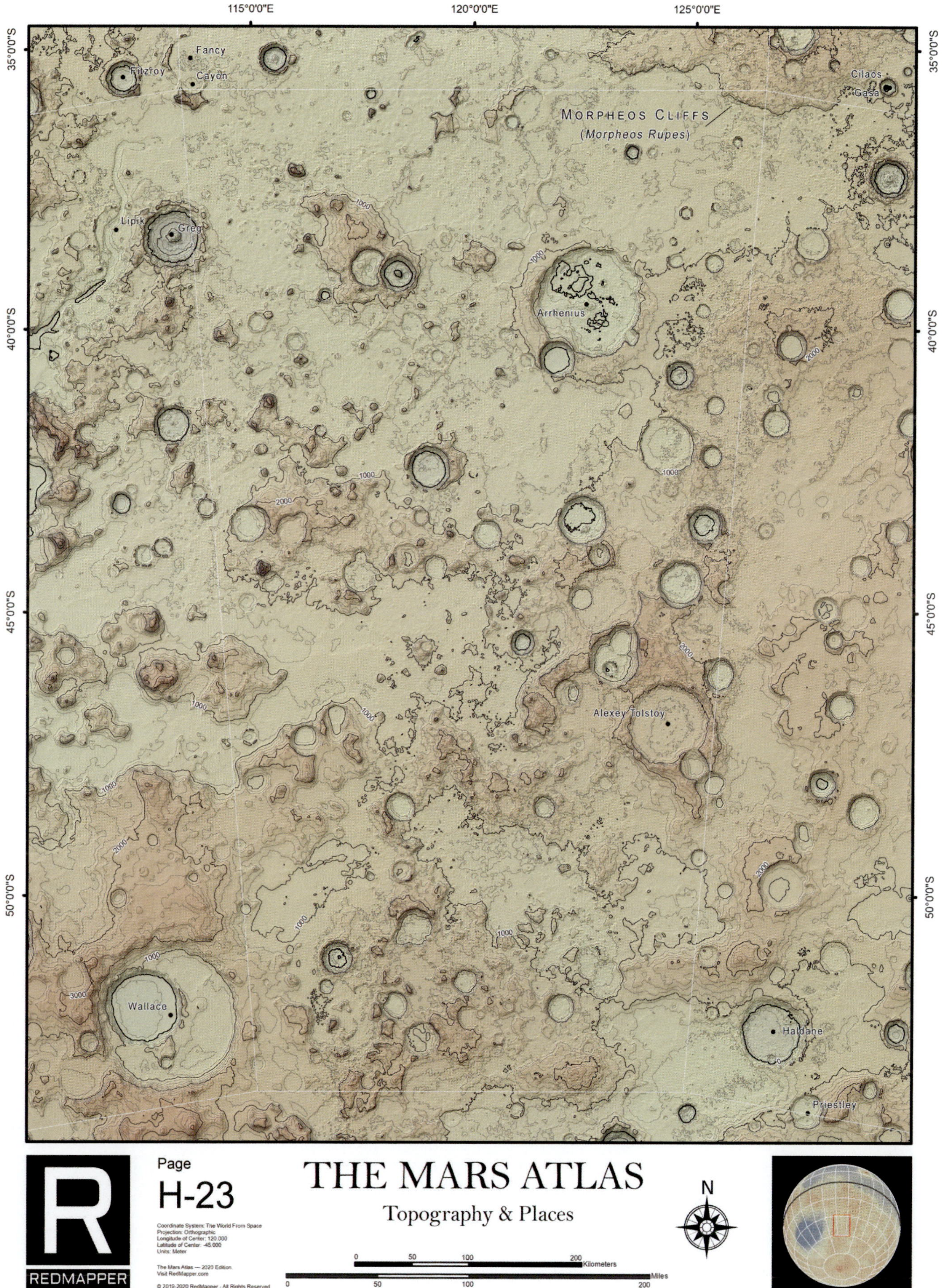

115°0'0"E
120°0'0"E
125°0'0"E
35°0'0"S
40°0'0"S
45°0'0"S
50°0'0"S
Fancy
Fitzroy
Cayon
Cilaos
Gasa
MORPHEOS CLIFFS
(Morpheos Rupes)
Lipik
Greg
Arrhenius
Alexey Tolstoy
Wallace
Haldane
Priestley
1000
2000
3000
REDMAPPER
Page
H-23
Coordinate System: The World From Space
Projection: Orthographic
Longitude of Center: 120.000
Latitude of Center: -45.000
Units: Meter
The Mars Atlas — 2020 Edition.
Visit RedMapper.com
© 2019-2020 RedMapper - All Rights Reserved
THE MARS ATLAS
Topography & Places
N
0
50
100
200
Kilometers
Miles

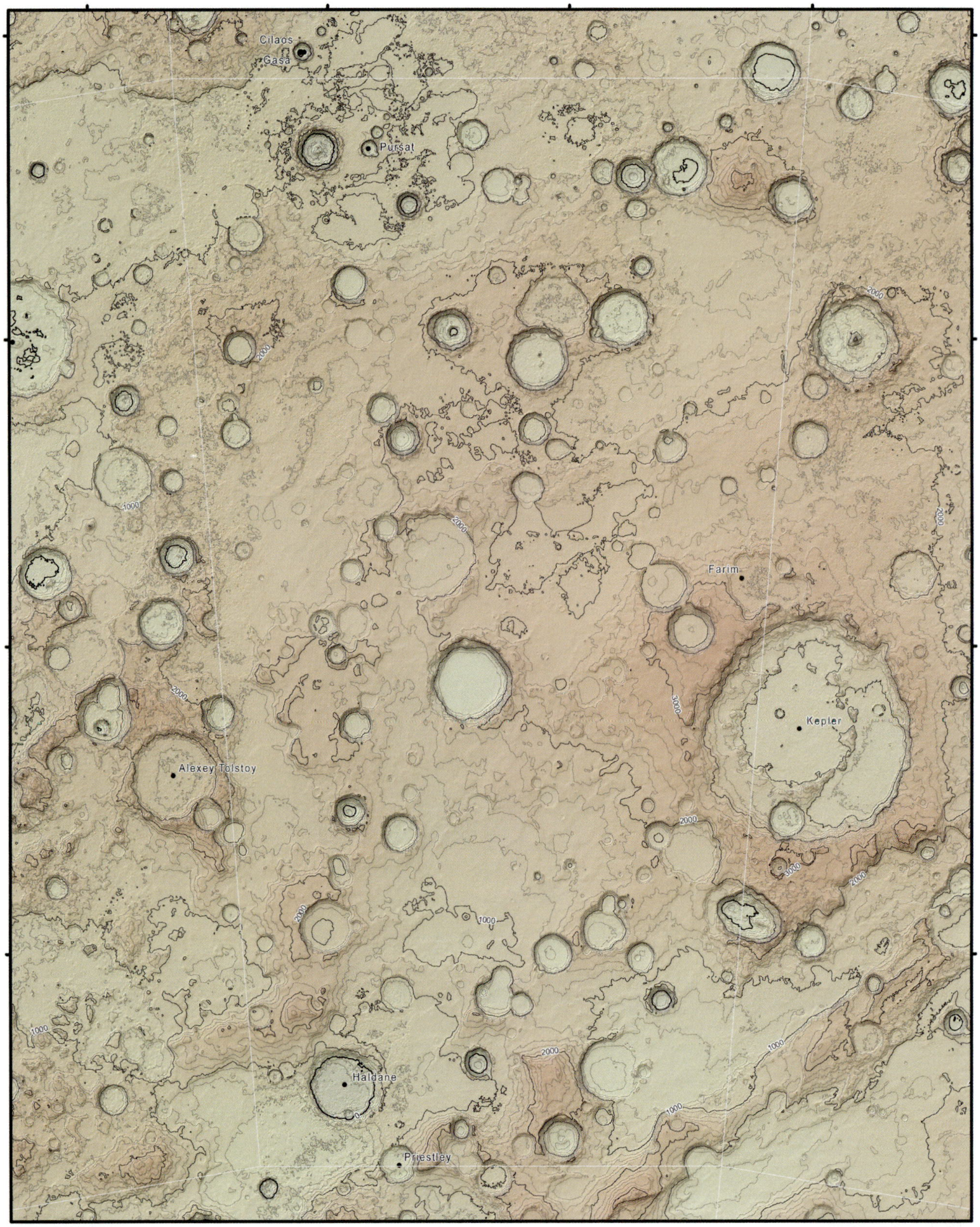

125°0'0"E
130°0'0"E
135°0'0"E
140°0'0"E

35°0'0"S
40°0'0"S
45°0'0"S
50°0'0"S

Cilaos
Gasa
Pursat
Farim
Kepler
Alexey Tolstoy
Haldane
Priestley

Page

H-24

Coordinate System: The World From Space
Projection: Orthographic
Longitude of Center: 133.332
Latitude of Center: -45.000
Units: Meter

The Mars Atlas — 2020 Edition.
Visit RedMapper.com

THE MARS ATLAS

Topography & Places

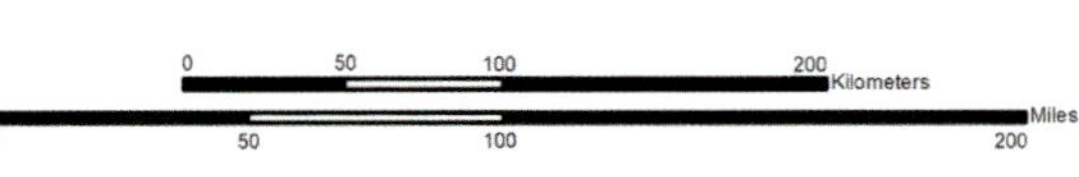

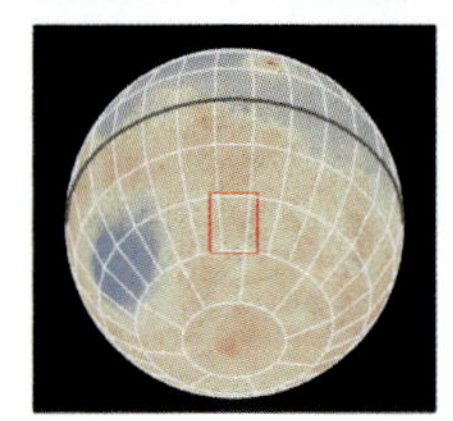

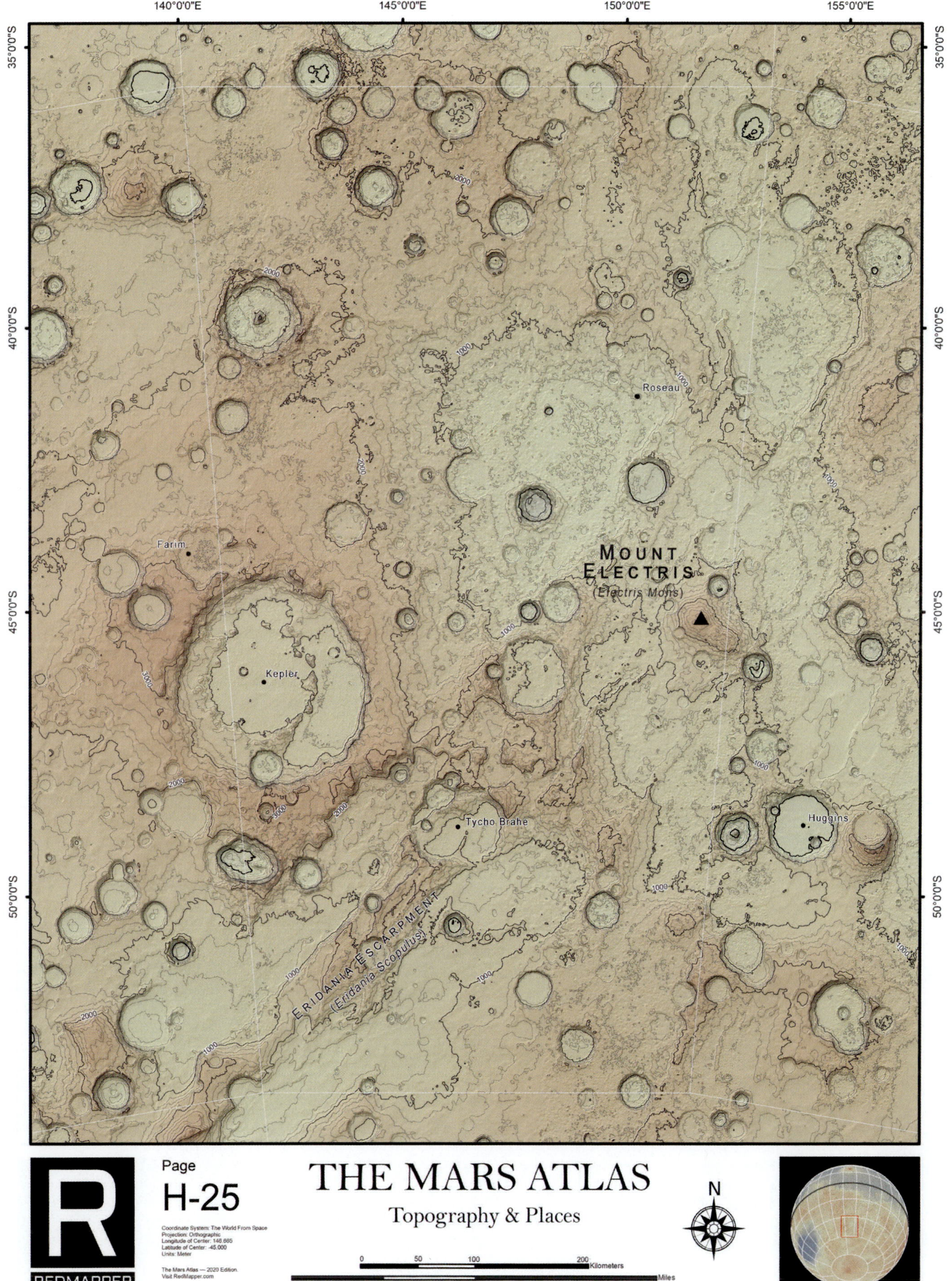

REDMAPPER

Page

H-25

Coordinate System: The World From Space
Projection: Orthographic
Longitude of Center: 146.665
Latitude of Center: -45.000
Units: Meter

The Mars Atlas — 2020 Edition.
Visit RedMapper.com

THE MARS ATLAS

Topography & Places

0 50 100 200 Kilometers

0 50 100 200 Miles

N

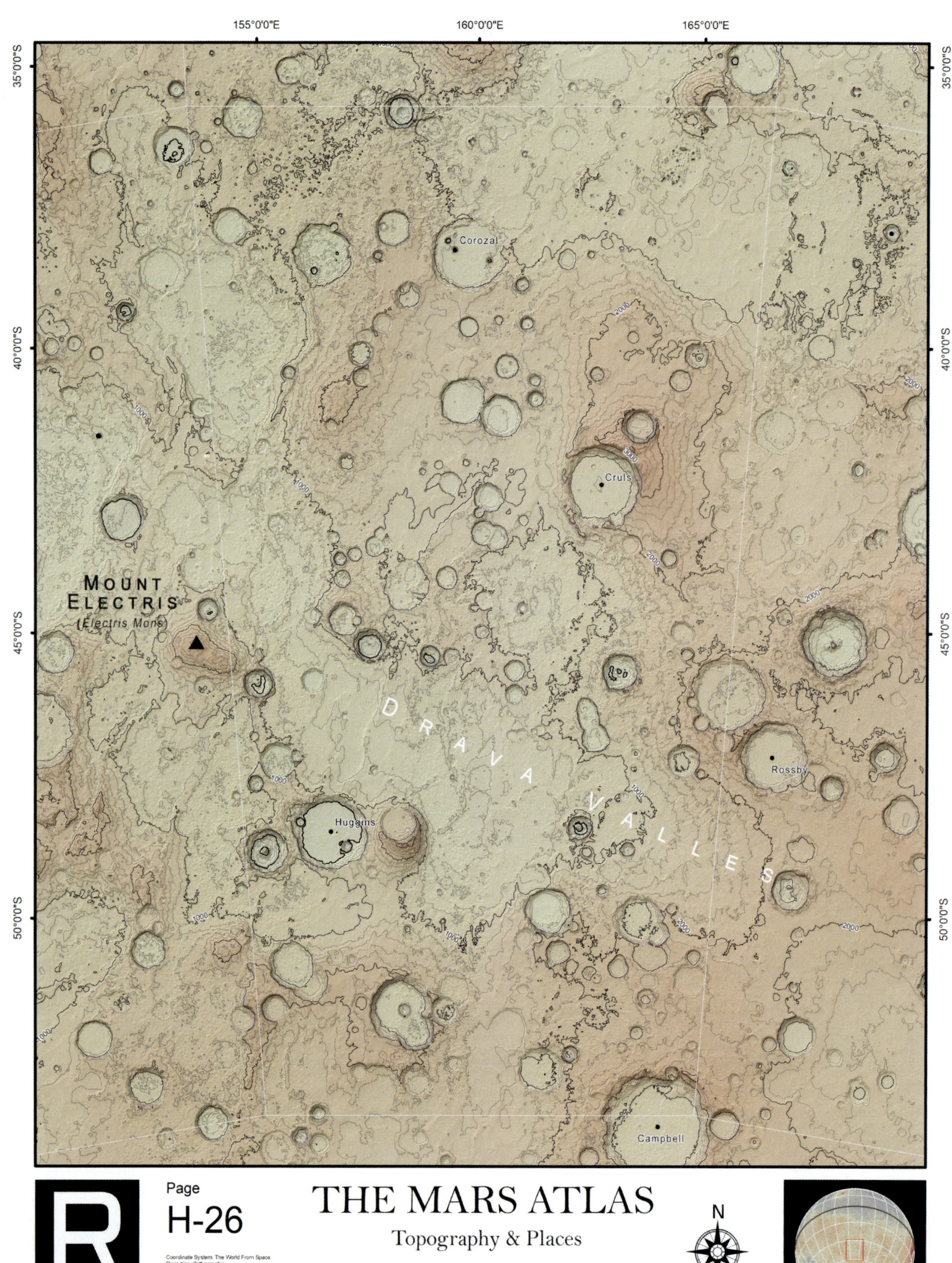

R
REDMAPPER

Page
H-26

Coordinate System: The World From Space
Projection: Orthographic
Longitude of Center: 160.000
Latitude of Center: -45.000

The Mars Atlas — 2020 Edition.
Visit RedMapper.com

THE MARS ATLAS

Topography & Places

N

0 50 100 200 Kilometers

0 50 100 200 Miles

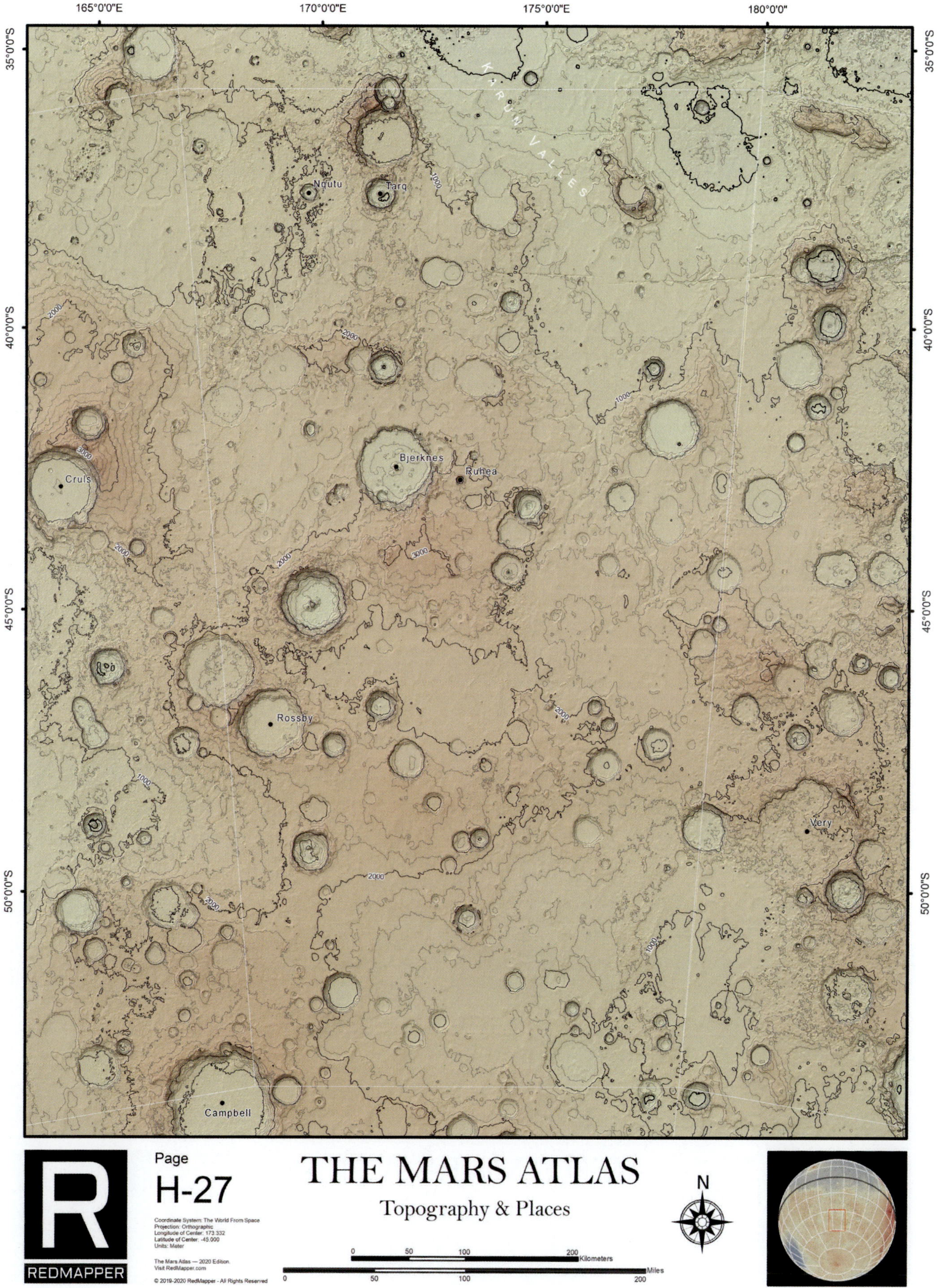
165°0'0"E
170°0'0"E
175°0'0"E
180°0'0"
35°0'0"S
40°0'0"S
45°0'0"S
50°0'0"S
KARUN VALLES
Ngutu
Tarq
Bjerknes
Ruhea
Cruls
Rossby
Very
Campbell
1000
2000
3000
REDMAPPER
Page
H-27
Coordinate System: The World From Space
Projection: Orthographic
Longitude of Center: 173.332
Latitude of Center: -45.000
Units: Meter
The Mars Atlas — 2020 Edition.
Visit RedMapper.com
© 2019-2020 RedMapper - All Rights Reserved
THE MARS ATLAS
Topography & Places
N
0 50 100 200 Kilometers
0 50 100 200 Miles

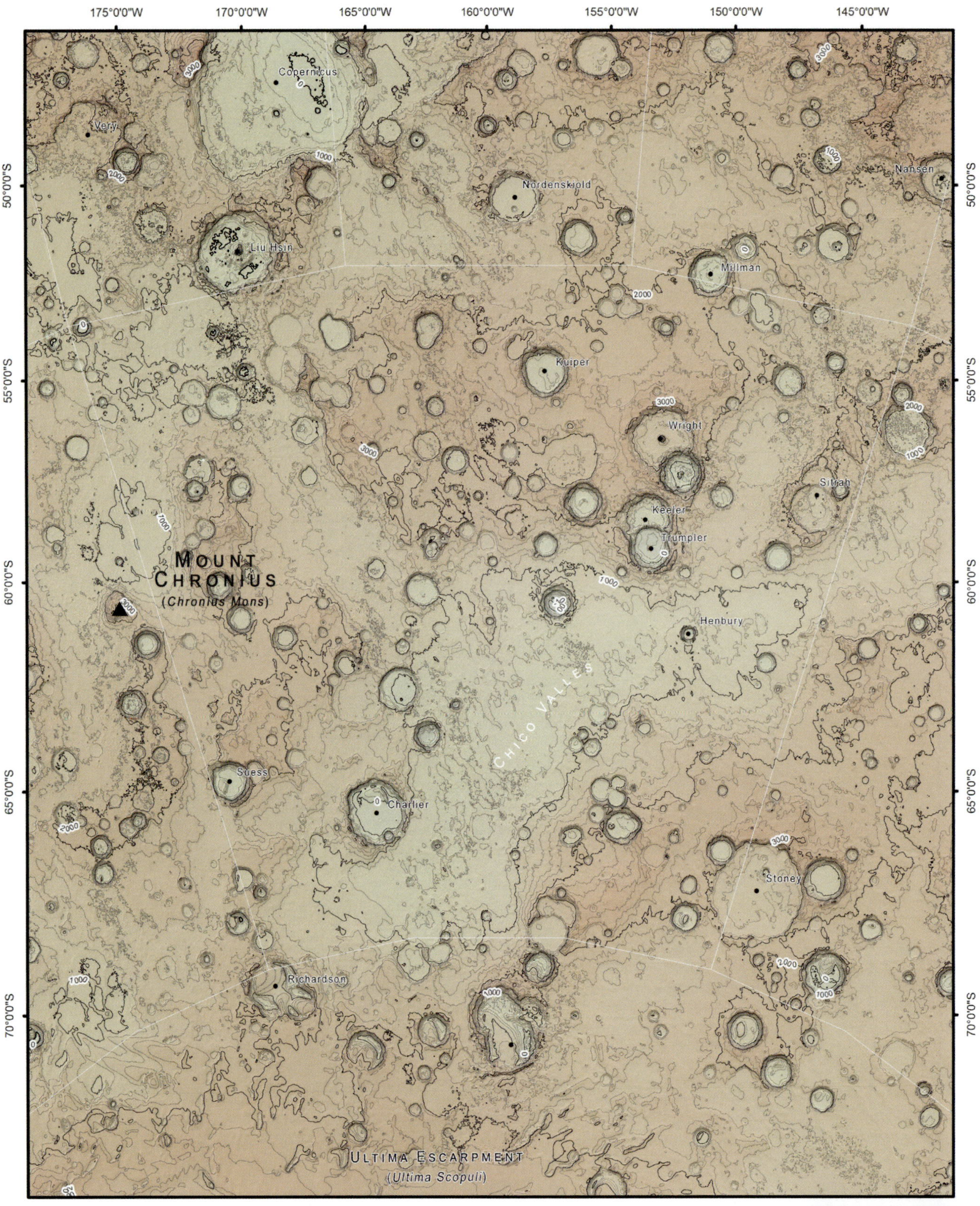

175°0'0"W 170°0'0"W 165°0'0"W 160°0'0"W 155°0'0"W 150°0'0"W 145°0'0"W

50°0'0"S
55°0'0"S
60°0'0"S
65°0'0"S
70°0'0"S

Copernicus
Very
Nordenskiold
Nansen
Liu Hsin
Millman
Kuiper
Wright
Sitrah
Keeler
Trumpler
MOUNT CHRONIUS
(*Chronius Mons*)
Henbury
CHICO VALLES
Suess
Charlier
Stoney
Richardson
ULTIMA ESCARPMENT
(*Ultima Scopuli*)

Page

I-1

Coordinate System: The World From Space
Projection: Orthographic
Longitude of Center: -160.000
Latitude of Center: -63.000
Units: Meter

The Mars Atlas — 2020 Edition.
Visit RedMapper.com

THE MARS ATLAS

Topography & Places

N

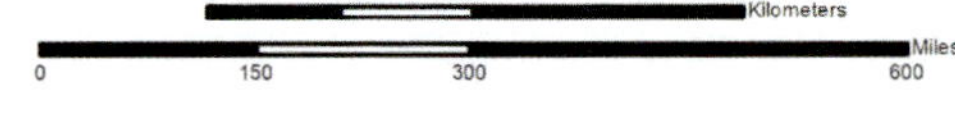

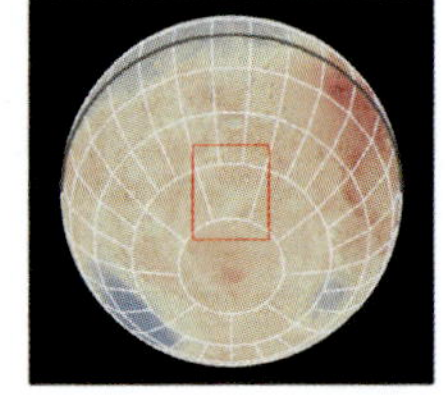

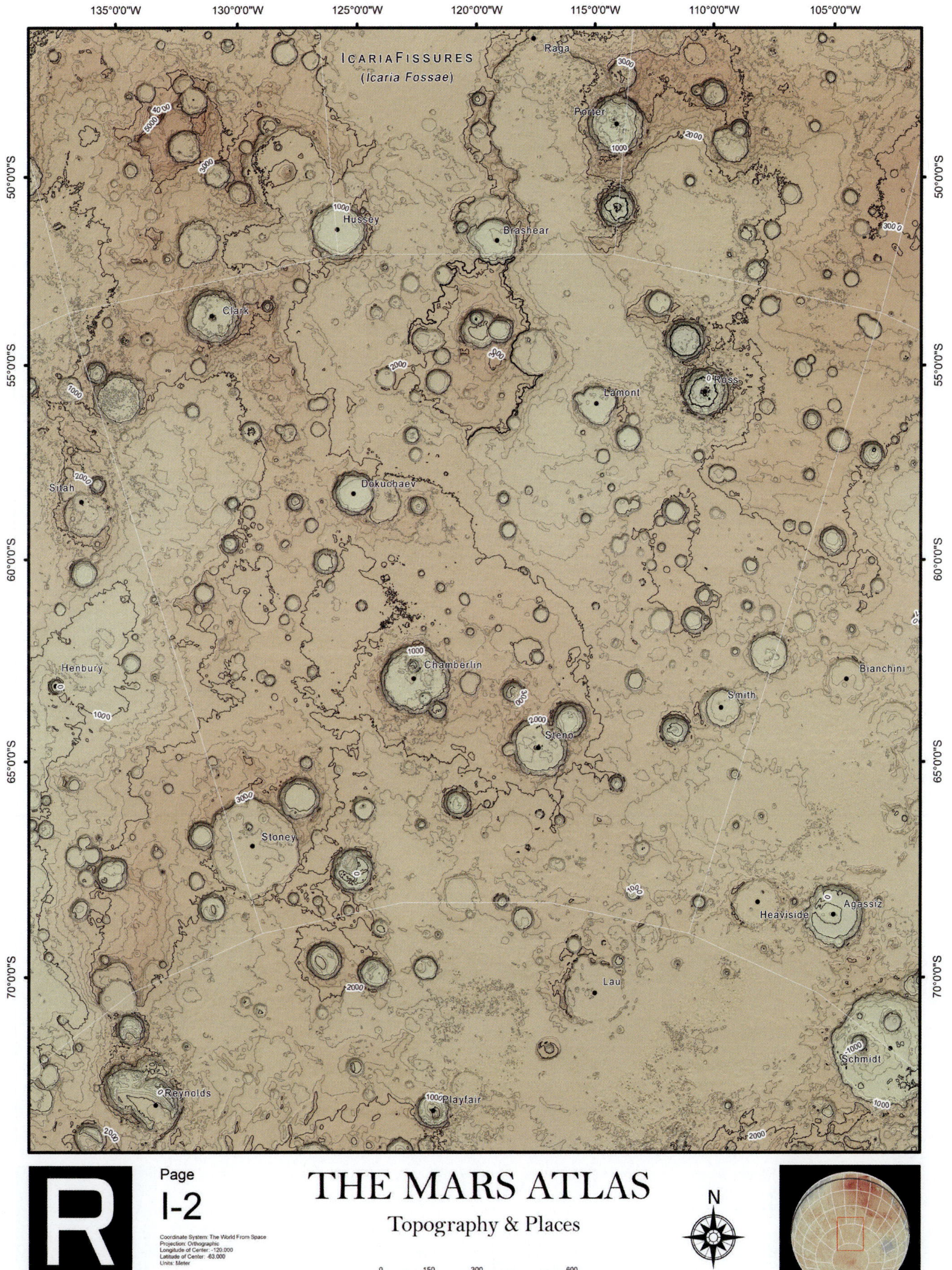

Page

I-2

THE MARS ATLAS

Topography & Places

Coordinate System: The World From Space
Projection: Orthographic
Longitude of Center: -120.000
Latitude of Center: -63.000
Units: Meter

The Mars Atlas — 2020 Edition.
Visit RedMapper.com

REDMAPPER

N

0 150 300 600 Kilometers

0 150 300 600 Miles

95°0'0"W 90°0'0"W 85°0'0"W 80°0'0"W 75°0'0"W 70°0'0"W 65°0'0"W

50°0'0"S 55°0'0"S 60°0'0"S 65°0'0"S 70°0'0"S

THAUMASIA FISSURES
(Thaumasia Fossae)

AONIA MONS
(Aonia Mons)

Lowell

Douglas

Coblentz

AONIA HILL
(Aonia Tholus)

AONIA PLANUM

ARGYRE RUPES
(Argyre Rupes)

Fontana

Bianchini

Von kármán

Smith

Agassiz

Heaviside

Steno

Schmidt

ARGENTIA PLANUM

Du Toit

Lau

ANGUSTI BASINS
(Cavi Angusti)

Joly

Sarn

3000 2000 1000 0 -1000

Page

I-3

Coordinate System: The World From Space
Projection: Orthographic
Longitude of Center: -80.000
Latitude of Center: -63.000
Units: Meter

The Mars Atlas — 2020 Edition.
Visit RedMapper.com

REDMAPPER

THE MARS ATLAS

Topography & Places

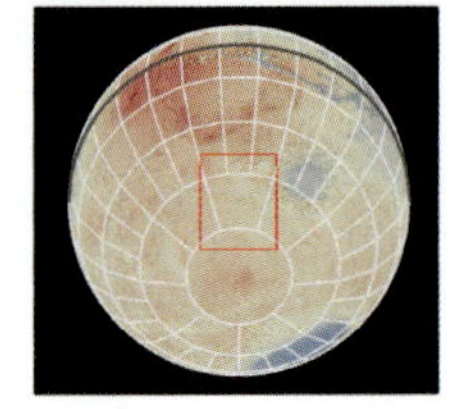

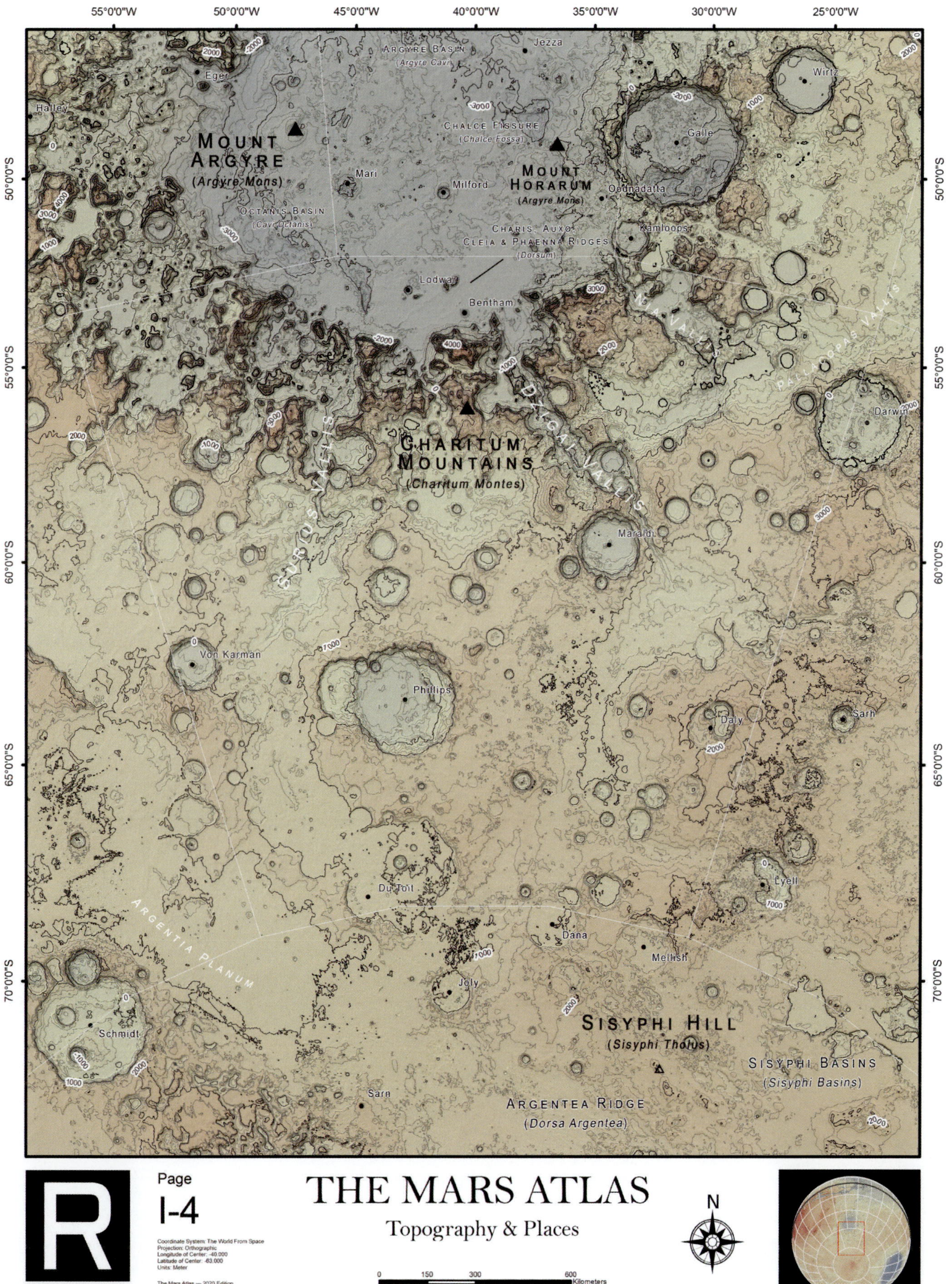

Page

I-4

THE MARS ATLAS

Topography & Places

Coordinate System: The World From Space
Projection: Orthographic
Longitude of Center: -40.000
Latitude of Center: -63.000
Units: Meter

The Mars Atlas — 2020 Edition
Visit RedMapper.com

REDMAPPER

N

0 150 300 600 Kilometers

0 150 300 600 Miles

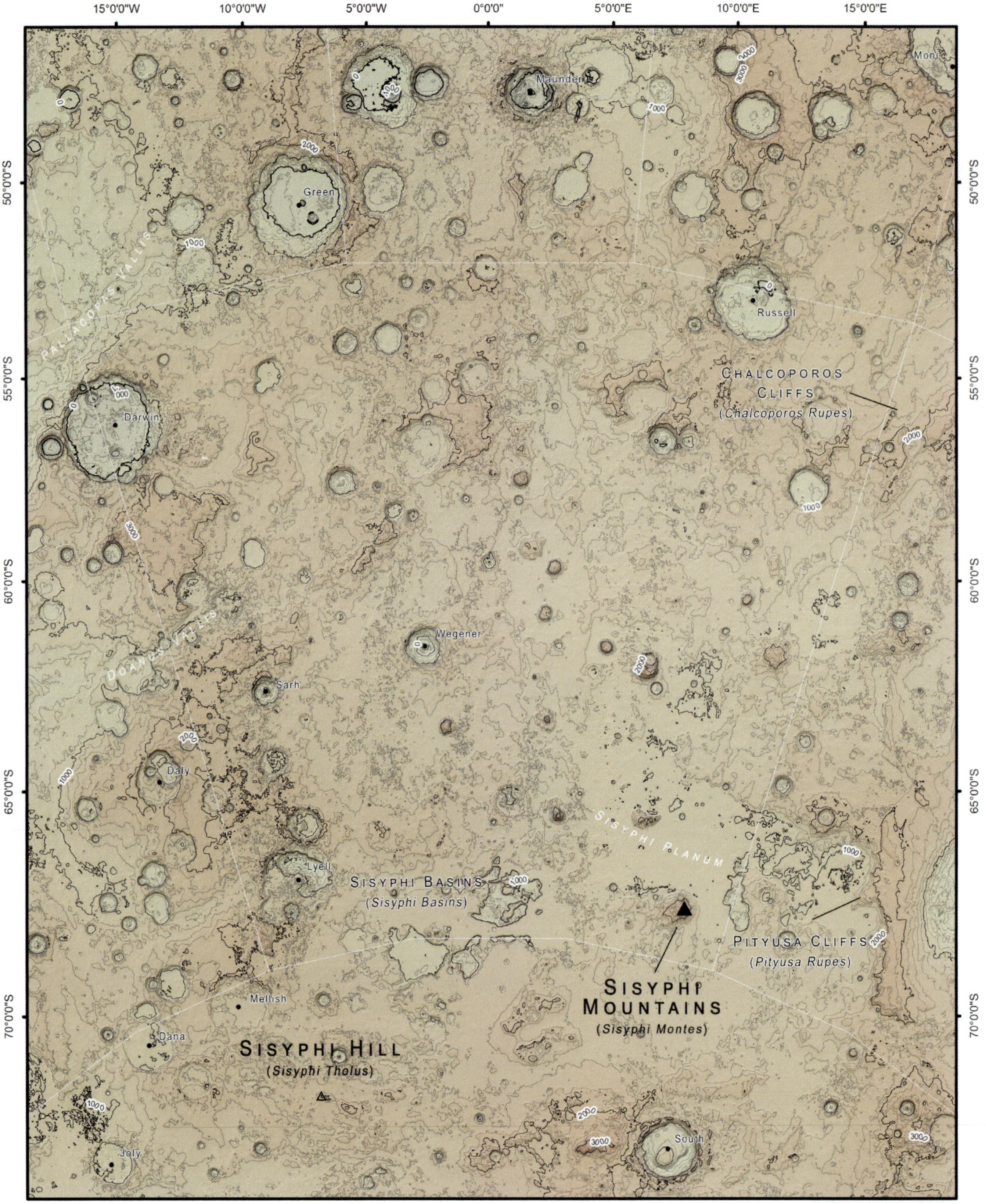

15°0'0"W
10°0'0"W
5°0'0"W
0°0'0"
5°0'0"E
10°0'0"E
15°0'0"E

50°0'0"S
55°0'0"S
60°0'0"S
65°0'0"S
70°0'0"S

Maunder
Moni
Green
Pallacopas Vallis
Russell
Chalcoporos Cliffs
(Chalcoporos Rupes)
Darwin
Wegener
Doanus Vallis
Sarh
Daly
Lyell
Sisyphi Basins
(Sisyphi Basins)
Sisyphi Planum
Pityusa Cliffs
(Pityusa Rupes)
Sisyphi Mountains
(Sisyphi Montes)
Mellish
Dana
Sisyphi Hill
(Sisyphi Tholus)
South
Joly

Page

I-5

Coordinate System: The World From Space
Projection: Orthographic
Longitude of Center: 0.000
Latitude of Center: -63.000
Units: Meter

The Mars Atlas — 2020 Edition.
Visit RedMapper.com

REDMAPPER

THE MARS ATLAS

Topography & Places

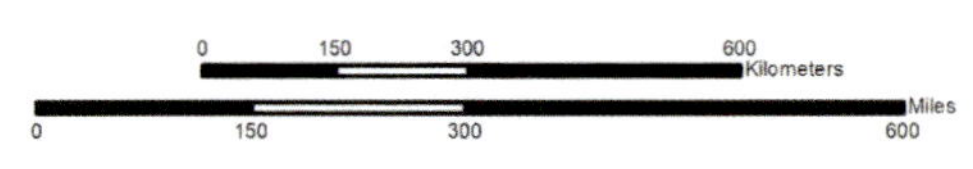

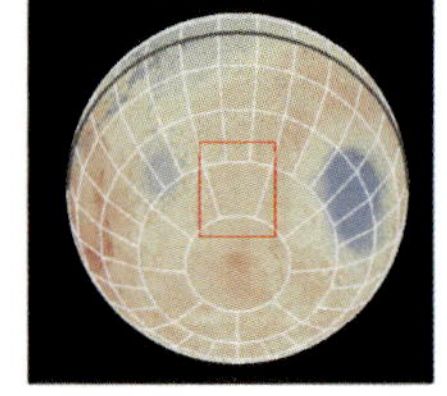

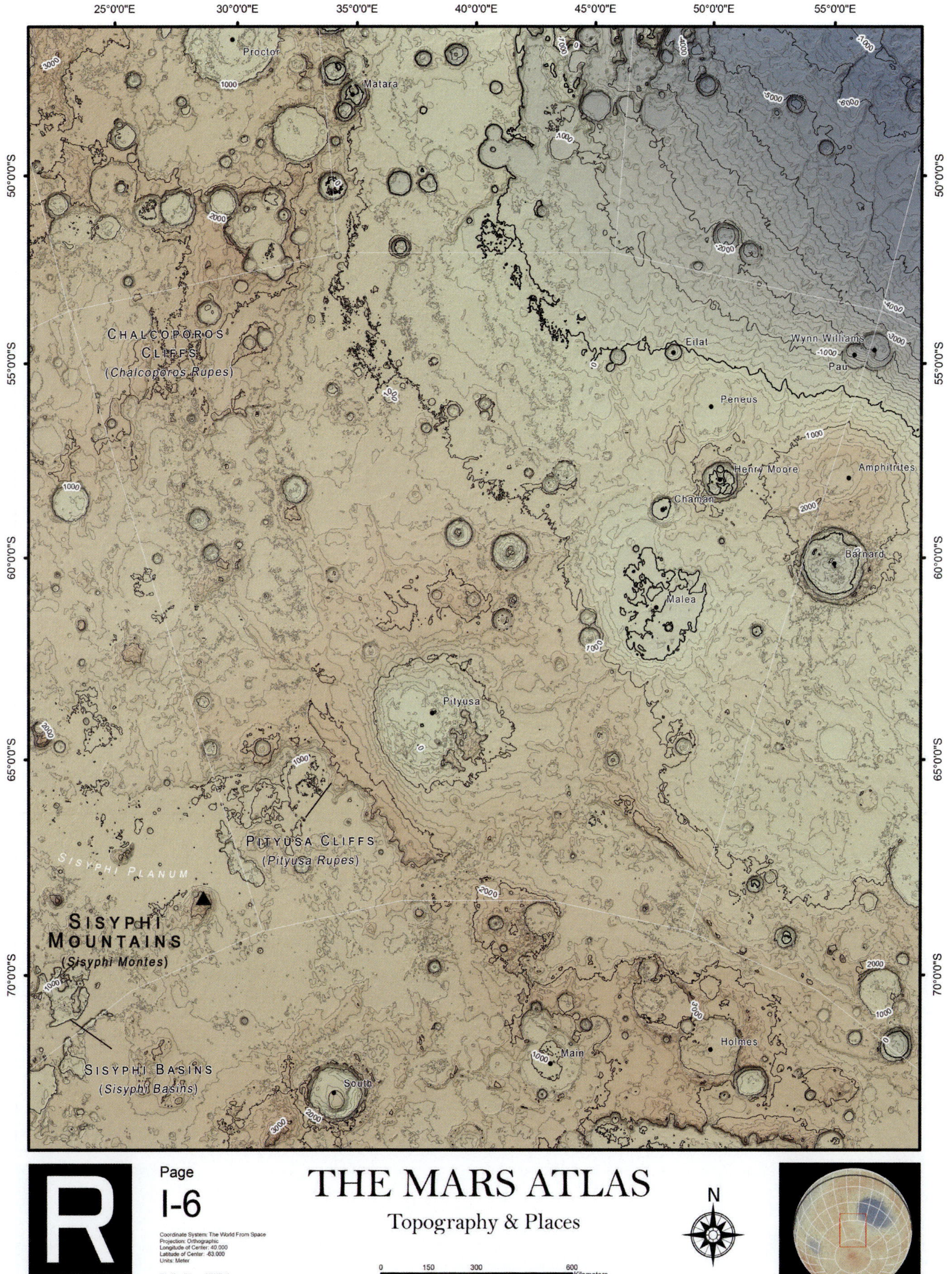

REDMAPPER

Page

I-6

Coordinate System: The World From Space
Projection: Orthographic
Longitude of Center: 40.000
Latitude of Center: -63.000
Units: Meter

The Mars Atlas — 2020 Edition.
Visit RedMapper.com

THE MARS ATLAS

Topography & Places

N

0 150 300 600 Kilometers

0 150 300 600 Miles

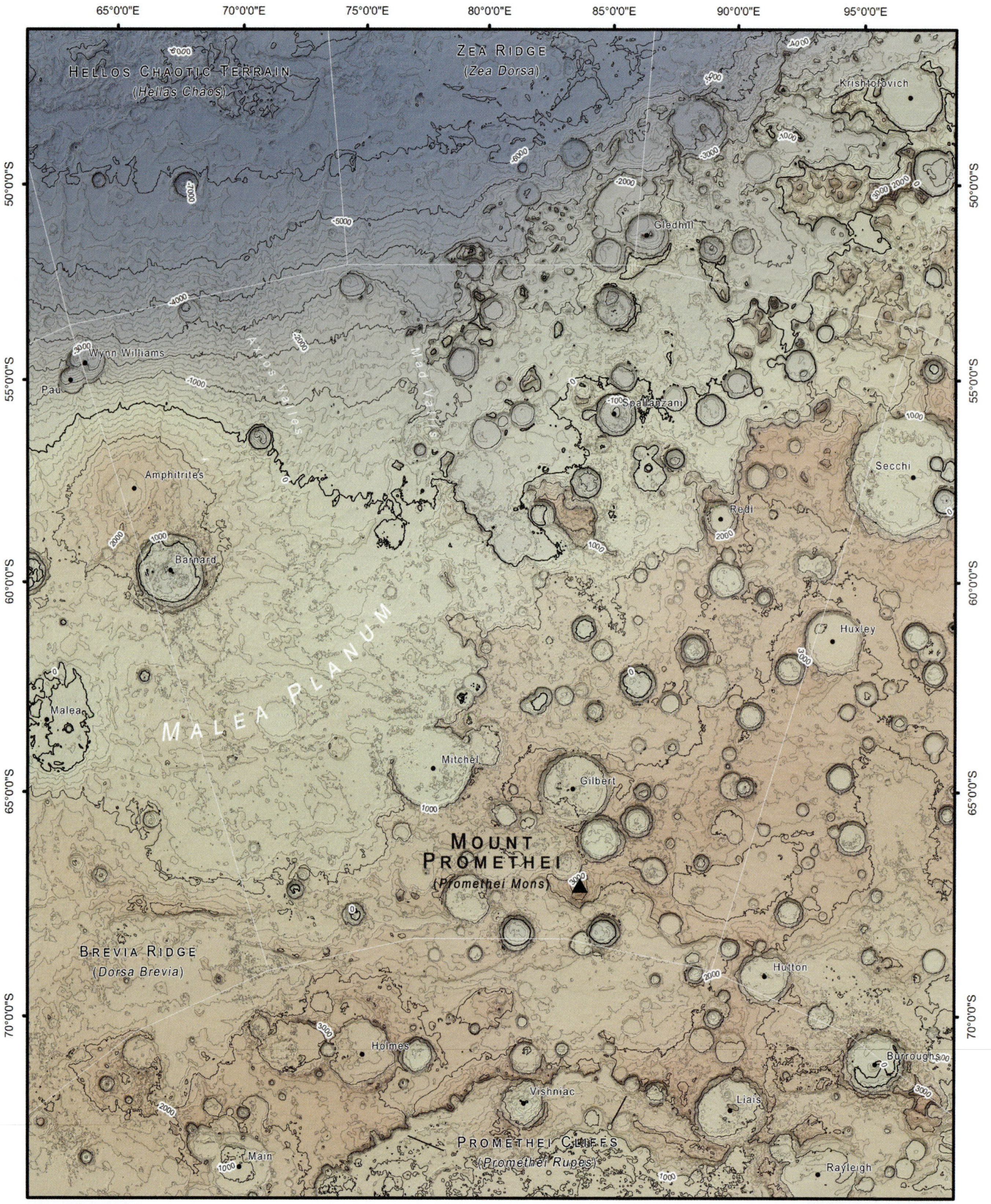

65°0'0"E 70°0'0"E 75°0'0"E 80°0'0"E 85°0'0"E 90°0'0"E 95°0'0"E

50°0'0"S 55°0'0"S 60°0'0"S 65°0'0"S 70°0'0"S

HELLAS CHAOTIC TERRAIN
(Hellas Chaos)
ZEA RIDGE
(Zea Dorsa)
Krishtofovich
Gledhill
Wynn Williams
Pau
Axius Valles
Mad Vallis
Spallanzani
Secchi
Amphitrites
Barnard
Redi
Huxley
MALEA PLANUM
Malea
Mitchel
Gilbert
MOUNT PROMETHEI
(Promethei Mons)
BREVIA RIDGE
(Dorsa Brevia)
Hutton
Holmes
Burroughs
Vishniac
Liais
PROMETHEI CLIFFS
(Promethei Rupes)
Main
Rayleigh

Page

I-7

Coordinate System: The World From Space
Projection: Orthographic
Longitude of Center: 80.000
Latitude of Center: -63.000
Units: Meter

The Mars Atlas — 2020 Edition.
Visit RedMapper.com

THE MARS ATLAS

Topography & Places

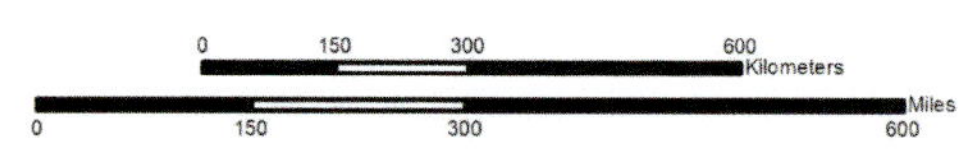

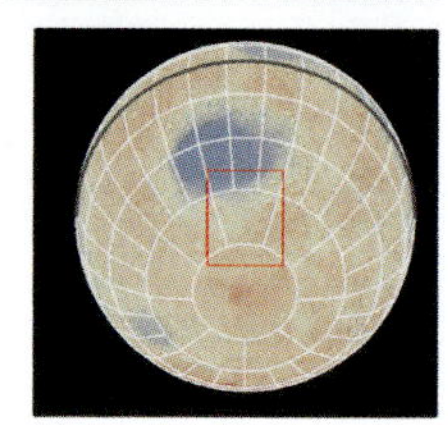

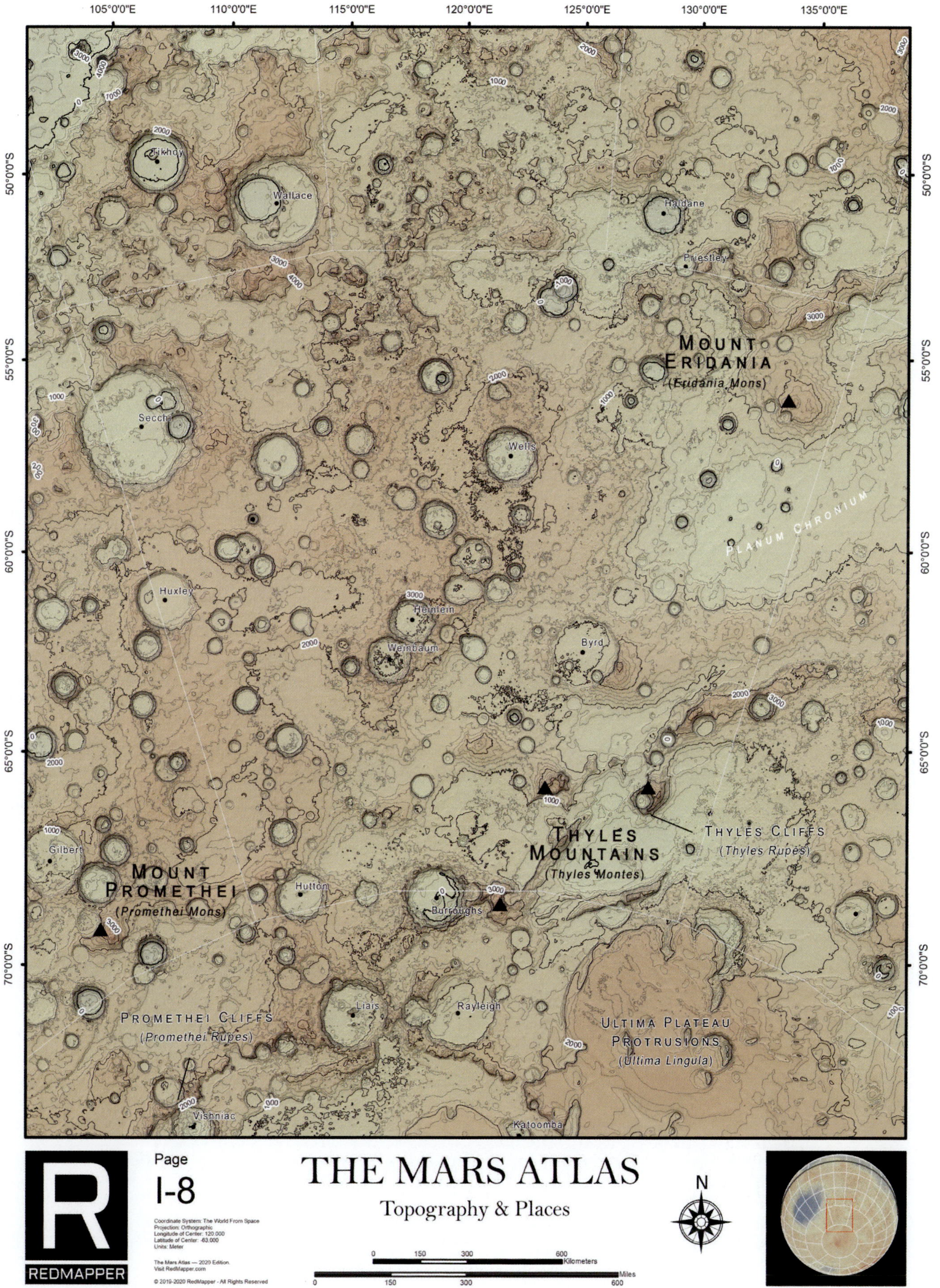

REDMAPPER

Page

I-8

Coordinate System: The World From Space
Projection: Orthographic
Longitude of Center: 120.000
Latitude of Center: -63.000
Units: Meter

The Mars Atlas — 2020 Edition.
Visit RedMapper.com

THE MARS ATLAS

Topography & Places

0 150 300 600 Kilometers

0 150 300 600 Miles

N

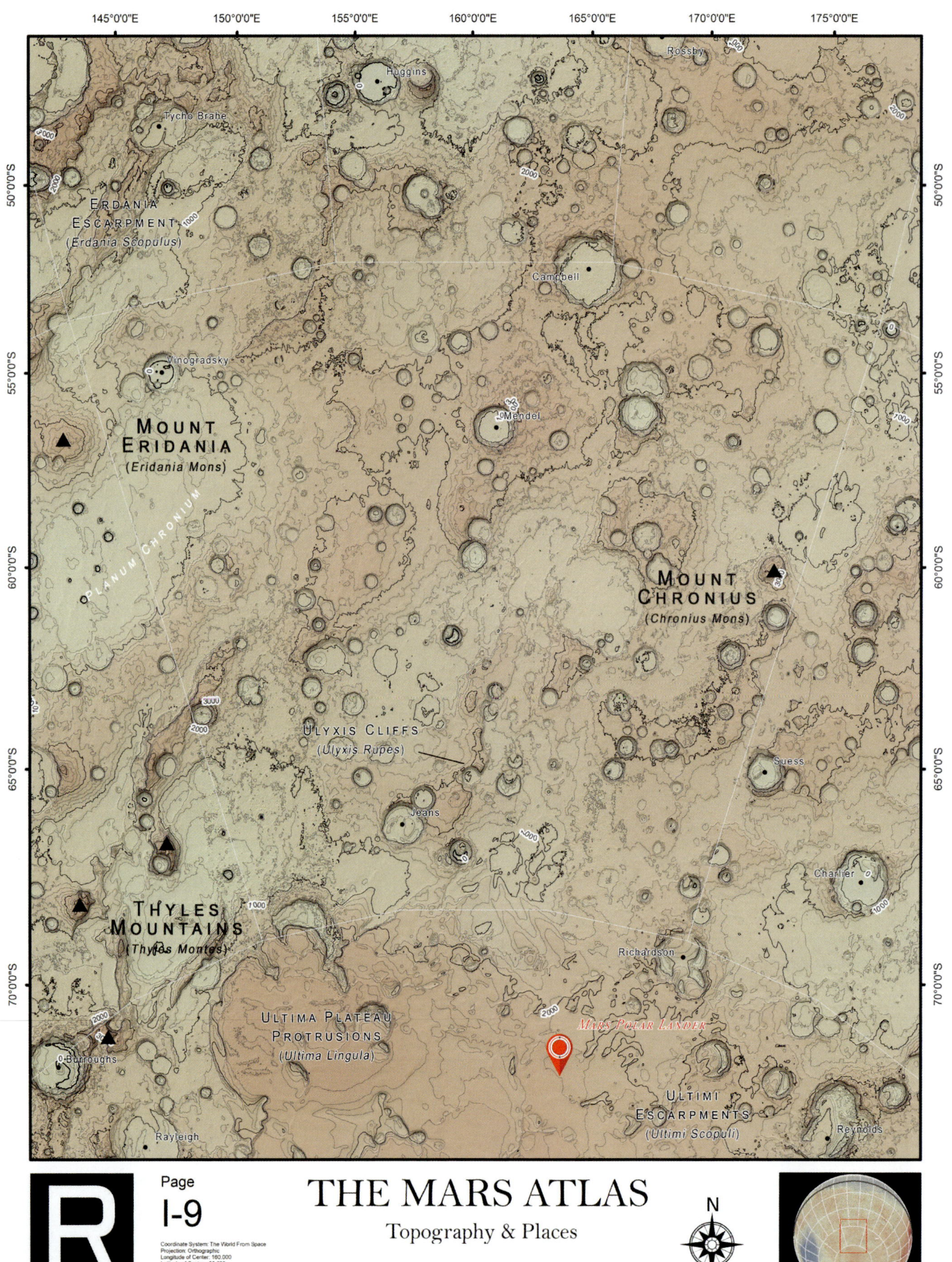

Page

I-9

Coordinate System: The World From Space
Projection: Orthographic
Longitude of Center: 160.000
Latitude of Center: -63.000
Units: Meter

The Mars Atlas — 2020 Edition.
Visit RedMapper.com

THE MARS ATLAS

Topography & Places

N

0 150 300 600 Kilometers

0 150 300 600 Miles

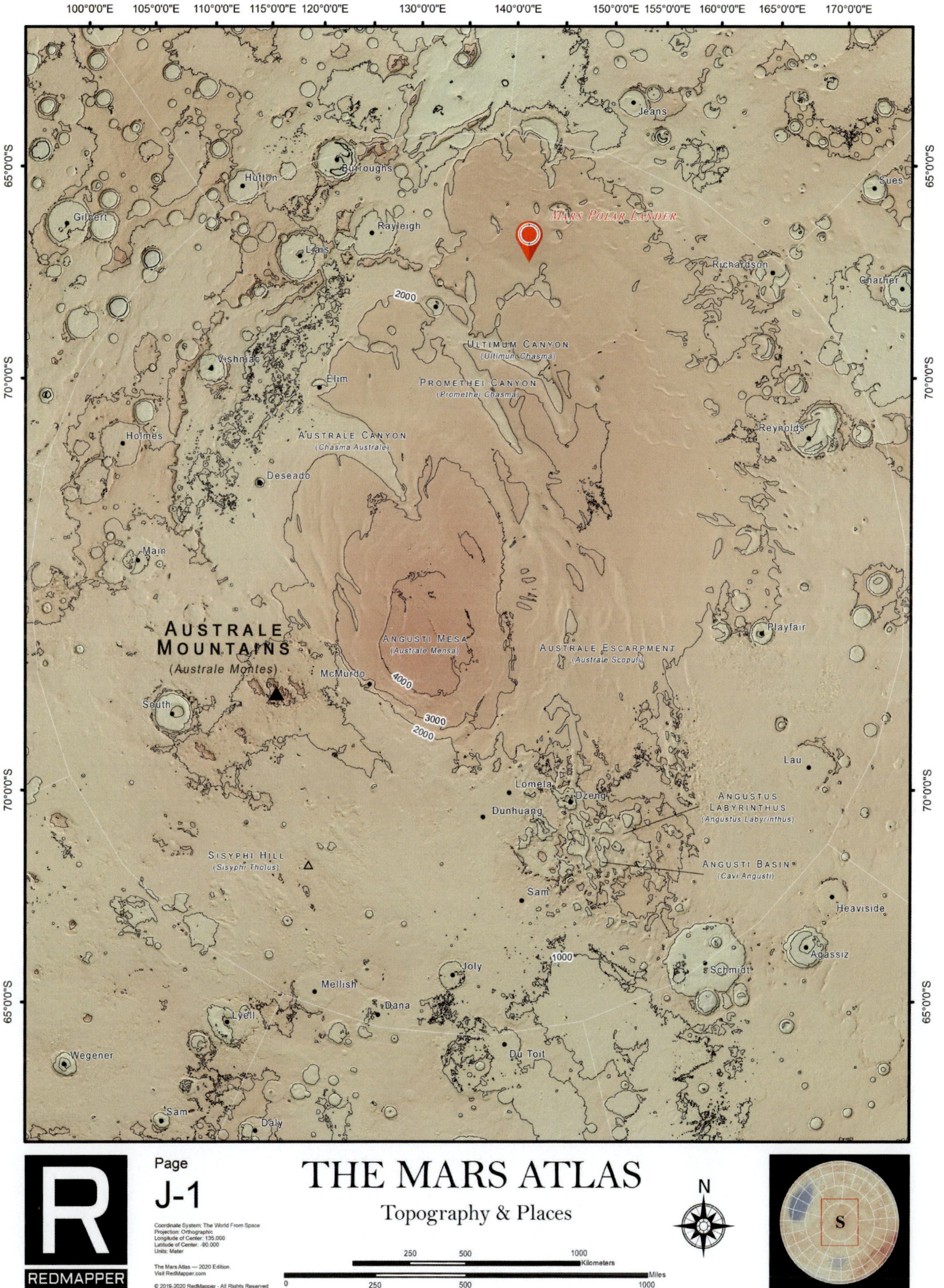

REDMAPPER

Page

J-1

Coordinate System: The World From Space
Projection: Orthographic
Longitude of Center: 135.000
Latitude of Center: -90.000
Units: Meter

The Mars Atlas — 2020 Edition
Visit RedMapper.com

THE MARS ATLAS

Topography & Places

250 500 1000 Kilometers

0 250 500 1000 Miles

N

S

INDEX OF PLACES

NAME	FEATURE TYPE	PAGE	LATITUDE	LONGITUDE
Bamba	Crater	F11	41° 35' 19.221" W	3° 21' 45.123" S
Bamberg	Crater	C14	3° 5' 59.698" W	39° 42' 26.039" N
Banes	Crater	E14	4° 19' 12.000" W	10° 45' 36.000" N
Banff	Crater	E12	30° 42' 27.779" W	17° 30' 35.071" N
Banh	Crater	D10	55° 29' 50.408" W	19° 25' 1.200" N
Baphyras Catena	Chain of craters	C8	84° 9' 29.852" W	38° 49' 48.214" N
Bar	Crater	G13	19° 29' 46.154" W	25° 14' 48.317" S
Barabashov	Crater	C9	68° 45' 5.416" W	47° 19' 30.659" N
Barnard	Crater	I7	61° 35' 21.819" E	61° 3' 46.315" S
Baro	Crater	G22	110° 41' 53.185" E	24° 48' 10.839" S
Barsukov	Crater	E12	29° 0' 56.738" W	7° 58' 21.612" N
Barth	Crater	E16	25° 40' 12.000" E	7° 26' 24.000" N
Bashkaus Valles	Valley	G14	3° 15' 48.687" W	25° 40' 39.359" S
Basin	Crater	E22	106° 59' 59.177" E	17° 49' 3.086" N
Batoka	Crater	F11	36° 39' 10.240" W	7° 33' 10.600" S
Batos	Crater	D12	29° 29' 57.578" W	21° 30' 13.726" N
Batson	Crater	G20	84° 9' 0.000" E	28° 54' 36.000" S
Baucau	Crater	G10	55° 6' 14.359" W	28° 22' 10.005" S
Baum	Crater	G16	28° 18' 2.813" E	24° 43' 9.258" S
Baykonyr	Crater	C24	132° 40' 44.123" E	46° 24' 35.691" N
Bazas	Crater	G21	93° 22' 50.886" E	27° 46' 36.303" S
Beagle 2	Historical Land. Site	E21	90° 25' 50.084" E	11° 31' 49.230" N
Becquerel	Crater	D13	7° 56' 34.800" W	21° 53' 31.920" N
Beer	Crater	F13	8° 10' 8.400" W	14° 28' 20.280" S
Beloha	Crater	H18	56° 42' 41.040" E	39° 34' 30.000" S
Beltra	Crater	D22	102° 24' 14.882" E	18° 0' 40.919" N
Belyov	Crater	H2	158° 0' 36.000" W	45° 1' 12.000" S
Belz	Crater	D11	43° 13' 40.702" W	21° 34' 27.130" N
Bend	Crater	G12	27° 44' 6.398" W	22° 24' 6.480" S
Bentham	Crater	I4	40° 33' 10.367" W	55° 46' 45.639" S
Bentong	Crater	G13	19° 2' 35.624" W	22° 18' 29.858" S
Bernard	Crater	G2	154° 12' 19.184" W	23° 14' 38.163" S
Berseba	Crater	F11	37° 36' 7.046" W	4° 23' 45.091" S
Beruri	Crater	E20	81° 14' 7.294" E	5° 15' 56.232" N
Betio	Crater	G8	78° 38' 57.122" W	23° 8' 4.058" S
Bhor	Crater	C24	134° 31' 41.062" E	41° 44' 50.793" N
Bianchini	Crater	I3	95° 17' 24.000" W	63° 51' 10.080" S
Biblis Patera	Crater	E5	123° 49' 3.728" W	2° 21' 20.758" N
Biblis Tholus	Mountain / Hill	E5	124° 22' 57.651" W	2° 31' 24.937" N
Bigbee	Crater	G11	34° 45' 2.605" W	24° 46' 40.302" S
Bira	Crater	D11	45° 32' 27.277" W	25° 5' 49.039" N
Bise	Crater	D10	56° 49' 55.273" W	20° 13' 27.666" N
Bison	Crater	G12	29° 8' 53.083" W	26° 18' 18.113" S
Bjerknes	Crater	H27	171° 28' 58.981" E	43° 0' 45.203" S
Bland	Crater	D22	108° 45' 21.158" E	18° 17' 42.607" N
Bled	Crater	D12	31° 27' 22.972" W	21° 35' 5.091" N
Blitta	Crater	G12	20° 57' 30.594" W	25° 53' 42.714" S
Blois	Crater	D10	55° 50' 55.862" W	23° 35' 42.753" N
Bluff	Crater	D22	110° 1' 46.511" E	23° 27' 57.570" N
Blunck	Crater	G11	36° 53' 49.321" W	27° 13' 41.667" S
Boeddicker	Crater	F26	162° 29' 14.401" E	14° 49' 17.879" S
Bogia	Crater	H20	83° 16' 4.080" E	44° 18' 42.120" S
Bogra	Crater	G12	28° 48' 16.679" W	24° 9' 51.078" S
Bok	Crater	D12	31° 36' 12.220" W	20° 34' 46.566" N
Bole	Crater	D10	53° 59' 40.720" W	25° 21' 15.845" N
Bombala	Crater	G22	106° 8' 2.400" E	27° 36' 7.560" S
Bond	Crater	G11	35° 56' 21.854" W	32° 47' 40.560" S
Bonestell	Crater	C10	30° 23' 34.423" W	41° 59' 52.326" N
Boola	Crater	A1	105° 11' 38.781" W	81° 15' 52.825" N
Bopolu	Crater	F14	6° 17' 50.574" W	2° 57' 46.761" S
Bor	Crater	D11	33° 40' 56.964" W	18° 10' 5.427" N
Bordeaux	Crater	D10	48° 53' 31.954" W	23° 7' 57.712" N
Boreales Scopuli	Escarpment	A1	90° 9' 36.000" W	88° 52' 48.000" N
Boreum Cavus	Depression / Basin	A1	20° 9' 0.000" W	84° 38' 24.000" N
Boru	Crater	G12	27° 52' 13.880" W	24° 20' 35.261" S
Bosporos Planum	Plain	G9	64° 29' 37.192" W	33° 52' 19.869" S
Bosporos Rupes	Cliff	H10	57° 33' 13.100" W	42° 44' 40.834" S
Bouguer	Crater	G16	27° 15' 59.836" E	18° 27' 40.101" S
Boulia	Crater	G22	111° 19' 7.979" E	22° 53' 7.900" S
Bozkir	Crater	H12	32° 10' 44.638" W	44° 8' 18.163" S
Bradbury	Crater	E20	85° 47' 43.916" E	2° 34' 54.434" N
Bradbury Land. Site	Historical Land. Site	F24	137° 26' 30.131" E	4° 35' 22.216" S
Brashear	Crater	H5	119° 1' 40.800" W	53° 48' 53.640" S
Brazos Valles	Valley	F15	18° 42' 11.569" E	6° 4' 44.610" S
Bree	Crater	C25	149° 37' 40.876" E	37° 38' 11.089" N
Bremerhaven	Crater	D10	48° 38' 31.200" W	23° 41' 52.865" N
Briault	Crater	F21	89° 40' 56.640" E	9° 58' 42.356" S
Bridgetown	Crater	D10	47° 4' 42.965" W	21° 53' 45.242" N
Bristol	Crater	D10	46° 55' 55.388" W	22° 5' 25.442" N
Broach	Crater	D10	56° 53' 40.185" W	23° 30' 50.228" N
Bronkhorst	Crater	F10	55° 12' 21.181" W	10° 42' 12.786" S
Brush	Crater	D22	111° 20' 18.092" E	21° 42' 0.211" N
Bulhar	Crater	C24	134° 31' 7.404" E	50° 21' 28.911" N

INDEX OF PLACES

NAME	FEATURE TYPE	PAGE	LATITUDE	LONGITUDE
Bunge	Crater	G10	48° 35' 31.200" W	33° 49' 1.560" S
Bunnik	Crater	H3	142° 6' 0.000" W	38° 4' 12.000" S
Burroughs	Crater	J1	117° 5' 57.170" E	72° 17' 23.417" S
Burton	Crater	F2	156° 19' 51.600" W	13° 52' 49.800" S
Buta	Crater	G12	32° 24' 43.013" W	23° 14' 55.340" S
Butte	Crater	F11	38° 54' 51.204" W	5° 4' 57.554" S
Buvinda Vallis	Valley	D25	151° 57' 50.689" E	33° 10' 5.721" N
Byala	Crater	G9	66° 28' 5.193" W	25° 44' 2.761" S
Byrd	Crater	I8	127° 49' 48.000" E	65° 13' 17.760" S
Byske	Crater	F11	33° 57' 5.784" W	4° 58' 23.117" S
Cadiz	Crater	D10	49° 1' 40.372" W	23° 8' 46.974" N
Cagli	Crater	E14	3° 33' 0.000" W	4° 43' 48.000" N
Cairns	Crater	D10	47° 27' 26.155" W	23° 33' 52.942" N
Calahorra	Crater	D11	38° 39' 3.455" W	26° 27' 24.342" N
Calamar	Crater	D10	54° 52' 23.666" W	18° 16' 29.783" N
Calbe	Crater	G12	28° 52' 22.551" W	25° 8' 11.413" S
Calydon Fossa	Depression / Basin	F7	87° 58' 53.745" W	7° 26' 2.758" S
Camargo	Crater	E22	109° 38' 30.329" E	17° 41' 57.332" N
Camichel	Crater	E10	51° 36' 34.739" W	2° 15' 42.517" N
Camiling	Crater	F11	37° 59' 55.818" W	0° 42' 50.052" S
Camiri	Crater	H11	42° 10' 19.200" W	44° 39' 16.920" S
Campbell	Crater	I9	165° 34' 51.600" E	54° 15' 14.760" S
Campos	Crater	G12	27° 48' 51.074" W	21° 47' 53.099" S
Can	Crater	C13	14° 35' 6.509" W	48° 12' 23.550" N
Canala	Crater	D8	80° 4' 35.097" W	24° 21' 14.985" N
Canas	Crater	G21	89° 51' 29.225" E	31° 11' 15.085" S
Canaveral	Crater	C24	135° 49' 50.490" E	46° 49' 46.430" N
Canberra	Crater	C24	132° 39' 43.134" E	47° 12' 11.528" N
Candor Chaos	Chaotic terrain	F9	72° 35' 5.902" W	6° 56' 12.020" S
Candor Chasma	Canyon / Chasm	F9	70° 46' 45.652" W	6° 31' 58.198" S
Candor Colles	Mountain / Hill	F8	75° 33' 58.553" W	6° 37' 59.294" S
Candor Labes	Chaotic terrain	F8	75° 59' 41.678" W	4° 47' 26.496" S
Candor Mensa	Mesa	F8	73° 31' 19.636" W	6° 15' 51.603" S
Cangwu	Crater	C7	89° 35' 28.843" W	41° 50' 44.018" N
Canillo	Crater	E23	116° 28' 31.959" E	10° 13' 40.768" N
Cankuzo	Crater	G18	52° 2' 2.764" E	19° 25' 18.120" S
Canso	Crater	D9	60° 37' 13.320" W	21° 21' 50.881" N
Cantoura	Crater	E10	51° 42' 57.977" W	14° 50' 18.861" N
Capen	Crater	E15	14° 18' 46.800" E	6° 34' 50.338" N
Capri Chasma	Canyon / Chasm	F11	42° 3' 57.384" W	8° 16' 25.492" S
Capri Mensa	Mesa	F10	47° 11' 16.629" W	13° 43' 49.290" S
Caralis Chaos	Chaotic terrain	H27	178° 36' 17.348" E	37° 12' 1.379" S
Cardona	Crater	G12	31° 59' 4.189" W	19° 38' 49.156" S
Cartago	Crater	G13	17° 58' 20.543" W	23° 15' 9.720" S
Cassini	Crater	D16	32° 6' 37.080" E	23° 21' 4.680" N
Castril	Crater	F27	175° 18' 0.709" E	14° 41' 57.181" S
Catota	Crater	C10	26° 58' 30.278" W	51° 40' 4.851" N
Cave	Crater	D11	35° 38' 20.610" W	21° 36' 51.566" N
Cavi Angusti	Depression / Basin	J1	74° 44' 50.348" W	78° 9' 48.947" S
Caxias	Crater	G6	100° 40' 49.925" W	28° 57' 15.318" S
Cayon	Crater	G23	113° 36' 58.359" E	35° 55' 33.527" S
Cefalu	Crater	D11	38° 53' 18.393" W	23° 38' 9.470" N
Centauri Montes	Mountain / Hill	H21	95° 31' 18.692" E	38° 40' 16.469" S
Ceraunius Catena	Chain of craters	C6	108° 5' 40.768" W	37° 6' 1.635" N
Ceraunius Fossae	Depression / Basin	D6	110° 8' 52.118" W	27° 0' 13.194" N
Ceraunius Tholus	Mountain / Hill	D7	97° 15' 4.916" W	24° 0' 7.478" N
Cerberus Dorsa	Ridge	F22	105° 17' 26.796" E	13° 44' 22.519" S
Cerberus Fossae	Depression / Basin	E26	166° 22' 25.786" E	11° 16' 35.800" N
Cerberus Palus	Plain	E25	148° 8' 42.633" E	5° 46' 35.577" N
Cerberus Tholi	Mountain / Hill	E26	164° 24' 50.343" E	4° 28' 53.375" N
Cerulli	Crater	D16	22° 6' 57.960" E	32° 12' 6.840" N
Ceti Chasma	Canyon / Chasm	F9	68° 22' 18.083" W	5° 1' 40.001" S
Ceti Labes	Chaotic terrain	F8	75° 43' 59.101" W	6° 46' 39.396" S
Ceti Mensa	Mesa	F8	76° 1' 19.308" W	5° 53' 37.952" S
Chafe	Crater	E22	102° 24' 35.277" E	15° 6' 4.702" N
Chalce Fossa	Depression / Basin	H11	39° 35' 13.802" W	51° 39' 55.496" S
Chalce Montes	Mountain / Hill	H11	37° 39' 8.897" W	53° 43' 24.163" S
Chalcoporos Rupes	Cliff	I6	20° 33' 57.200" E	55° 38' 17.679" S
Chaman	Crater	I6	50° 57' 46.001" E	60° 51' 51.234" S
Chamberlin	Crater	I2	124° 17' 28.108" W	65° 50' 38.718" S
Changsong	Crater	D10	57° 20' 22.185" W	23° 28' 13.298" N
Chapais	Crater	G12	20° 33' 14.400" W	22° 20' 44.520" S
Charis Dorsum	Ridge	I4	41° 27' 58.738" W	55° 51' 45.564" S
Charitum Montes	Mountain / Hill	I4	40° 17' 31.617" W	58° 5' 49.512" S
Charleston	Crater	D10	47° 47' 48.618" W	22° 37' 34.661" N
Charlier	Crater	I1	168° 28' 4.061" W	68° 33' 38.166" S
Charlieu	Crater	C8	83° 59' 13.801" W	38° 9' 15.985" N
Charybdis Scopulus	Escarpment	G16	20° 4' 34.465" E	24° 8' 28.409" S
Chasma Australe	Canyon / Chasm	J1	95° 1' 53.405" E	82° 21' 16.696" S
Chasma Boreale	Canyon / Chasm	A1	47° 38' 25.232" W	82° 32' 28.478" N
Chatturat	Crater	D7	94° 56' 35.714" W	35° 22' 48.381" N
Chauk	Crater	D10	55° 54' 34.028" W	23° 21' 11.437" N
Cheb	Crater	G13	19° 26' 35.779" W	24° 11' 46.163" S

INDEX OF PLACES

NAME	FEATURE TYPE	PAGE	LATITUDE	LONGITUDE
Chefu	Crater	G22	112° 14' 29.850" E	22° 54' 43.830" S
Chekalin	Crater	G12	26° 48' 30.736" W	24° 16' 31.081" S
Chia	Crater	E10	59° 39' 10.685" W	1° 34' 15.626" N
Chico Valles	Valley	I1	152° 13' 56.280" W	66° 46' 7.578" S
Chimbote	Crater	F11	39° 40' 58.545" W	1° 25' 23.933" S
Chincoteague	Crater	C23	124° 6' 58.810" E	41° 12' 16.315" N
Chinju	Crater	F11	42° 8' 35.933" W	4° 30' 58.463" S
Chinook	Crater	D10	55° 27' 32.352" W	22° 30' 8.016" N
Chive	Crater	D10	56° 0' 47.905" W	21° 40' 54.987" N
Choctaw	Crater	H11	37° 14' 33.307" W	41° 11' 8.324" S
Chom	Crater	C14	2° 30' 54.186" W	38° 34' 26.760" N
Choyr	Crater	G15	18° 43' 17.498" E	32° 25' 44.019" S
Chronius Mons	Mountain / Hill	I9	178° 0' 44.691" E	61° 29' 36.497" S
Chrysas Mensa	Mesa	F9	70° 5' 51.226" W	6° 45' 58.169" S
Chryse Chaos	Chaotic terrain	E11	37° 11' 22.434" W	9° 51' 52.469" N
Chryse Colles	Mountain / Hill	E11	41° 51' 50.718" W	8° 9' 2.510" N
Chryse Planitia	Plain	D11	40° 18' 26.281" W	28° 25' 59.719" N
Chukhung	Crater	C9	72° 25' 12.000" W	38° 28' 12.000" N
Chupadero	Crater	E20	83° 26' 1.937" E	6° 7' 58.085" N
Chur	Crater	E12	29° 18' 20.631" W	16° 56' 3.982" N
Cilaos	Crater	G24	129° 28' 57.668" E	35° 42' 49.438" S
Circle	Crater	G12	25° 31' 31.781" W	22° 10' 18.141" S
Clallam Canyon	Canyon / Chasm	F11	34° 13' 33.281" W	3° 4' 36.523" S
Clanis Valles	Valley	D18	58° 27' 58.154" E	33° 14' 24.575" N
Claritas Fossae	Depression / Basin	G6	104° 14' 15.899" W	27° 53' 14.436" S
Claritas Rupes	Cliff	G6	105° 15' 38.449" W	25° 2' 13.053" S
Clark	Crater	I2	133° 12' 14.276" W	55° 8' 37.338" S
Clasia Vallis	Valley	D18	57° 2' 7.776" E	33° 46' 23.757" N
Cleia Dorsum	Ridge	I4	45° 59' 14.091" W	54° 51' 45.150" S
Clogh	Crater	D10	47° 40' 4.702" W	20° 33' 47.821" N
Clota Vallis	Valley	G12	20° 30' 8.232" W	25° 35' 28.823" S
Clova	Crater	D10	52° 2' 42.530" W	21° 27' 57.105" N
Cluny	Crater	G12	27° 18' 7.835" W	23° 51' 52.597" S
Cobalt	Crater	G12	27° 2' 0.167" W	25° 47' 15.910" S
Coblentz	Crater	I3	90° 18' 29.075" W	54° 53' 59.643" S
Cobres	Crater	F2	153° 35' 52.800" W	11° 41' 47.760" S
Coimbra	Crater	E14	5° 18' 50.463" W	4° 10' 35.123" N
Colles Nili	Mountain / Hill	C19	62° 52' 34.262" E	38° 43' 13.305" N
Coloe Fossae	Depression / Basin	C18	56° 46' 49.525" E	36° 39' 0.265" N
Colon	Crater	D10	47° 3' 57.475" W	22° 44' 56.973" N
Columbia Mem. Station	Historical Land. Site	F27	175° 28' 31.585" E	14° 34' 6.365" S
Columbia Valles	Valley	F11	42° 54' 5.303" W	9° 26' 20.697" S
Columbus	Crater	G2	165° 58' 55.383" W	29° 17' 35.786" S
Comas Sola	Crater	G2	158° 30' 43.200" W	19° 35' 24.000" S
Conches	Crater	F11	34° 12' 9.947" W	4° 13' 25.243" S
Concord	Crater	E11	34° 1' 13.322" W	16° 31' 30.467" N
Coogoon Valles	Valley	E12	21° 44' 20.858" W	17° 11' 19.561" N
Cooma	Crater	G6	108° 20' 59.781" W	23° 41' 9.675" S
Copernicus	Crater	H1	168° 49' 40.800" W	48° 50' 40.560" S
Coprates Catena	Chain of craters	F9	62° 5' 14.312" W	15° 0' 14.072" S
Coprates Chasma	Canyon / Chasm	F9	60° 44' 18.422" W	13° 21' 55.310" S
Coprates Labes	Chaotic terrain	F9	67° 47' 24.809" W	11° 49' 12.997" S
Coprates Mensa	Mesa	F9	71° 9' 48.673" W	12° 12' 1.416" S
Coprates Montes	Mountain / Hill	F9	65° 11' 8.730" W	13° 7' 6.328" S
Coracis Fossae	Depression / Basin	G8	80° 51' 46.729" W	35° 48' 57.296" S
Corby	Crater	C24	137° 33' 41.265" E	42° 52' 31.965" N
Corinto	Crater	E25	141° 42' 25.841" E	16° 56' 47.974" N
Coronae Montes	Mountain / Hill	G20	86° 6' 33.890" E	34° 18' 33.902" S
Coronae Planum	Plain	G19	65° 24' 33.272" E	32° 40' 55.968" S
Coronae Scopulus	Escarpment	G19	64° 56' 36.981" E	33° 15' 21.405" S
Corozal	Crater	H26	159° 25' 21.509" E	38° 47' 18.899" S
Cost	Crater	E22	104° 1' 11.072" E	14° 58' 43.251" N
Cray	Crater	C13	16° 7' 5.070" W	44° 5' 44.820" N
Creel	Crater	F11	38° 51' 5.876" W	6° 2' 50.253" S
Crewe	Crater	G13	19° 31' 36.728" W	24° 50' 27.127" S
Crivitz	Crater	F27	174° 47' 20.253" E	14° 33' 2.377" S
Crommelin	Crater	E13	10° 8' 13.200" W	5° 4' 39.396" N
Cross	Crater	G2	157° 41' 28.449" W	30° 11' 46.395" S
Crotone	Crater	A1	69° 18' 39.768" W	82° 12' 45.891" N
Cruls	Crater	H26	163° 1' 32.321" E	42° 54' 45.048" S
Cruz	Crater	C14	1° 57' 58.691" W	38° 27' 48.207" N
Cue	Crater	G21	93° 13' 42.248" E	35° 50' 7.352" S
Culter	Crater	F10	53° 56' 2.389" W	8° 50' 34.074" S
Curie	Crater	D14	4° 45' 2.883" W	28° 46' 59.377" N
Curiosity Rover	Rover Location	F24	137° 21' 12.438" E	4° 43' 15.549" S
Cusus Valles	Valley	E18	50° 22' 20.476" E	14° 2' 42.539" N
Cyane Catena	Chain of craters	C5	118° 18' 11.718" W	36° 15' 15.914" N
Cyane Fossae	Depression / Basin	D5	121° 10' 17.116" W	31° 15' 13.684" N
Cydnus Rupes	Cliff	C22	112° 12' 33.232" E	52° 31' 53.513" N
Cydonia Colles	Mountain / Hill	C13	12° 13' 15.054" W	39° 4' 27.987" N
Cydonia Labyrinthus	Labyrinth terrain	C13	12° 3' 23.864" W	41° 17' 26.589" N
Cydonia Mensae	Mesa	D13	12° 19' 59.844" W	34° 33' 42.498" N
Cypress	Crater	H10	47° 21' 10.225" W	47° 16' 49.756" S

INDEX OF PLACES

NAME	FEATURE TYPE	PAGE	LATITUDE	LONGITUDE
Da Vinci	Crater	E11	39° 15' 52.968" W	1° 28' 0.477" N
Daan	Crater	H21	91° 34' 40.778" E	40° 28' 35.027" S
Daedalia Planum	Plain	G5	125° 56' 58.153" W	18° 20' 51.587" S
Daet	Crater	F11	41° 47' 45.055" W	7° 17' 16.594" S
Daga Vallis	Valley	F11	42° 25' 15.583" W	12° 4' 13.122" S
Dalu Cavus	Depression / Basin	F7	99° 0' 0.000" W	6° 54' 0.000" S
Daly	Crater	I4	23° 7' 0.012" W	66° 17' 40.241" S
Dana	Crater	J1	32° 47' 38.618" W	72° 29' 32.419" S
Danielson	Crater	E13	7° 3' 4.222" W	7° 58' 26.053" N
Dank	Crater	D22	106° 59' 49.919" E	21° 57' 21.538" N
Dao Vallis	Valley	H21	88° 53' 12.301" E	37° 36' 45.273" S
Darvel	Crater	E10	51° 0' 36.539" W	17° 46' 58.913" N
Darwin	Crater	I5	19° 9' 3.835" W	56° 58' 14.014" S
Davies	Crater	C14	0° 5' 26.499" E	45° 57' 38.681" N
Dawes	Crater	F17	38° 3' 30.240" E	9° 6' 22.320" S
de Vaucouleurs	Crater	F27	170° 59' 45.828" E	13° 18' 45.608" S
Deba	Crater	G13	17° 17' 59.515" W	23° 57' 17.482" S
Dechu	Crater	H2	157° 59' 24.000" W	42° 15' 0.000" S
Degana	Crater	G11	45° 29' 44.729" W	23° 43' 29.912" S
Dein	Crater	C14	2° 30' 48.586" W	38° 12' 27.357" N
Dejnev	Crater	G2	164° 38' 27.600" W	25° 8' 12.120" S
Delta	Crater	H11	39° 10' 24.302" W	45° 57' 36.339" S
Denning	Crater	F17	33° 31' 28.943" E	17° 25' 39.720" S
Dersu	Crater	D10	51° 53' 23.720" W	22° 38' 8.298" N
Dese	Crater	H12	30° 36' 46.891" W	45° 24' 43.759" S
Deseado	Crater	J1	70° 17' 15.357" E	80° 37' 25.486" S
Dessau	Crater	H10	53° 7' 32.506" W	42° 45' 34.531" S
Deuteronilus Colles	Mountain / Hill	C16	21° 41' 49.603" E	41° 56' 55.889" N
Deuteronilus Mensae	Mesa	C16	23° 55' 6.221" E	45° 6' 34.262" N
Deva Vallis	Valley	F2	156° 51' 58.432" W	7° 40' 27.565" S
Dia Cau	Crater	F11	42° 39' 53.191" W	0° 21' 46.839" S
Dilly	Crater	E26	157° 13' 19.067" E	13° 15' 58.334" N
Dingo	Crater	G13	17° 29' 37.457" W	23° 42' 20.605" S
Dinorwic	Crater	G6	101° 27' 48.358" W	30° 1' 57.859" S
Dionysus Patera	Crater	E4	133° 13' 48.000" W	17° 57' 36.000" N
Dison	Crater	G13	16° 30' 4.791" W	25° 1' 56.028" S
Dittaino Valles	Valley	F9	66° 51' 47.579" W	1° 25' 40.944" S
Dixie	Crater	E10	55° 54' 51.608" W	17° 46' 36.608" N
Doanus Vallis	Valley	I4	25° 35' 25.046" W	63° 1' 27.830" S
Doba	Crater	E23	119° 37' 14.209" E	10° 55' 3.229" N
Dogana	Crater	F10	53° 39' 57.009" W	10° 0' 37.792" S
Dokka	Crater	A1	145° 45' 19.644" W	77° 10' 16.632" N
Dokuchaev	Crater	I2	127° 5' 4.362" W	60° 37' 26.008" S
Dollfus	Crater	G14	4° 15' 45.703" W	21° 35' 17.285" S
Domoni	Crater	C5	125° 36' 34.958" W	51° 23' 3.948" N
Doon	Crater	D22	109° 30' 27.883" E	23° 31' 47.566" N
Dorsa Argentea	Ridge	J1	33° 23' 13.824" W	77° 37' 42.532" S
Dorsa Brevia	Ridge	I7	63° 10' 49.917" E	71° 2' 49.558" S
Douglass	Crater	H9	70° 32' 9.600" W	51° 20' 40.560" S
Dowa	Crater	G22	110° 14' 18.053" E	31° 39' 27.752" S
Downe	Crater	F27	175° 46' 58.800" E	15° 58' 39.000" S
Drava Valles	Valley	H26	165° 59' 12.734" E	48° 51' 50.952" S
Drilon Vallis	Valley	E10	52° 20' 9.465" W	7° 10' 18.079" N
Dromore	Crater	D10	49° 34' 56.564" W	19° 52' 48.140" N
Du Martheray	Crater	E21	93° 34' 51.240" E	5° 27' 15.552" N
Du Toit	Crater	I4	49° 35' 47.453" W	71° 37' 20.879" S
Dubis Vallis	Valley	F3	148° 7' 42.209" W	5° 9' 27.384" S
Dubki	Crater	G10	55° 12' 3.040" W	34° 58' 7.310" S
Dukhan	Crater	E11	39° 8' 30.040" W	7° 45' 20.380" N
Dulce Vallis	Valley	F24	136° 32' 22.053" E	4° 49' 6.725" S
Dulovo	Crater	E20	84° 33' 31.932" E	3° 37' 18.176" N
Dungeness Point	Plateau protrusions	F11	34° 16' 35.310" W	3° 21' 1.620" S
Dunhuang	Crater	J1	48° 31' 39.578" W	80° 50' 22.507" S
Dunkassa	Crater	H4	137° 3' 42.693" W	37° 29' 36.929" S
Durius Valles	Valley	F27	171° 58' 34.788" E	17° 18' 2.986" S
Dush	Crater	D10	54° 1' 5.552" W	22° 29' 39.292" N
Dzeng	Crater	J1	70° 28' 21.490" W	80° 30' 27.195" S
Dzigai Vallis	Valley	I4	36° 35' 34.056" W	58° 6' 6.100" S
E Mareotis Tholus	Mountain / Hill	D8	85° 7' 59.846" W	35° 54' 59.206" N
Eads	Crater	G12	29° 54' 33.402" W	28° 29' 0.038" S
Eagle	Crater	C13	8° 9' 55.198" W	43° 48' 30.415" N
East Crater	Crater	F11	35° 49' 55.372" W	2° 46' 35.058" S
Eberswalde	Crater	G12	33° 17' 49.434" W	23° 58' 31.129" S
Echt	Crater	G12	28° 11' 14.533" W	21° 58' 22.407" S
Echus Chaos	Chaotic terrain	E8	74° 43' 8.482" W	10° 47' 16.477" N
Echus Chasma	Canyon / Chasm	E8	79° 57' 42.456" W	2° 28' 17.748" N
Echus Fossae	Depression / Basin	E8	76° 44' 47.279" W	2° 36' 40.194" N
Echus Montes	Mountain / Hill	E8	78° 8' 5.810" W	7° 34' 16.988" N
Echus Palus	Plain	E8	77° 16' 12.000" W	12° 17' 24.000" N
Edam	Crater	G12	20° 2' 19.143" W	26° 16' 50.642" S
Eddie	Crater	E25	142° 11' 44.886" E	12° 19' 25.772" N
Eden Patera	Crater	D13	11° 3' 26.754" W	33° 46' 9.220" N
Eger	Crater	H10	51° 52' 4.800" W	48° 17' 40.560" S

INDEX OF PLACES

NAME	FEATURE TYPE	PAGE	LATITUDE	LONGITUDE
Hegemone Dorsum	Ridge	I4	44° 54' 12.239" W	54° 43' 13.878" S
Heimdal	Crater	B2	124° 33' 38.179" W	68° 19' 31.610" N
Heinlein	Crater	I8	116° 18' 34.457" E	64° 28' 53.354" S
Hellas Chaos	Chaotic terrain	H19	64° 24' 29.104" E	47° 7' 16.337" S
Hellas Chasma	Canyon / Chasm	G19	65° 27' 57.588" E	34° 38' 14.048" S
Hellas Montes	Mountain / Hill	H21	97° 36' 31.470" E	37° 38' 2.635" S
Hellas Planitia	Plain	H19	70° 30' 9.008" E	42° 25' 48.400" S
Hellespontus Montes	Mountain / Hill	H17	42° 45' 35.485" E	44° 22' 5.599" S
Helmholtz	Crater	H12	21° 16' 22.800" W	45° 23' 50.640" S
Henbury	Crater	I1	147° 43' 35.137" W	63° 29' 40.462" S
Henry	Crater	E16	23° 26' 47.400" E	10° 47' 23.640" N
Henry Moore	Crater	I6	53° 53' 43.334" E	59° 43' 24.664" S
Hephaestus Fossae	Depression / Basin	D23	122° 50' 49.281" E	20° 50' 8.750" N
Hephaestus Rupes	Cliff	D23	114° 54' 9.946" E	23° 32' 18.750" N
Her Desher Vallis	Valley	G10	47° 56' 1.223" W	25° 4' 30.266" S
Hera Patera	Crater	D4	133° 46' 12.000" W	18° 27' 0.000" N
Herculaneum	Crater	D10	58° 39' 14.980" W	19° 18' 24.614" N
Hermes Patera	Crater	D4	133° 25' 48.000" W	18° 19' 48.000" N
Hermus Vallis	Valley	F3	147° 48' 22.713" W	5° 18' 59.274" S
Herschel	Crater	F24	129° 53' 34.800" E	14° 28' 48.000" S
Hesperia Dorsa	Ridge	G22	113° 9' 19.842" E	22° 48' 5.159" S
Hesperia Planum	Plain	G22	109° 53' 38.402" E	21° 25' 21.473" S
Hibes Montes	Mountain / Hill	E27	171° 20' 34.392" E	3° 47' 27.060" N
Hiddekel Cavus	Depression / Basin	D15	16° 14' 24.000" E	29° 25' 48.000" N
Hiddekel Rupes	Cliff	E15	16° 28' 12.000" E	16° 46' 12.000" N
Himera Valles	Valley	G12	22° 39' 24.209" W	21° 32' 24.019" S
Hipparchus	Crater	H3	151° 12' 17.461" W	44° 27' 8.408" S
Hit	Crater	C24	138° 20' 56.722" E	47° 3' 31.706" N
Holden	Crater	G11	34° 1' 8.400" W	26° 2' 24.000" S
Holmes	Crater	J1	66° 32' 58.920" E	74° 51' 18.720" S
Honda	Crater	G13	16° 23' 42.212" W	22° 23' 51.275" S
Hooke	Crater	H11	44° 23' 45.600" W	44° 55' 14.520" S
Hope	Crater	C13	10° 17' 46.464" W	44° 50' 7.562" N
Horarum Mons	Mountain / Hill	H11	36° 33' 41.863" W	51° 3' 12.512" S
Horowitz	Crater	G25	140° 45' 11.468" E	32° 3' 35.666" S
Houston	Crater	C24	135° 56' 49.482" E	48° 13' 32.286" N
Hrad Vallis	Valley	C24	135° 54' 47.266" E	38° 10' 5.158" N
Hsuanch eng	Crater	C24	132° 41' 19.426" E	46° 43' 13.933" N
Huallaga Vallis	Valley	G20	79° 4' 28.726" E	26° 39' 59.634" S
Huancayo	Crater	F11	39° 45' 57.120" W	3° 38' 14.195" S
Huggins	Crater	H26	155° 50' 38.400" E	49° 2' 9.960" S
Humboldt Plateau	Plateau protrusions	F11	34° 54' 25.316" W	4° 45' 47.718" S
Hunten	Crater	H16	23° 41' 29.282" E	39° 10' 38.086" S
Huo Hsing Vallis	Valley	D19	66° 36' 18.564" E	30° 11' 14.711" N
Hussey	Crater	H5	126° 35' 8.710" W	53° 19' 4.882" S
Hutton	Crater	I8	104° 36' 0.721" E	71° 38' 5.535" S
Huxley	Crater	I8	100° 46' 12.000" E	62° 40' 25.367" S
Huygens	Crater	F18	55° 34' 54.120" E	13° 52' 54.840" S
Hyblaeus Catena	Chain of craters	D25	140° 37' 6.298" E	21° 35' 48.215" N
Hyblaeus Chasma	Canyon / Chasm	D25	141° 15' 41.446" E	21° 58' 31.879" N
Hyblaeus Dorsa	Ridge	E24	130° 19' 23.342" E	13° 9' 46.356" N
Hyblaeus Fossae	Depression / Basin	D24	137° 3' 20.387" E	21° 26' 15.544" N
Hydaspis Chaos	Chaotic terrain	E12	26° 55' 44.238" W	3° 5' 14.917" N
Hydrae Cavus	Depression / Basin	F9	61° 18' 27.540" W	7° 55' 30.823" S
Hydrae Chaos	Chaotic terrain	F10	59° 58' 25.741" W	5° 54' 8.442" S
Hydrae Chasma	Canyon / Chasm	F9	62° 0' 25.338" W	6° 45' 2.465" S
Hydraotes Chaos	Chaotic terrain	E11	35° 17' 29.288" W	1° 7' 4.614" N
Hydraotes Colles	Mountain / Hill	F11	33° 40' 43.437" W	0° 1' 13.438" S
Hypanis Valles	Valley	E11	46° 24' 56.522" W	9° 27' 45.630" N
Hyperborea Lingula	Plateau protrusions	A1	53° 32' 17.478" W	80° 19' 4.657" N
Hyperboreae Undae	Dunes	A1	49° 29' 7.885" W	79° 57' 53.906" N
Hyperborei Cavi	Depression / Basin	A1	49° 41' 50.315" W	79° 54' 52.076" N
Hyperboreus Labyrinthus	Labyrinth terrain	A1	59° 45' 3.279" W	80° 17' 1.672" N
Hypsas Vallis	Valley	D18	57° 59' 11.598" E	33° 37' 54.860" N
Iamuna Chaos	Chaotic terrain	F11	40° 36' 37.263" W	0° 16' 56.879" S
Iamuna Dorsa	Ridge	D10	50° 23' 58.841" W	20° 58' 15.163" N
Iani Chaos	Chaotic terrain	F13	17° 2' 31.192" W	2° 11' 31.398" S
Iazu	Crater	F14	5° 10' 37.686" W	2° 42' 18.717" S
Iberus Vallis	Valley	D25	152° 4' 19.480" E	21° 14' 53.954" N
Ibragimov	Crater	G10	59° 34' 12.000" W	25° 25' 53.760" S
Icaria Fossae	Depression / Basin	H5	125° 9' 42.114" W	48° 5' 19.558" S
Icaria Planum	Plain	H6	106° 2' 8.088" W	43° 16' 13.093" S
Icaria Rupes	Cliff	G4	139° 40' 48.000" W	35° 47' 24.000" S
Idaeus Fossae	Depression / Basin	C10	51° 11' 44.382" W	37° 19' 54.206" N
Igal	Crater	G22	110° 53' 48.935" E	20° 5' 29.251" S
Ikej	Crater	D22	112° 30' 2.603" E	20° 57' 19.596" N
Imgr	Crater	D22	111° 11' 1.032" E	19° 7' 23.723" N
Indus Vallis	Valley	D17	38° 52' 39.552" E	18° 56' 56.845" N
Innsbruck	Crater	F11	39° 57' 49.072" W	6° 23' 26.111" S
Ins	Crater	D22	108° 53' 59.599" E	24° 29' 16.941" N
Insight Lander	Historical Land. Site	E24	4° 30' 7.199" N	135° 37' 22.799" E
Inuvik	Crater	A1	28° 19' 6.078" W	78° 35' 15.648" N
Irbit	Crater	G12	24° 54' 40.582" W	24° 20' 35.062" S

INDEX OF PLACES

NAME	FEATURE TYPE	PAGE	LATITUDE	LONGITUDE
Irharen	Crater	D25	140° 49' 19.200" E	34° 29' 25.360" N
Isara Valles	Valley	F3	146° 24' 57.554" W	5° 18' 21.736" S
Isidis Dorsa	Ridge	E21	88° 12' 43.727" E	12° 55' 26.209" N
Isidis Planitia	Plain	E21	88° 22' 37.957" E	13° 56' 8.354" N
Isil	Crater	G21	87° 55' 58.440" E	27° 1' 2.280" S
Ismenia Patera	Crater	C14	1° 47' 57.932" E	38° 33' 3.468" N
Ismeniae Fossae	Depression / Basin	C17	38° 21' 9.494" E	41° 18' 52.205" N
Ismenius Cavus	Depression / Basin	D15	17° 4' 30.072" E	33° 53' 53.337" N
Issedon Paterae	Crater	C7	90° 15' 7.347" W	38° 7' 34.084" N
Issedon Tholus	Mountain / Hill	C7	94° 49' 39.265" W	36° 3' 16.641" N
Ister Chaos	Chaotic terrain	E10	56° 33' 46.451" W	12° 57' 12.934" N
Istok	Crater	H8	85° 48' 57.579" W	45° 6' 0.901" S
Ituxi Vallis	Valley	D25	153° 19' 7.000" E	25° 26' 54.080" N
Ius Chasma	Canyon / Chasm	F8	84° 23' 20.675" W	7° 17' 21.847" S
Ius Labes	Chaotic terrain	F8	78° 27' 45.003" W	7° 28' 25.131" S
Ius Mensa	Mesa	F8	75° 19' 4.440" W	8° 36' 9.465" S
Izamal	Crater	E14	1° 37' 12.000" E	3° 46' 48.000" N
Izendy	Crater	G6	101° 26' 17.609" W	28° 52' 57.073" S
Jaisalmer	Crater	D20	84° 8' 24.000" E	33° 29' 24.000" N
Jal	Crater	G12	28° 45' 48.380" W	26° 14' 58.157" S
Jama	Crater	D10	53° 10' 48.191" W	21° 23' 40.514" N
Jampur	Crater	C8	81° 32' 57.610" W	38° 42' 51.002" N
Janssen	Crater	E17	37° 36' 33.701" E	2° 41' 34.188" N
Jarry Desloges	Crater	F20	83° 51' 11.880" E	9° 22' 29.604" S
Jeans	Crater	I9	154° 10' 54.698" E	69° 38' 15.808" S
Jen	Crater	C13	10° 34' 27.546" W	39° 53' 0.152" N
Jezero	Crater	D20	77° 41' 14.457" E	18° 24' 29.520" N
Jezza	Crater	H11	37° 55' 18.264" W	48° 25' 12.636" S
Jiji	Crater	E14	1° 44' 24.000" W	8° 46' 12.000" N
Jijiga	Crater	D10	53° 57' 8.427" W	25° 6' 26.185" N
Jodrell	Crater	C24	132° 18' 14.587" E	47° 28' 8.207" N
Johannesburg	Crater	C24	133° 11' 12.599" E	47° 55' 18.441" N
Johnstown	Crater	F10	51° 3' 56.222" W	9° 48' 14.617" S
Jojutla	Crater	A1	169° 48' 10.179" W	81° 35' 21.036" N
Joly	Crater	J1	42° 41' 8.265" W	74° 30' 2.031" S
Jones	Crater	G13	19° 49' 30.250" W	18° 53' 2.450" S
Jori	Crater	G20	83° 24' 0.000" E	28° 22' 48.000" S
Jorn	Crater	G20	76° 25' 31.546" E	27° 11' 23.835" S
Jovis Fossae	Depression / Basin	D5	115° 49' 33.424" W	19° 45' 59.408" N
Jovis Tholus	Mountain / Hill	D5	117° 24' 52.492" W	18° 12' 14.301" N
Jumla	Crater	G20	86° 26' 27.960" E	21° 17' 34.080" S
Juventae Cavi	Depression / Basin	F10	58° 9' 34.407" W	3° 54' 32.163" S
Juventae Chasma	Canyon / Chasm	F9	61° 23' 13.191" W	3° 21' 58.313" S
Juventae Dorsa	Ridge	E9	71° 1' 5.906" W	0° 23' 18.452" N
Juventae Mensa	Mesa	F9	65° 37' 31.929" W	7° 55' 47.849" S
Kachug	Crater	D22	107° 35' 19.366" E	18° 8' 57.623" N
Kagoshima	Crater	C24	135° 43' 45.390" E	47° 19' 8.058" N
Kagul	Crater	G13	19° 1' 37.651" W	23° 44' 0.686" S
Kaid	Crater	F11	44° 42' 0.174" W	4° 27' 52.317" S
Kaiser	Crater	H15	19° 6' 30.977" E	46° 11' 38.375" S
Kaj	Crater	G12	29° 23' 10.141" W	27° 2' 42.222" S
Kakori	Crater	H12	29° 51' 7.488" W	41° 29' 24.360" S
Kalba	Crater	F26	154° 49' 57.070" E	5° 53' 6.859" S
Kaliningrad	Crater	C24	134° 57' 46.041" E	48° 28' 28.012" N
Kalocsa	Crater	E13	6° 56' 56.778" W	6° 55' 22.669" N
Kalpin	Crater	E25	141° 16' 12.000" E	8° 55' 48.000" N
Kamativi	Crater	G21	99° 59' 36.240" E	20° 30' 13.320" S
Kamloops	Crater	H12	32° 36' 17.680" W	53° 27' 4.010" S
Kamnik	Crater	H2	161° 47' 22.311" W	37° 12' 21.175" S
Kampot	Crater	H11	45° 35' 32.429" W	41° 46' 58.820" S
Kanab	Crater	G13	19° 0' 11.820" W	27° 12' 4.612" S
Kandi	Crater	G23	122° 5' 49.475" E	32° 44' 44.842" S
Kankossa	Crater	F10	55° 32' 24.000" W	11° 49' 48.000" S
Kansk	Crater	G13	17° 15' 57.150" W	20° 31' 28.082" S
Kantang	Crater	G13	17° 34' 33.639" W	24° 26' 26.212" S
Kaporo	Crater	F15	14° 21' 0.000" E	0° 7' 12.000" S
Karpinsk	Crater	H12	32° 9' 0.240" W	45° 34' 23.280" S
Karshi	Crater	G13	19° 19' 13.300" W	23° 16' 39.111" S
Kartabo	Crater	H10	52° 27' 23.905" W	40° 50' 55.043" S
Karun Valles	Valley	G27	174° 5' 57.030" E	35° 56' 24.144" S
Karzok	Crater	D4	131° 44' 13.549" W	18° 24' 12.698" N
Kasabi	Crater	G21	89° 3' 31.744" E	27° 46' 17.400" S
Kasei Valles	Valley	D9	62° 52' 42.232" W	25° 8' 8.389" N
Kashira	Crater	G13	18° 18' 24.260" W	27° 5' 13.247" S
Kasimov	Crater	G12	22° 56' 23.278" W	24° 37' 48.000" S
Kasra	Crater	D22	103° 37' 53.648" E	21° 58' 32.166" N
Katoomba	Crater	J1	127° 48' 32.251" E	79° 0' 39.343" S
Kaup	Crater	D12	33° 9' 48.706" W	22° 38' 5.640" N
Kaw	Crater	E22	104° 16' 54.814" E	16° 24' 5.922" N
Kayne	Crater	F27	173° 33' 34.570" E	15° 29' 42.345" S
Keeler	Crater	I1	151° 14' 40.491" W	60° 41' 30.619" S
Kem	Crater	H12	32° 57' 59.332" W	44° 56' 10.326" S
Kenge	Crater	F22	102° 57' 0.000" E	16° 21' 36.000" S

INDEX OF PLACES

INDEX OF PLACES

NAME	FEATURE TYPE	PAGE	LATITUDE	LONGITUDE
Margaritifer Chaos	Chaotic terrain	F12	21° 42' 16.740" W	9° 18' 5.569" S
Margaritifer Terra	Land mass	F12	24° 55' 20.320" W	1° 50' 58.781" S
Mari	Crater	H11	45° 53' 2.851" W	52° 0' 41.400" S
Maricourt	Crater	C9	71° 10' 4.773" W	53° 20' 17.596" N
Marikh Vallis	Valley	G14	4° 19' 8.664" E	19° 9' 29.965" S
Mariner	Crater	G2	164° 14' 6.000" W	34° 40' 54.120" S
Mars 2 Lander	Historical Land. Site	E10	47° 0' 0.122" W	3° 59' 59.963" N
Mars 3 Lander	Historical Land. Site	H2	157° 59' 59.614" W	44° 59' 59.977" S
Mars 6 Lander	Historical Land. Site	G13	19° 24' 0.090" W	23° 53' 59.873" S
Mars Pathfinder Lander	Historical Land. Site	D12	33° 15' 2.646" W	19° 5' 53.293" N
Mars Polar Lander	Historical Land. Site	J1	164° 42' 12.961" E	76° 41' 36.686" S
Marte Vallis	Valley	E1	177° 5' 45.058" W	14° 4' 36.863" N
Marth	Crater	E14	3° 26' 49.670" W	12° 56' 11.828" N
Martin	Crater	G9	69° 15' 3.660" W	21° 20' 41.378" S
Martynov	Crater	G11	36° 24' 48.098" W	30° 21' 43.528" S
Martz	Crater	G25	144° 10' 36.971" E	34° 54' 53.259" S
Masursky	Crater	E12	32° 18' 18.000" W	12° 4' 2.280" N
Matara	Crater	H17	34° 35' 36.167" E	49° 36' 36.863" S
Matrona Vallis	Valley	F27	176° 11' 24.408" E	7° 39' 34.570" S
Maumee Valles	Valley	D10	52° 51' 11.333" W	19° 30' 25.060" N
Maunder	Crater	H14	1° 45' 5.004" E	49° 36' 14.040" S
Mawrth Vallis	Valley	D13	16° 58' 21.443" W	22° 25' 49.577" N
Mazamba	Crater	G9	69° 40' 22.800" W	27° 31' 54.840" S
McCauley	Crater	G20	83° 13' 48.000" E	27° 21' 0.000" S
McLaughlin	Crater	D12	22° 22' 1.200" W	21° 53' 47.760" N
McMurdo	Crater	J1	0° 35' 15.799" E	84° 23' 0.485" S
Medrissa	Crater	D10	56° 34' 6.636" W	18° 38' 30.468" N
Medusae Fossae	Depression / Basin	F2	164° 11' 44.263" W	2° 9' 58.623" S
Mega	Crater	F11	36° 54' 17.400" W	1° 25' 49.860" S
Meget	Crater	D22	107° 18' 46.087" E	18° 51' 53.651" N
Melas Chasma	Canyon / Chasm	F9	72° 32' 19.487" W	10° 31' 3.946" S
Melas Dorsa	Ridge	G9	72° 5' 45.732" W	18° 55' 7.414" S
Melas Fossae	Depression / Basin	G9	71° 31' 19.232" W	26° 16' 58.754" S
Melas Labes	Chaotic terrain	F9	71° 42' 12.919" W	8° 31' 54.339" S
Melas Mensa	Mesa	F8	74° 9' 7.802" W	10° 43' 23.624" S
Mellish	Crater	J1	23° 44' 27.600" W	72° 37' 38.640" S
Mellit	Crater	E14	1° 44' 3.014" W	7° 7' 25.752" N
Memnonia Fossae	Depression / Basin	G2	153° 49' 15.844" W	23° 37' 36.960" S
Mena	Crater	G13	18° 45' 25.325" W	32° 6' 34.934" S
Mendel	Crater	I9	161° 14' 55.848" E	58° 47' 1.130" S
Mendota	Crater	D24	138° 19' 54.913" E	35° 49' 39.624" N
Meridiani Land. Site	Historical Land. Site	F14	5° 31' 16.663" W	1° 56' 43.775" S
Meridiani Planum	Plain	F14	3° 8' 24.000" W	0° 2' 24.000" S
Meridiani Serpentes	Ridge	E15	8° 6' 0.000" E	1° 18' 0.000" N
Meroe Patera	Crater	E19	68° 45' 56.426" E	6° 58' 52.458" N
Micoud	Crater	C15	16° 20' 17.700" E	50° 33' 47.663" N
Mie	Crater	C24	139° 39' 17.448" E	48° 9' 52.447" N
Mila	Crater	G12	20° 45' 4.531" W	27° 9' 48.678" S
Milankovic	Crater	B1	146° 34' 43.037" W	54° 27' 41.991" N
Milford	Crater	H11	41° 29' 26.039" W	52° 24' 44.215" S
Millman	Crater	H3	149° 38' 20.400" W	53° 56' 58.920" S
Millochau	Crater	G20	85° 6' 17.640" E	21° 11' 21.840" S
Milna	Crater	G13	12° 14' 19.701" W	23° 27' 47.160" S
Minio Vallis	Valley	F3	151° 40' 17.772" W	4° 22' 46.908" S
Mirtos	Crater	D10	51° 45' 20.012" W	22° 7' 29.446" N
Mistretta	Crater	G6	109° 7' 45.141" W	24° 40' 50.922" S
Mitchel	Crater	I7	76° 0' 27.714" E	67° 31' 33.769" S
Miyamoto	Crater	F13	6° 56' 59.593" W	2° 52' 4.656" S
Mliba	Crater	H21	87° 59' 3.781" E	39° 36' 25.284" S
Moa Valles	Valley	D10	54° 41' 58.396" W	35° 37' 13.521" N
Moanda	Crater	G11	39° 56' 52.079" W	35° 55' 35.999" S
Mohawk	Crater	C14	5° 21' 4.712" W	42° 53' 16.735" N
Mojave	Crater	E12	32° 59' 10.924" W	7° 29' 2.779" N
Molesworth	Crater	G25	149° 16' 13.665" E	27° 29' 42.668" S
Moni	Crater	H15	18° 45' 58.173" E	47° 0' 35.321" S
Montevallo	Crater	E10	54° 15' 54.494" W	15° 15' 5.499" N
Morava Valles	Valley	F12	24° 12' 14.845" W	13° 34' 0.857" S
Morella	Crater	F10	51° 23' 13.284" W	9° 34' 47.735" S
Moreux	Crater	C17	44° 32' 38.040" E	41° 47' 31.920" N
Moroz	Crater	G12	20° 34' 22.800" W	23° 46' 16.680" S
Morpheos Rupes	Cliff	H23	125° 34' 35.404" E	35° 59' 58.957" S
Mosa Vallis	Valley	F16	22° 12' 15.775" E	15° 5' 29.421" S
Moss	Crater	D22	109° 29' 16.641" E	19° 14' 0.654" N
Muara	Crater	D13	19° 18' 49.778" W	24° 19' 16.778" N
Muller	Crater	G24	127° 53' 11.408" E	25° 44' 8.344" S
Munda Vallis	Valley	F3	146° 10' 16.084" W	5° 22' 5.156" S
Murgoo	Crater	G12	22° 26' 44.048" W	23° 38' 13.682" S
Murray	Crater	G16	28° 3' 46.538" E	23° 17' 37.544" S
Mut	Crater	D11	35° 45' 43.282" W	22° 21' 22.435" N
Mutch	Crater	E10	55° 12' 25.200" W	0° 35' 43.548" N
N Mareotis Tholus	Mountain / Hill	C8	86° 12' 22.823" W	36° 22' 47.799" N
Naar	Crater	D11	42° 7' 47.510" W	22° 54' 49.484" N
Naic	Crater	D22	107° 28' 21.770" E	24° 27' 11.043" N

INDEX OF PLACES

NAME	FEATURE TYPE	PAGE	LATITUDE	LONGITUDE
Nain	Crater	C24	126° 50' 34.060" E	41° 28' 3.000" N
Naju	Crater	C23	122° 51' 23.844" E	44° 59' 13.584" N
Nako	Crater	G20	82° 55' 48.000" E	29° 40' 12.000" S
Naktong Vallis	Valley	E17	33° 23' 35.343" E	4° 53' 14.473" N
Nakusp	Crater	D11	35° 26' 48.728" W	24° 43' 53.959" N
Nan	Crater	G13	19° 56' 15.429" W	26° 41' 20.897" S
Nanedi Valles	Valley	E10	48° 36' 58.578" W	5° 2' 55.202" N
Nansen	Crater	H3	140° 25' 18.820" W	49° 55' 5.604" S
Napo Vallis	Valley	G20	78° 1' 32.368" E	25° 58' 8.159" S
Nardo	Crater	G12	32° 50' 28.943" W	27° 30' 19.177" S
Naro Vallis	Valley	F19	60° 42' 41.575" E	3° 59' 49.136" S
Naruko	Crater	H2	161° 41' 59.408" W	36° 14' 7.255" S
Naryn	Crater	E23	123° 17' 43.725" E	14° 53' 20.580" N
Nat Cavus	Depression / Basin	F7	96° 54' 0.000" W	11° 54' 0.000" S
Naukan	Crater	D12	30° 34' 33.204" W	21° 15' 16.236" N
Navan	Crater	G12	23° 30' 10.350" W	25° 53' 9.275" S
Navua Valles	Valley	G20	82° 41' 1.615" E	33° 56' 6.200" S
Nazca	Crater	G21	93° 40' 15.633" E	31° 37' 36.349" S
Nectaris Fossae	Depression / Basin	G10	57° 9' 24.340" W	23° 5' 6.151" S
Nectaris Montes	Mountain / Hill	F10	54° 38' 51.518" W	14° 38' 18.576" S
Negele	Crater	G21	95° 59' 48.674" E	35° 47' 59.193" S
Negril	Crater	D19	69° 25' 58.960" E	20° 11' 34.336" N
Neive	Crater	D22	107° 4' 26.138" E	23° 10' 57.491" N
Nema	Crater	D10	52° 7' 40.016" W	20° 41' 43.507" N
Nepa	Crater	G13	19° 39' 55.990" W	24° 58' 24.831" S
Nepenthes Mensae	Mesa	E23	119° 25' 14.302" E	9° 11' 31.862" N
Nepenthes Planum	Plain	E23	113° 47' 31.358" E	14° 0' 35.099" N
Nereidum Montes	Mountain / Hill	H11	43° 12' 40.677" W	37° 34' 23.468" S
Nestus Valles	Valley	F2	158° 28' 33.880" W	7° 1' 41.942" S
Neukum	Crater	H16	28° 24' 0.000" E	44° 54' 0.000" S
Never	Crater	D22	105° 46' 20.843" E	23° 30' 16.431" N
Neves	Crater	F25	151° 18' 27.697" E	3° 23' 33.776" S
New Bern	Crater	D10	49° 9' 7.917" W	21° 32' 0.369" N
New Haven	Crater	D10	49° 15' 43.842" W	22° 4' 30.738" N
New Plymouth	Crater	F27	175° 52' 7.630" E	15° 46' 35.799" S
Newcomb	Crater	G14	1° 2' 37.597" E	24° 16' 25.801" S
Newport	Crater	D10	48° 57' 26.567" W	22° 14' 16.711" N
Newton	Crater	H2	158° 1' 43.691" W	40° 30' 7.309" S
Nhill	Crater	G6	103° 19' 50.105" W	28° 40' 43.121" S
Nia Chaos	Chaotic terrain	F9	67° 22' 34.715" W	6° 44' 29.261" S
Nia Fossae	Depression / Basin	F9	71° 46' 28.225" W	14° 43' 59.179" S
Nia Mensa	Mesa	F9	67° 18' 54.598" W	7° 43' 18.674" S
Nia Tholus	Mountain / Hill	F8	74° 56' 51.363" W	6° 35' 11.585" S
Nia Vallis	Valley	H11	34° 48' 26.708" W	53° 31' 40.256" S
Nicer Vallis	Valley	F2	158° 11' 25.806" W	6° 57' 51.509" S
Nicholson	Crater	E2	164° 25' 49.811" W	0° 12' 39.969" N
Nier	Crater	C22	106° 6' 32.400" E	42° 47' 15.360" N
Niesten	Crater	G18	57° 44' 43.906" E	27° 59' 58.433" S
Nif	Crater	D10	56° 14' 6.024" W	19° 54' 20.831" N
Niger Vallis	Valley	G21	92° 34' 5.405" E	34° 57' 34.172" S
Nili Fossae	Depression / Basin	D20	76° 41' 41.382" E	22° 1' 5.358" N
Nili Patera	Crater	E19	67° 10' 2.391" E	8° 58' 7.763" N
Nili Tholus	Mountain / Hill	E19	67° 20' 58.723" E	9° 8' 53.927" N
Nilokeras Fossa	Depression / Basin	D10	57° 49' 42.763" W	24° 35' 40.595" N
Nilokeras Mensae	Mesa	D10	51° 56' 55.192" W	30° 28' 50.294" N
Nilokeras Scopulus	Escarpment	D10	55° 51' 6.875" W	31° 42' 55.158" N
Nilosyrtis Mensae	Mesa	D19	68° 28' 13.279" E	34° 46' 29.531" N
Nilus Chaos	Chaotic terrain	D8	76° 57' 17.040" W	25° 23' 13.735" N
Nilus Dorsa	Ridge	D8	79° 3' 52.299" W	20° 40' 57.665" N
Nilus Mensae	Mesa	D9	72° 13' 37.543" W	22° 12' 0.669" N
Nipigon	Crater	D8	81° 50' 29.090" W	33° 45' 32.226" N
Niquero	Crater	H2	165° 58' 17.171" W	38° 47' 12.600" S
Nirgal Vallis	Valley	G11	41° 41' 5.123" W	28° 9' 34.630" S
Nitro	Crater	G12	24° 0' 15.954" W	21° 15' 44.787" S
Njesko	Crater	G20	85° 6' 30.795" E	35° 14' 59.345" S
Noachis Terra	Land mass	H14	5° 9' 24.284" W	50° 24' 26.042" S
Noc Cavus	Depression / Basin	F7	96° 42' 0.000" W	13° 24' 0.000" S
Noctis Fossae	Depression / Basin	F7	98° 50' 53.425" W	2° 41' 36.384" S
Noctis Labyrinthus	Labyrinth terrain	F6	101° 11' 19.920" W	6° 21' 44.940" S
Noctis Tholus	Mountain / Hill	F7	98° 12' 0.000" W	12° 18' 0.000" S
Noma	Crater	G12	24° 18' 20.005" W	25° 25' 42.205" S
Noord	Crater	G13	11° 15' 59.912" W	19° 16' 19.693" S
Nordenskiold	Crater	H2	158° 45' 33.210" W	52° 21' 58.961" S
Northport	Crater	D10	54° 28' 40.908" W	18° 31' 24.779" N
Novara	Crater	G13	10° 41' 16.043" W	24° 53' 52.588" S
Nqutu	Crater	H27	169° 33' 0.000" E	38° 2' 24.000" S
Nune	Crater	E11	38° 45' 35.809" W	17° 33' 0.282" N
Nutak	Crater	E12	30° 15' 22.276" W	17° 24' 52.619" N
Nybyen	Crater	H13	16° 39' 22.198" W	37° 1' 28.605" S
Obock	Crater	F25	150° 31' 39.097" E	2° 0' 26.669" S
Ocampo	Crater	D24	138° 18' 14.400" E	32° 39' 54.000" N
Oceanidum Fossa	Depression / Basin	I4	29° 30' 41.142" W	61° 34' 46.874" S
Oceanidum Mons	Mountain / Hill	I4	41° 13' 33.198" W	54° 55' 33.598" S

INDEX OF PLACES

INDEX OF PLACES

CONTRIBUTOR'S PAGE

First 200 backers of The Mars Atlas

#	Name	#	Name	#	Name	#	Name
1	Joanna K. Johnson	51	Max Jäger	101	Frank	151	William Lorenzo
2	Mae Shephard	52	T. Stengler	102	Dr. Bernd Hanewinckel	152	Lewis coppard
3	Stanley & Florence Ray	53	Oliver Grieser	103	Marc Rouleau	153	Benjamin Chapman
4	Gareth Brede	54	Philipp Mahn	104	Angus Abranson	154	Stijn Calders
5	Paul Weller	55	Markus Klimt	105	Alexander Hoare	155	Jared Coplin
6	Lauren E. Ray	56	Bernd Schumacher	106	Walter Obendrauf	156	Matt D'Addona
7	Tyler J. Hawks	57	Phil Taprogge	107	Christopher Drew	157	Robert L. Oakes
8	Victoria & Rafaella Mery	58	Heiko Dormann	108	David & Valerie Herda	158	Rachel Montgomery
9	Scott & Denise Ray	59	Michael Rother	109	Robert Carr	159	Rob Lucas
10	Lumm	60	Stefan "Misel" Misch	110	Ryan Blanger	160	Jonathan J. Gallegos
11	Ross Mogridge	61	Jakob Tomasi	111	David Ryack	161	Jamie Hyde
12	Benjamin Fabian	62	Tom J. Wunschick	112	Timo Walter	162	Felipe Carvallo
13	Scott R. Ilsley	63	Stefan Polk	113	William Slater	163	Sean McNeely
14	Rex G. Forrester	64	Benoît Tremblay	114	Heinz-Josef Winkels	164	Vincent Catanzaro
15	Craig Stevig, M(ascp)	65	Ulisse Dusci	115	Alexander Müller	165	Calvin Brindley
16	Jean-Baptiste Tune	66	Gjon Preci	116	Kenneth Bladt	166	Stephen Favetti
17	Danielle Williamson	67	Andreas Janker	117	Gregory Landegger	167	Garald D. Michaud
18	James L. Burk	68	Tim Winslow	118	Chris Haines	168	Garrett Kingman
19	Gregory Frank	69	Joseph Smith	119	Dr. E. Erik Bender	169	Kate Greiner
20	Andrew Gallo	70	Zachrisson Gisbert	120	Shane Calvert	170	Roland Amacher
21	John Peterson	71	Martin Kropp	121	Gobind Basmall	171	Carl Hall
22	Donald Hladiuk	72	Paul & Karen Herkes	122	Fabian Buß	172	Skyler & Heather Young
23	Andrew Rohl	73	Olaf Leidinger	123	Steven S. Long	173	N. Trevor Brierly
24	Erin Valenciano	74	Ahmed Mirza	124	CJ Rohrich	174	Chris Beck
25	Marc Hellmann	75	Thomas Hörmannn	125	Brian Wilson	175	Scott Edmondson
26	Dmytro Vasylyev	76	Lucie E. Niestroj	126	Allen Lum	176	Neal Johnson
27	Jed Scott	77	Stanko Cirovic	127	Yumyog	177	David Hedberg
28	Petra, Ela, David & Martin Kucerovi	78	Guido Adam	128	Martin Hans	178	Elise Gesting
29	Niklas Wendel	79	Uwe Gogolin	129	Adrian Widera	179	Cecil Keebler
30	Clint Roesbeke	80	C. Senges	130	Florian Kugler	180	Steve Bolton
31	Ilyana et Cyril ANGER	81	Ole Marggraf	131	Sándor Barics	181	Marco Nehrkorn
32	Kenneth H. Fichtler	82	Aaron K. Clark	132	Ron Beiswanger	182	Johnathen J. Everett
33	Philippe van Nedervelde	83	Florian Voigt	133	Kelsey Wierenga	183	Wendy L. Schultz
34	Ang Zuan Kee	84	Caspar Williams	134	Ian Ramsden	184	Roland E. Miller, III
35	A. B.	85	Jörg Leu	135	Andrew Miazga	185	Simon Scott T Stromberg
36	Gerardo Guevara	86	Sina Schwetje	136	Gerd Bordon	186	Steven Lewis
37	Juan Miguel Suay Belenguer	87	Jesse & Carri Renaud	137	Tina Mammoser	187	Thomas McGarry
38	Ben Brandt	88	Peter Jenisch	138	Trevor Hullender	188	May
39	Marcel-Jan Krijgsman	89	Fabian Wittich	139	Michael L. Jaegers	189	Raymond Kenny
40	Kohl125	90	Fabian Haselwarter	140	Brandon Porter	190	Oliver Ockenden
41	Matt Norman	91	Kit Larkin & Mandy Farneman	141	Steve Hayhurst	191	David Greenfield
42	Nick Larter	92	David J. Shephard	142	Evelynn S. Barringer	192	Raffaele D'Angelo
43	Fam. Barrenstein	93	William Leung	143	James Boys	193	John R. Shibley
44	Johannes Blume	94	Jagoba Velasco	144	Osian G. Williams	194	Stephen Eickhoff
45	Samuel John Phillip Lee Holmes	95	David Sprague	145	Connor Barber	195	Taegon
46	David Mittag	96	Adam Howell	146	Steffen Jauch	196	David H. Smith
47	Richard Woeber	97	Tiziano Ammann	147	Danny Ventsias	197	James Kendall
48	Thomas Peter from jtr.ch	98	Steffan Mensikov	148	L. Andrew Williams	198	Simon T.
49	Michael "Mila" Langhans	99	Christian Meyer	149	Ryan McDermott	199	Lawrence McGlynn
50	Julian S. Dieckmann	100	Patric Götz	150	Russell Sagraves	200	Scott A. OBrien

A special thank you goes out to the these individuals for their support of RedMapper's Kickstarter campaign. Because of their contribution and enthusiasm, The Mars Atlas is now in print and available to anyone with a passion for Mars and space exploration. These 200 intrepid souls represent the first wave of backers of this project and deserve special recognition.